AF832882

SINTERING – METHODS AND PRODUCTS

Edited by **Volodymyr Shatokha**

Sintering – Methods and Products
Edited by Volodymyr Shatokha

Published by InTech

Janeza Trdine 9, 51000 Rijeka, Croatia

Copyright © 2012 InTech

All chapters are distributed under the Creative Commons Attribution 3.0 license, which allows users to download, copy and build upon published articles even for commercial purposes, as long as the author and publisher are properly credited, which ensures maximum dissemination and a wider impact of our publications. After this work has been published by InTech, authors have the right to republish it, in whole or part, in any publication of which they are the author, and to make other personal use of the work. Any republication, referencing or personal use of the work must explicitly identify the original source.

As for readers, this license allows users to download, copy and build upon published chapters even for commercial purposes, as long as the author and publisher are properly credited, which ensures maximum dissemination and a wider impact of our publications.

Notice

Statements and opinions expressed in the chapters are these of the individual contributors and not necessarily those of the editors or publisher. No responsibility is accepted for the accuracy of information contained in the published chapters. The publisher assumes no responsibility for any damage or injury to persons or property arising out of the use of any materials, instructions, methods or ideas contained in the book.

Publishing Process Manager Sasa Leporic

Technical Editor Teodora Smiljanic

Cover Designer InTech Design Team

Additional hard copies can be obtained from orders@intechopen.com

Sintering – Methods and Products, Edited by Volodymyr Shatokha
p. cm.
ISBN 978-953-51-0371-4

We are IntechOpen, the world's leading publisher of Open Access books

Built by scientists, for scientists

4,100+
Open access books available

116,000+
International authors and editors

120M+
Downloads

151
Countries delivered to

Our authors are among the
Top 1%
most cited scientists

12.2%
Contributors from top 500 universities

CLARIVATE ANALYTICS
BOOK CITATION INDEX
INDEXED

WEB OF SCIENCE™

Selection of our books indexed in the Book Citation Index
in Web of Science™ Core Collection (BKCI)

Interested in publishing with us?
Contact book.department@intechopen.com

Numbers displayed above are based on latest data collected.
For more information visit www.intechopen.com

For EU product safety concerns:
IN TECH d.o.o., Prolaz Marije Krucifikse Kozulić 3, 51000 Rijeka, Croatia,
info@intechopen.com or visit our website at intechopen.com.

Meet the editor

Dr Volodymyr Shatokha graduated in 1982 obtaining the title of a Ferrous Metallurgy Engineer from the Dnipropetrovsk Metallurgical Institute, Ukraine. His further career is tied up with his Alma Mater (since 1999 - National Metallurgical Academy of Ukraine). In 1985 he received his PhD degree. His further research was mainly in the field of iron ore sinter and pellets properties, and the slag forming processes in ironmaking. He has been teaching students since 1992. In 1998 he obtained the title of Professor at the Ironmaking Chair. During 1999-2002 Dr. Shatokha was the Dean of the Metallurgical Faculty. Since 2002 he has been working as the Vice Rector for Research and Pedagogy. His current research mainly deals with recycling methods for the ferrous wastes, in particular through iron ore sintering. He has authored over 130 research papers, 7 books and 6 patents. He holds the title of Honorary Professor at the Inner-Mongolia University of Science and Technology, China and is a member of the Advisory Board for the Iron and Steel Institute of Japan International Journal. He is a Laureate of the State Prize of Ukraine in Science and Technology (2010).

Contents

Preface XI

Part 1 Sintering Technologies and Methods 1

Chapter 1 **Latest Generation Sinter Process Optimization Systems 3**
Thomas Kronberger, Martin Schaler and Christoph Schönegger

Chapter 2 **Modeling Sintering Process of Iron Ore 23**
Jose Adilson de Castro

Chapter 3 **Development of Sintered MCrAlY Alloys for Aeronautical Applications 47**
Fernando Juárez López and Ricardo Cuenca Alvarez

Chapter 4 **Metal Laser Sintering for Rapid Tooling in Application to Tyre Tread Pattern Mould 73**
Jelena Milovanovic, Milos Stojkovic and Miroslav Trajanovic

Chapter 5 **Sintering of Supported Metal Catalysts 91**
José Luis Contreras and Gustavo A. Fuentes

Chapter 6 **Sintering Characteristics of Injection Moulded 316L Component Using Palm-Based Biopolymer Binder 127**
Mohd Afian Omar and Istikamah Subuki

Chapter 7 **Liquid Phase Sintering of Fe-Cu-Sn-Pb System for Tribological Applications 147**
Cristina Teisanu

Chapter 8 **Photonic Sintering of Silver Nanoparticles: Comparison of Experiment and Theory 173**
Jeff West, Michael Carter, Steve Smith and James Sears

Chapter 9 **Microwave-Induced Combustion Synthesis of Luminescent Aluminate Powders 189**
A. Potdevin, N. Pradal, M.-L. François, G. Chadeyron, D. Boyer and R. Mahiou

Part 2 Characterisation and Properties of Sintered Materials 213

Chapter 10 **On the Application of 3D X-Ray Microtomography for Studies in the Field of Iron Ore Sintering Technology 215**
Volodymyr Shatokha, Iurii Korobeynikov and Eric Maire

Chapter 11 **Effect of the Additives of Nanosized Nb and Ta Carbides on Microstructure and Properties of Sintered Stainless Steel 233**
Uílame Umbelino Gomes, José Ferreira da Silva Jr. and Gislâine Bezerra Pinto Ferreira

Chapter 12 **Properties and Structure of Sintered Boron Containing Carbon Steels 249**
G. Bagliuk

Chapter 13 **Influence of Sintering Temperature on Magnetotransport Behavior of Some Nanocrystalline Manganites 267**
G. Venkataiah , Y. Kalyana Lakshmi and P. Venugopal Reddy

Chapter 14 **The Quantification of Crystalline Phases in Materials: Applications of Rietveld Method 293**
Cláudia T. Kniess, João Cardoso de Lima and Patrícia B. Prates

Preface

Sintering refers to the science and technology of production objects from fines or powders. Being a very complex phenomenon, sintering covers a broad subject field and combines chemistry, physics and mechanics. Sintering deals with various material systems such as metals, non-metals, ceramics, polymers, and their combinations. Processes of sintering may occur in a wide temperature range - in a solid state as well as with participation of a liquid phase.

As a material processing method, sintering embraces extremely wide group of technologies used to obtain such different products as for example iron ore agglomerate and luminescent films. In recent times some novel methods emerge such as for example laser sintering, electric current assisted sintering, photonic sintering, microwave sintering, and spark plasma sintering.

Modern science and technology of sintering aim at the energy-efficient production of materials with intended properties – whether it is mechanical strength, heat resistivity, magnetotransport characteristics, tribological behaviour and so on. These properties are associated with structure and morphology, therefore characterisation methods of sintered product are of special attention.

The book, *Sintering - Methods and Products,* assembles various aspects of sintering in chapters written by the experienced scientists from all over the world. It comprises two sections focusing on *Sintering technologies and methods*, and *Characterisation and properties of sintered materials*. Sharing their results and expertise with the reading audience, the authors have dedicated their time and energy to make this book happen. I am very thankful to all the authors for their contribution.

I thank sincerely *InTech - Open Access Publisher* for the initiative of publishing this book. Reading audience of researchers and students dealing with or interested in sintering has now a really unique opportunity to find comprehensive, cutting edge information covering a wide range of interdisciplinary issues in the domain of science and technology of sintering in a single book.

Dr. Volodymyr Shatokha
National Metallurgical Academy of Ukraine,
Ukraine

Part 1

Sintering Technologies and Methods

1

Latest Generation Sinter Process Optimization Systems

Thomas Kronberger, Martin Schaler and Christoph Schönegger
Siemens VAI Metals Technologies GmbH (Siemens VAI) Linz, Austria

1. Introduction

SIMETAL Sinter VAiron is an advanced process optimization system which covers the sinter production process from ore preparation in the blending yards and sinter plant up to the blast furnace. It was developed in a close cooperation between the Austrian steel producer voestalpine Stahl and the engineering and plant-building company Siemens VAI. The overall target of this system is to achieve stable process conditions at a high productivity level with a uniform sinter quality at low production costs. This is achieved through the application of a number of sophisticated tracking, diagnosis and control models and systems which are bundled within an overall expert system.

2. System objectives

In the sintering process, chemical and physical parameters such as basicity and product diameters must satisfy pre-set target values within defined standard deviations in order to meet the quality requirements of the blast furnace. Sinter quality begins with the selection and mixing of the raw materials in the blending yard and dosing plant which are integrated in a common control model of the Sinter Process. The chemical properties are homogenized by an automatic adaptation of the raw material mix. An enhanced burn-through-point control system which takes into account physical and chemical properties of the sinter mix is incorporated in the system. The system has to counteract changes caused by fluctuations, which is achieved by a closed-loop control of the process.

The main targets of the SIMETAL Sinter VAiron[1] process control system are summarized as follows:

- Minimizing fuel consumption – The fuel rate is a key factor in production costs.
- Avoidance of heavy control actions – If only minor control actions are necessary, the sinter machine performance is stabilized significantly.
- Avoidance of critical process situations – The sooner the system reacts to critical process situations, such as an inhomogeneous mixture, poor surface ignition or incomplete burn-through of mix, the smoother the overall sintering process is, resulting in a more uniform product quality.

---

[1]Later on in this document we will briefly call the system '*VAiron Sinter*'.

- Synoptical operational decisions throughout all shifts - Constant operating conditions throughout all shifts will increase the lifetime of the equipment and reduce production costs.
- Reduction of emissions - With the closed-loop operation mode of the VAiron Sinter Expert System, the production parameters can be optimized within the environmental emission limits, in particular, SO_2 emissions.

3. System structure and technological controls

A reliable and well proven basis automation system is the backbone of modern sinter plant operation. VAiron Sinter is characterized by a modular system structure (Fig. 1).

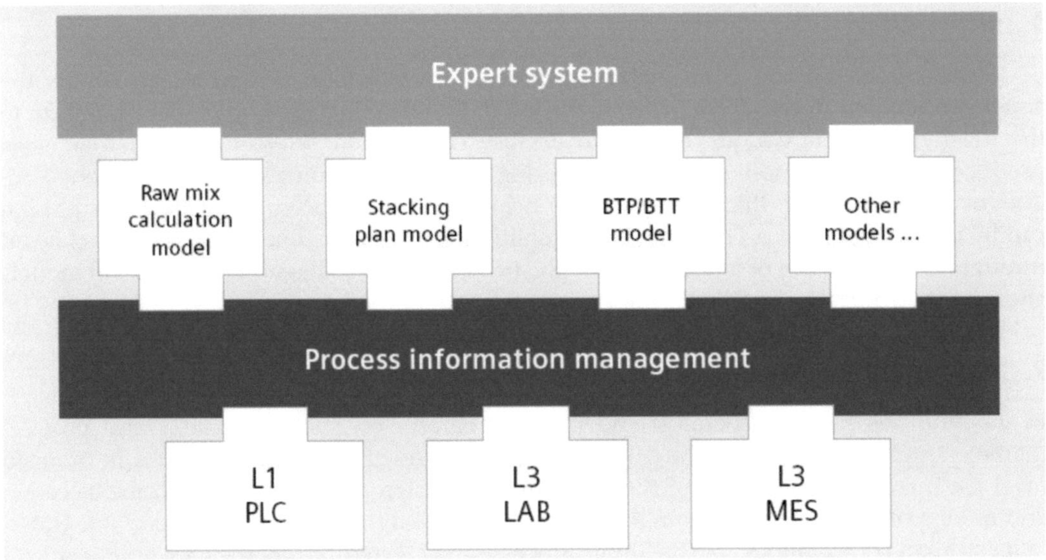

Fig. 1. System Structure.

In addition to basic functions such as data acquisition and set-point execution, the technological controls (main control loops) are implemented in the basis automation system. These include raw-mix-ratio control, raw-mix-feed control, moisture control, surge-hopper-level control, drum-feeder control, ignition-hood control, exhaust-gas-cooler control and cooler control. The focus of these basic control functions is to assure a smooth and reliable sintering process and to enable a continued process optimization.

3.1 VAironment – Process information and data-management system

A multithreaded, three-tier, client-server real-time application is the basis for the hardware and software configuration in VAiron Sinter. The data acquisition function pre-processes the data from a broad spectrum of raw data sources (front-end signals, material weights, laboratory data, events, model results and cost data, etc.) before storing these in the plant database (Fig. 2).

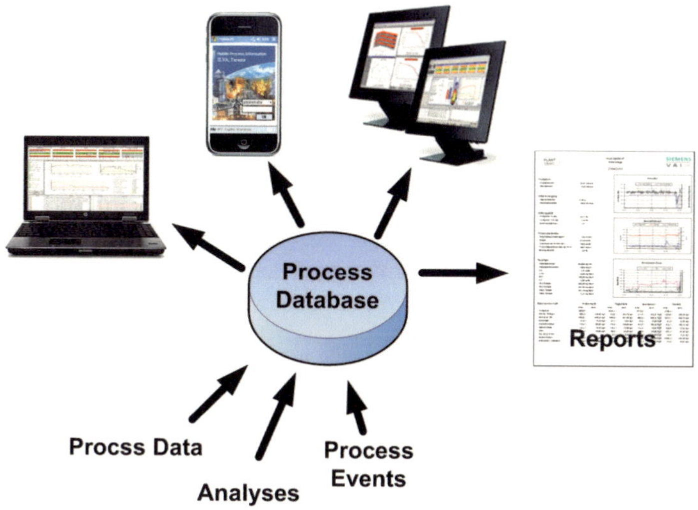

Fig. 2. VAironment Process Information and Data Management System.

The process information management system provides a flexible and powerful database for the continuous improvement of process knowledge. VAiron Sinter interprets process data, performs model calculations and visualizes the results in Windows- or web-based graphical user interfaces. Additional data analysis, interpretation and visualization tools can connect to VAiron easily (COM, ODBC). The module-based system is highly configurable in order to allow for adaptations related to modifications in the plant setup or operational philosophy.

Data handling encompasses the chemical and physical data of the sinter strand as well as the process history. The raw sinter mix and the production process is monitored in detail from ore preparation to the blast furnace. Unfavorable conditions can be detected and eliminated (**Fig. 3**).

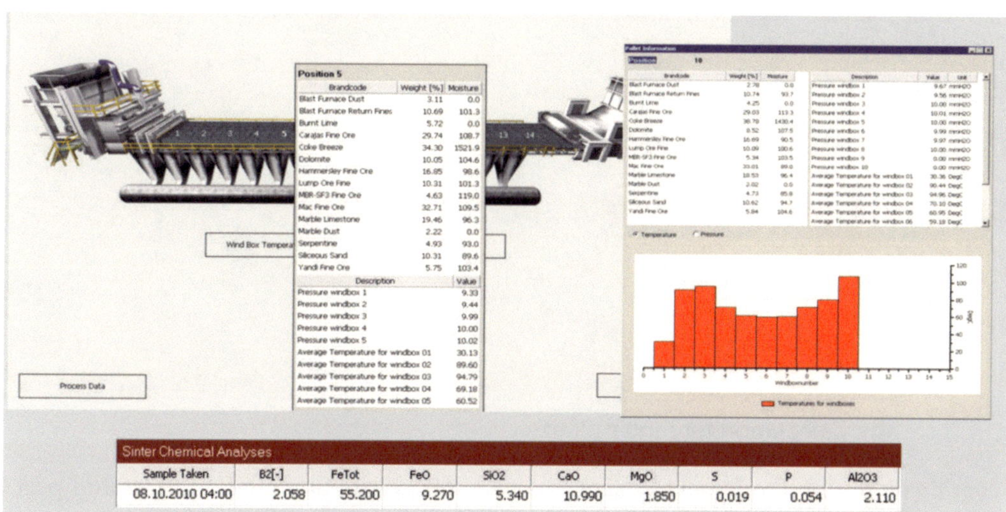

Fig. 3. Detailed Tracking of Material Packages.

3.2 VAiron process models

A number of process models are available in the VAiron Sinter automation package as outlined in the following.

3.2.1 Raw mix calculation model

Producing target quality requires accurate charging of the raw materials (ores, coke, additives, etc.). To modify the raw mix recipe, the coke addition, sinter basicity, raw material analyses and their influence on sinter parameters must be taken into consideration. This procedure is complex and requires computer assistance.

The purpose of the raw mix calculation model is to establish a raw mix composition, in order to automatically achieve the assigned target values for coke addition, sinter basicity, Fe_{tot}, SiO_2, etc.

Up to four variable materials may be chosen - one for the basicity equation, one for Fe_{tot} balance, and one for SiO_2 balance as well as one for the MgO balance.

The results of the raw mix calculation can be activated manually to run on Level 1.

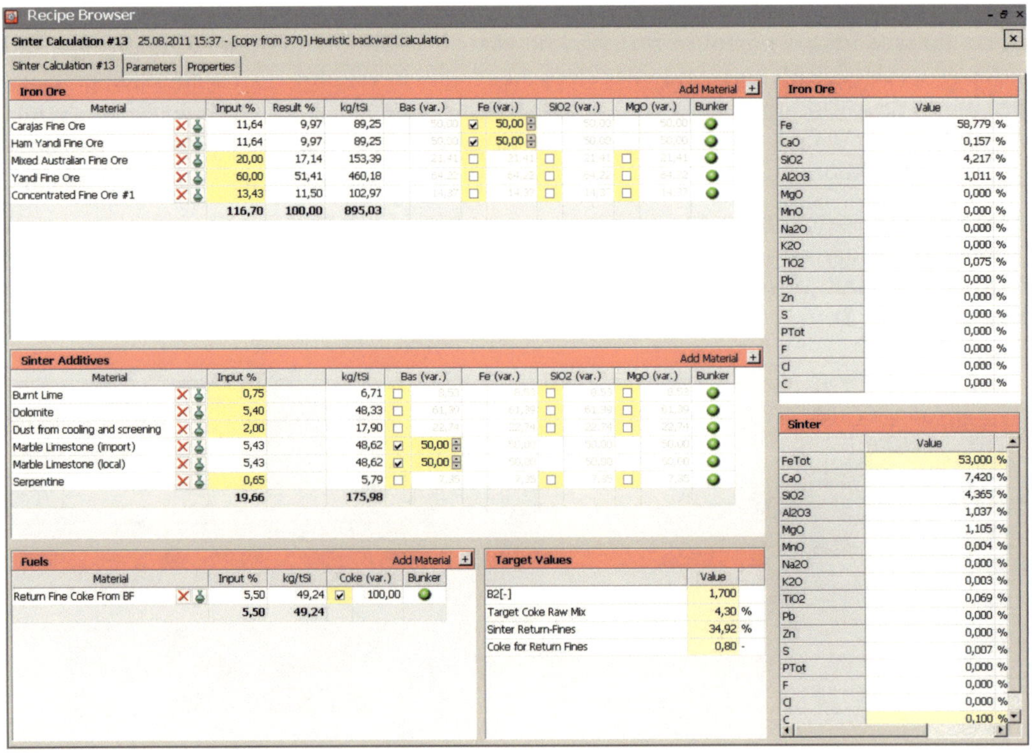

Fig. 4. Raw Mix Calculation for Sinter Plant.

In combination with the Expert System, the raw mix calculation model is a central part of the closed-loop operation, which is a unique highlight of the VAiron Sinter automation solution. A screen of the user interface of the raw mix calculation model is shown in Fig. 4.

3.2.1.1 Calculation of material ratios (backward calculation)

The aim of this calculation mode is to calculate the set-points for the raw material system (material ratios) in order to reach the chemical sinter composition with the desired results.

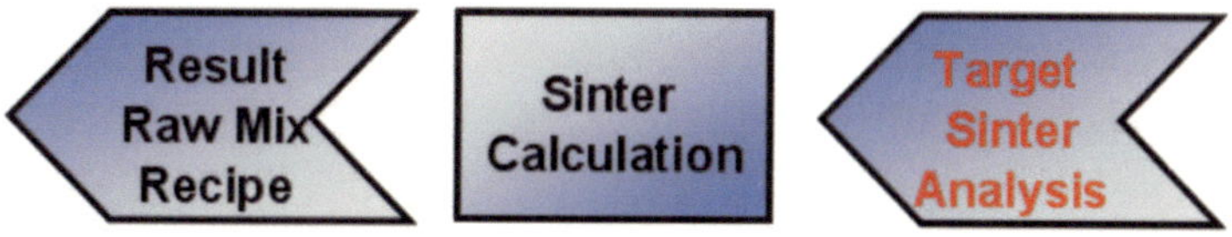

Fig. 5. Backward Calculation Procedure.

Several calculation options are available:

- To aim at a certain basicity
- To aim at a certain Fe_{tot}
- To aim at a certain SiO_2
- To aim at a certain MgO
- Calculation of the set-point for fuel materials

The advantage of the backward calculation (Fig. 5) is to reach and keep the product quality as stable as possible and optimize the material costs.

3.2.1.2 Calculation of sinter composition (forward calculation)

The target of this calculation is the opposite of the calculation of the raw material ratios, i.e. to calculate the chemical analysis of sinter using the composition of the raw materials as an input.

Using this calculation option, the model calculates the theoretic chemical composition of the sinter product based on fixed material ratios input by process engineers or operators.

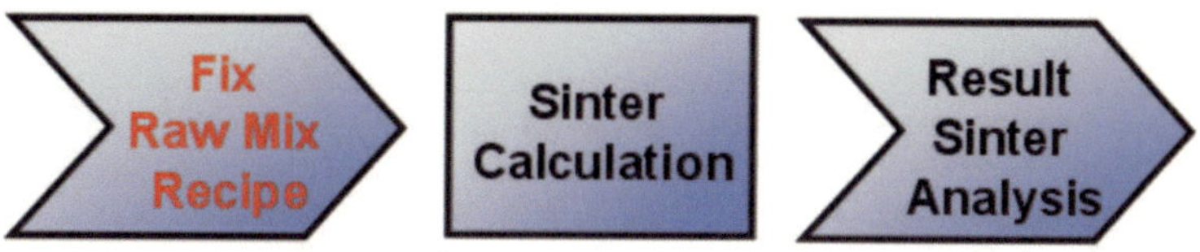

Fig. 6. Forward Calculation Procedure.

3.2.1.3 Online calculation of sinter composition (forward calculation)

This calculation is identical to the previously described calculation of the sinter composition with the difference that the actual charged material ratios from Level 1 are used as input for the calculation. Therefore, the model is started automatically in the background when a new Level 1 recipe is detected and it calculates the actual sinter composition using the actual raw mix data.

3.2.2 Stacking plan for blending ore bed

The model calculates a stacking plan for blending ore beds based on the raw mix composition calculated by the corresponding raw mix calculation. After considering the availability of materials and intermediate bunkers, the stacking process is organized in several stages. The model calculates the material flow rate during stacking so that a homogeneous blending ore bed with constant chemical properties is achieved. The stacking process is monitored and deviations from the plan are compensated automatically. A screen of the stack plan model for the blending ore bed user interface is shown in Fig. 7.

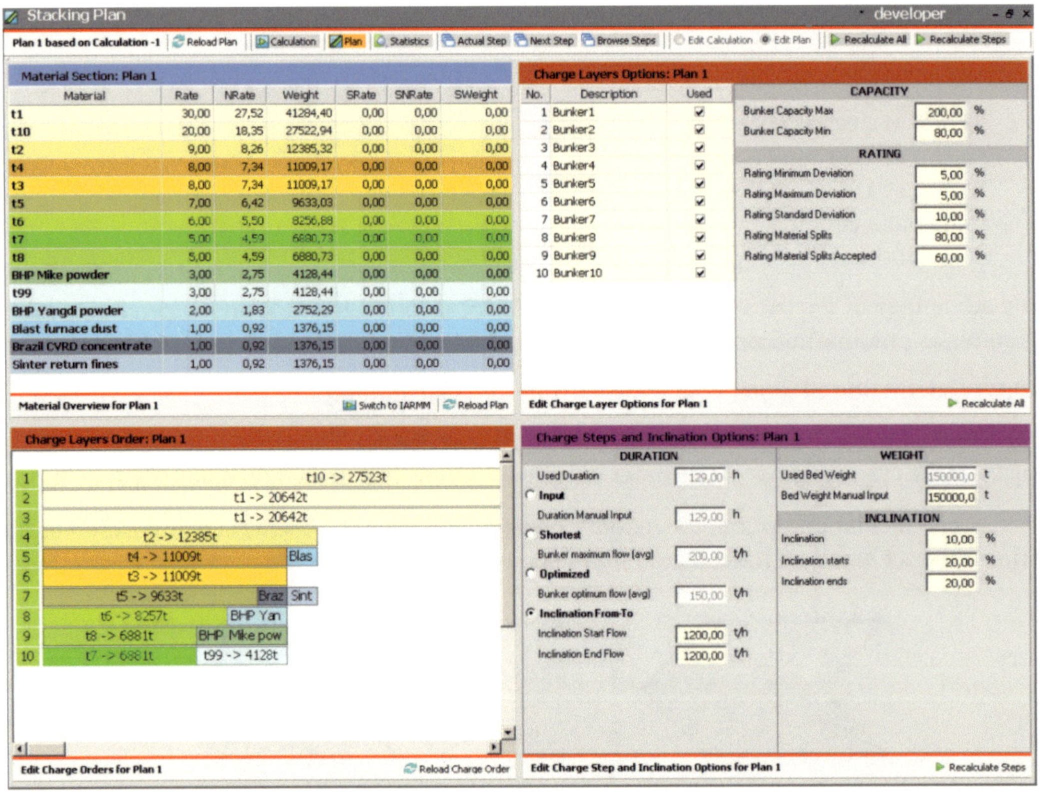

Fig. 7. Stacking Plan User Interface.

3.2.3 Blending ore bed distribution model

This model simulates the 3D geometry of the blending ore bed by calculating the volume of the material mixture per stacking step. For this calculation, the bulk densities and the angles of repose for each material type are required (Fig. 8). Furthermore, the spatial distribution of analysis data such as Fe and S are calculated. In the offline mode, the model calculates the geometry of the bed based on the stacking plan. In the online mode, the model builds up the bed using actual material data, the exact position of the charging device, the brands of the materials and the material quantities.

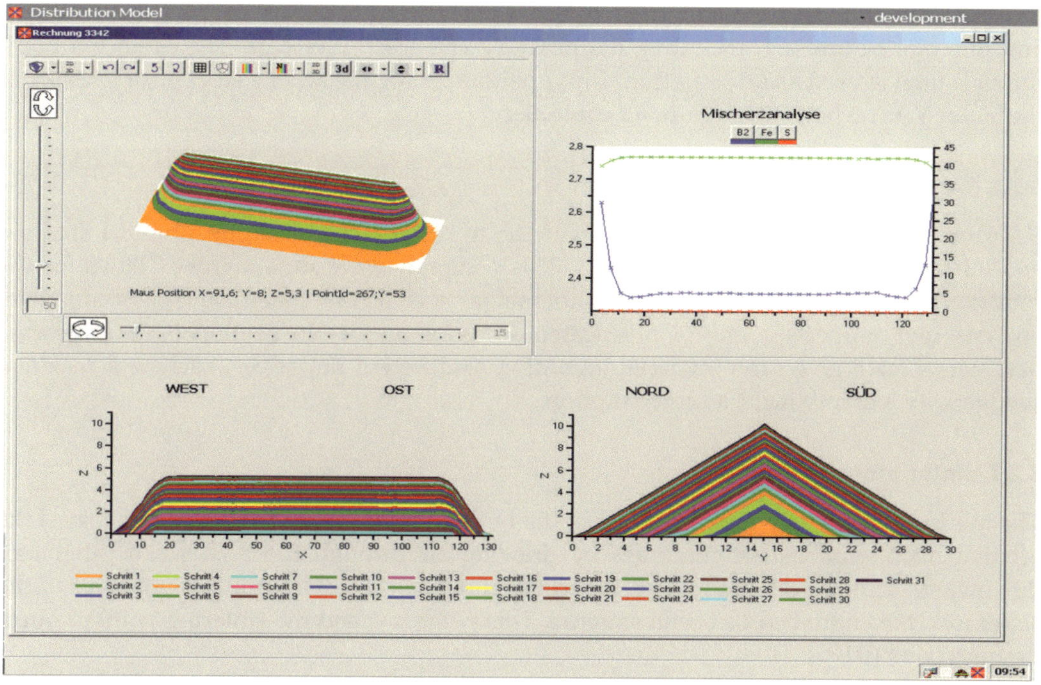

Fig. 8. Visualization of Ore Blending Bed Distribution Model.

3.2.4 Sinter process supervision models

In addition to the complex process models that are subsequently described, many auxiliary calculations are performed. These calculations include filters for suppressing short-term fluctuations. Examples are:

- Raw mix permeability is derived from ignition hood data, taking into account the pressure drop, bed height and waste-gas flow under the ignition hood
- Moisture calculation of the raw mix
- Average particle size of the raw materials
- The harmonic diameter of the sinter product (calculated from the grain-size distribution of the sinter analysis) as an important indicator of sinter quality[2]
- The actual burn-through point position which has a major influence on the control of the sinter strand velocity

A total of approximately 700 different model values are calculated.

---

[2]A particle size distribution with mass fractions $x_1, x_2, x_3, \ldots x_n$ corresponding to maximum diameters (measured as screen size) $d_1, d_2, \ldots d_n$ has a harmonic diameter
$D_h := 1 / \sum (x_i/d_i)$. The harmonic diameter is a good indicator of the permeability of the material (i.e., sinter in the blast furnace). A low fraction of material with small diameters leads to high D_h and thus to good permeability.

3.2.5 Burn-through time prediction model

The model predicts the dynamic behaviour of the sintering based on the process conditions and raw mix parameters, including permeability and waste-gas data. The predicted burn-through time is used as an important input parameter for the advanced control strategy of the Sinter VAiron burn-through point controller.

3.2.6 Productivity analysis tool

VAironment allows for the long-term archiving of recipes, chemical and physical analyses and all kinds of measured process data. This comprehensive data archive allows for the retrospective analysis of best process conditions for specific raw materials. The productivity analysis tool supports highly sophisticated search strategies in finding optimal process parameters for a given raw material according to different objectives, such as a maximal productivity and minimal fuel consumption.

3.2.7 Sinter process model

The top layer of the raw mix is ignited in the ignition hood. After the sinter mix leaves the ignition hood, combustion continues by drawing air through the bed which progresses downwards through the entire bed. When the combustion reaches the bottom layer of the sinter mix, the entire bed has been sintered. This point is called the sintering point or burn-through point (BTP).

Proper control of the sinter strand speed aims at positioning the BTP close to the end of the strand. If the BTP is situated before that ideal position, the area after the BTP is only used for cooling the sinter. This leads to a diminution of the active sintering area and a productivity decrease. If it is not reached within the sinter strand, un-sintered sinter mix is discharged and has to be recycled as return fines – leading to a productivity decrease. Furthermore, this results in poor sinter quality. The position of the BTP is measured by thermocouples installed in the last suction boxes and is characterized by the maximum value of the exhaust gas temperature detected by the thermocouples within these suction boxes.

A paramount control goal in the sintering process is that the material must be completely sintered by the time it reaches the end of the sinter strand (minimized return fines) and that the BTP is as close as possible to the ideal position for maximum productivity.

The gas flow through the sinter strand is a function of the permeability of the raw mix. As the total sintering time depends on the total gas flow, a higher permeability will obviously lead to shorter burn-through times. However, it is also clear that a higher gas flow through one section of the sinter strand will slightly reduce the gas flow through the material in other zones along the sinter strand. Taking this into consideration, a permeability-based simulation of the sintering process can thus be applied for an improved sinter-process control to achieve higher productivity. This solution approach was implemented in the sinter plant control system.

For example, the pronounced rise in the BTP curve beginning at time interval 43 of Fig. 9 is a consequence of an increased permeability of the raw mix at this position. The temporary BTP drop immediately preceding this rise is a result of the reduced gas flow through the raw mix, as explained above. As soon as the permeability falls, the BTP curve also drops accordingly (time interval 54). The inexact response correlation between the flat trend of the permeability curve and the irregular trend of the BTP can be explained as being the result of the accumulated nonlinear effects of varying gas flows through different sections of the sinter strand on the BTP. On the right-hand side of the diagram (time intervals 80 to 113), the inverse situation for a reduced permeability is shown.

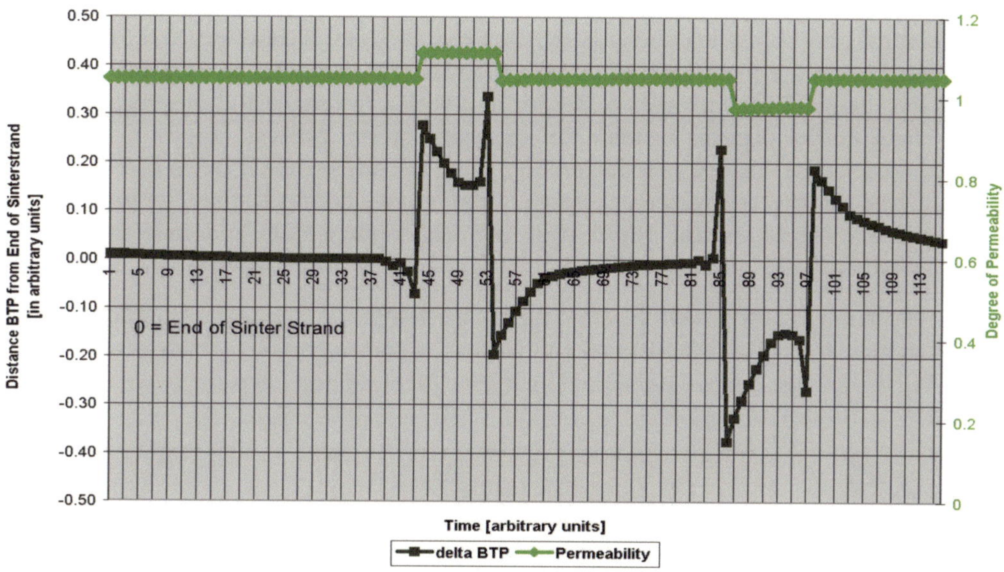

Fig. 9. Illustration of Simulated Dependency Between Permeability of Raw Mix and Distance of BTP from End of Sinter Strand; Plotted as Function Of Time (measured in arbitrary units).

If the sinter is not completely burned through before reaching the end of the strand, a decrease of the sinter strand speed is the logical control action. Choosing the proper speed reduction for the sinter at the end of the strand will result in an increased duration of the material spent at the beginning of the strand that is longer than ideal. This sinter will be burned through before reaching the end of the sinter strand, again necessitating an increase in the strand speed. This, in turn, means that the BTP of the following material will again be too close to the end of the sinter strand. For conventional automation solutions, e.g., PID (Proportional Integral Differential)-based control, this effect tends to lead to BTP and thus strand-speed oscillations when trying to compensate for the fluctuations. The sintering process model predicts the Burn-Through Time (BTT) as an indicator for the dynamic behavior of sintering, based on the process conditions and raw mix parameters, e.g., the material permeability. The compiled prediction of the BTT for discrete sinter strand segments is one of the important starting points for the calculation of the optimum sinter-strand speed by the Expert System.

3.3 VAiron sinter expert system

One of the most important factors for the control of the sinter process is to make operation of the sinter plant as smooth and steady as possible. The process can be disturbed, however, by changes in the properties of charged materials, failures in the process, human factors, process conditions, etc. Delays in corrective measures compensating for the interference factors may vary from minutes to hours, days and weeks. Correct timing of control actions and anticipation of disturbances is of utmost importance for maintaining high production rates and low production costs.

The knowledge of experienced sinter process engineers and operators on the process, the cause and effect relationships of process disturbances, metallurgical know-how, and the adopted control philosophy is modeled into the expert system. It monitors and forecasts the process status, gives alarms in case of process disturbances, suggests control measures, and describes the changes in the process in the form of verbal messages and graphical displays. With the help of the experts system, the expertise of the process control personnel is improved, the process control practice among the different shifts becomes more uniform, and, on the basis of forecasting, the operation of the process becomes smoother as compared to conventional process control. Monitoring of measurement data and indices based on those measurements becomes more efficient with the help of the expert system.

Therefore the sinter expert system has two main objectives. The first is the situation analysis of the phenomena called *diagnosis* and the second is the *therapy* in which proposals are presented to the process control personnel in order to achieve and keep stable conditions.

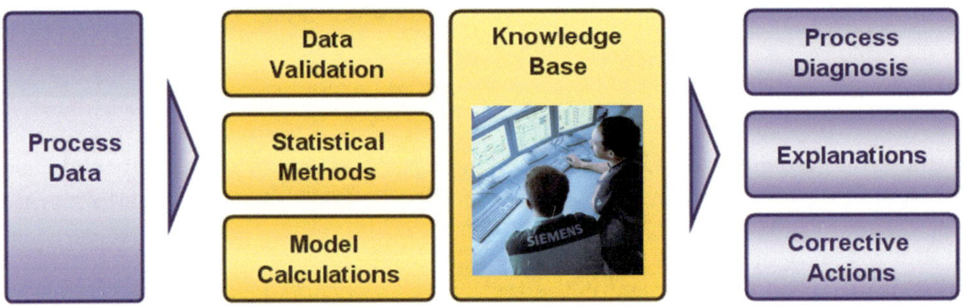

Fig. 10. Structure of VAiron Expert System.

The sinter expert system studies the occurrences of the phenomena around the sinter process by means of various technical calculations and makes conclusions based on those calculations. The calculations are based on results from process measurements.

The basic structure is shown in Fig. 10. The system diagnoses the overall sinter plant status and previous sintering conditions. The tasks and functions of the expert system are outlined in the following sections.

3.3.1 Life-phases of the expert system

At delivery time the system is prepared according to the rules defined by process experts of Siemens VAI and the customer. Experience gathered during the commissioning phase (with real time process data) will help to fine-tune the system. The expert system has to be maintained and enhanced during operation by the customer's personnel after take over. Thus the life-time of the expert system may be separated into three distinct phases:

Phase	Responsible	Description
Development of base system (based on Siemens VAI's experience and customer's know-how)	SVAI	Implementation of diagnosis structure, implementation of described diagnoses and actions, implementation of explanation capability, implementation of interfaces
Fine-tuning during commissioning (based on operational data)	SVAI / Customer	Fine-tuning of described diagnoses and actions
Maintenance and further enhancements (based on operational data)	Customer	Further adaptations, reaction on equipment changes, implementation of new diagnoses and/or actions

Table 1. Life-phases of the expert system.

Modifications of the expert system can be done by the customer's personnel according to the following topics:

- Fine-tuning of delivered diagnoses and rules - As described above the expert system has to be maintained throughout its lifetime. The simplest form of maintenance is fine tuning of the delivered application, i.e. modifications of existing but no addition of new parts. Customer's personnel may change tuning parameters stored outside the expert system.
- Addition of other diagnoses - If the expert system shall increase its diagnosing capability, one has to add new diagnoses.
- Addition of other therapies - If new suggestions of corrective actions shall be provided by the expert system, one has to add new rules.

The VAiron Sinter expert system includes a Metallurgical Model Toolbox and its own scripting language. This toolbox enables the modification of the logics of existing diagnosis and rules and the creation of new diagnosis and rules based on the specific requirements of operation.

3.3.2 Sinter plant control HMI – Expert system

Various process model results, control trends as well as the diagnoses are visualized in this HMI. It offers the following functionality:

- Checking of the performance of the various models using dedicated screens, i.e. tab folders in the Sinter Plant Control HMI.
- Modification of parameters for the individual models.
- Switching between closed-loop and semi-automatic mode for the individual controls.

The expert system HMI looks as follows:

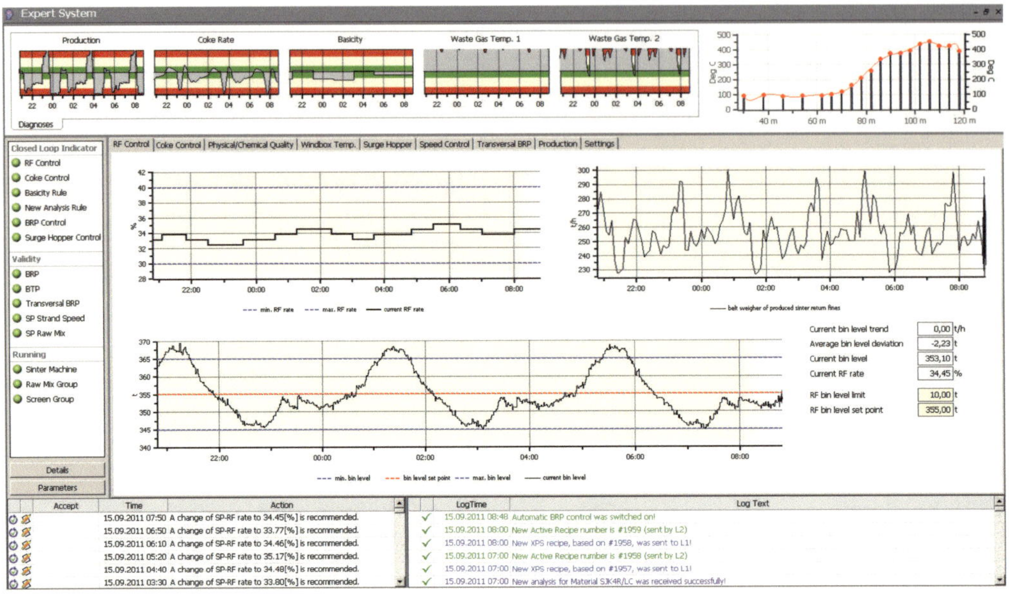

Fig. 11. Sinter Expert System HMI.

3.3.3 Diagnoses

The expert system studies the occurrence of phenomena in the sinter process using a variety of technical calculations and it draws conclusions derived from them. The calculations are based on a large amount of process measurement and analyses data that is collected continuously.

The following standard diagnoses are provided by the expert system:

Diagnosis	Description
Production	A moving average of the sinter production over a certain period of time (e.g. 1 hour) compared to the target sinter production.
Fuel Consumption	Moving average of the fuel consumption over a dedicated period of time (e.g. 1 hour) compared to target fuel consumption.
Chemical Quality	The basicity of the last received chemical sinter analysis is compared to the target basicity value.
Physical Quality	The harmonic diameter of the last received physical analysis is compared to the target value.
Environment	A moving average of the components of the waste gas analysis over a certain period of time (e.g. 1 hour) is compared to their target values.
Waste Gas Temperature	Moving average of temperature measurements in the main duct before the E.P. over a certain time period (e.g. 1/2 hours) is compared to the target value.

Table 2. Provided Expert System diagnoses.

3.3.4 Therapy

The expert system provides the following two types of user notifications regarding the therapy to achieve and keep a stable and smooth sinter process:

- Plausibility checks on measurements
- Corrective actions

3.3.4.1 Plausibility checks on measurements

A check action is created and presented to the user for all process variables that were found to be initially missing or invalid.

No further checks on the equipment are provided. This means that there are no further examinations, whether the measuring device is really faulty or not.

The check actions are given in textual form like

- The temperature of thermocouple in wind box #13 is unusual and therefore suspicious
- The analysis deviation (100%-∑elements) of the Sinter analysis exceeds 10%. Please check the analysis.

3.3.4.2 Corrective actions

Corrective actions are proposals for the operating personnel to change some process parameters (set-points). The expert system suggests at the same time one or more corrective actions out of a set of possible ones. Some of the corrective actions are provided qualitatively, that means the expert system suggests increasing or decreasing something instead of giving exact values to the user. Others are provided quantitatively, that means the new set-point is provided by the expert system.

Internally, the process of suggesting a corrective action is functionally divided into three groups based on the respective objectives as shown below:

- Situation Analysis (to judge the kind of process variations that have occurred)
- Phenomenon Recognition (to judge the kind of phenomenon expressed by that variation)
- Action Determination (to judge the action against the phenomenon)

Selected corrective actions have to be acknowledged by the operators. This is especially necessary if the expert system is in semi-automatic mode. The operators can enter a reason if they do not follow the expert system's suggestions and this action is suppressed until its status changes. This gives important information for tuning of the expert system.

The action execution can be separated into two operation modes which can be set for each rule and control individually as follows:

- Semi-Automatic Mode
- Closed-Loop Mode

The current operation mode for each rule/control is displayed in the expert system HMI (section) and indicated by green and red lights (see Fig. 11). The distinction between semi-automatic and closed-loop mode is described in detail in the section below.

Semi-automatic mode

If a rule or control is in semi-automatic mode, a detailed description of all recommended changes is provided to the operator by the expert system. During a configurable period of time (usually 10 - 15 minutes) the operator has the possibility to accept or decline the suggestion.

In case of rejecting the suggestion no further action will be executed. Additionally an input field is provided to key in a reason for refusing this suggestion.

If the operator has the opinion that the suggested set-point/recipe change is necessary to keep smooth process conditions, he has the possibility to accept the change. Afterwards the expert system is executing the changes and sends the new set-points/raw mix recipe to the Level 1 system automatically.

In case of neither rejecting nor accepting the recommended set-point/recipe change during the configurable period of time, the expert system will automatically reject the recommendation and no further actions will be executed as well.

Closed loop mode

For the recommendations of rules and controls switched to Level 2 closed loop mode a detailed description of necessary changes is provided as well. The pending time, in which the operator can accept or decline this suggestion, is also configurable (usually 5 - 10 minutes).

The only significant difference to the semi-automatic mode is the behavior after expiration of the configurable period of time without an operator action. In the closed loop case, contrary to the semi-automatic mode, the expert system will automatically accept the recommended changes and sends the new set-points/recipe to the Level 1 system.

3.3.4.3 Controls

Controls are actions that are executed continuously (e.g. every minute) for fast reactions to keep the process stable and usually run silently in the background. They are typically switched on or off on Level 1. Controls may also use a validity flag indicating its state to Level 1.

The controls shown in the following table are typically included in the expert system.

Corrective Action	Description
Change position of feeder gates	To ensure a homogeneous flame front in the transverse direction of the sinter strand, the transversal burn-through deviation control adjusts the packing degree. This leads to a constant burn-through of the sinter in transverse direction. The advantages for the sinter process are fewer and more stable sinter return fines.
Change sinter machine strand speed	The burn-through-point control maintains a target of the burn-through-position. This is achieved by modifying the speed of the sinter strand in accordance with the preset burn-through point position. The effect of this control leads to a maximized possible production of the used raw mix composition.

Table 3. Expert System Controls.

3.3.4.4 Rules

Rules are actions that are suggested at specific events (e.g. new sinter analysis) or they are triggered periodically. Since rules are slower controls or major changes in plant operation, a rule is always explained textually. So an operator can decide if he wants to follow the rule or not. In case of switching a rule to closed loop mode, it will be processed automatically after a certain period of time if the operator does not reject the suggestion with the expert system user interface.

The rules shown in the following table are evaluated by the expert system.

Corrective Action	Description
Change Basicity	Sinter properties represented by chemical parameters such as the basicity (CaO/SiO2) have to be kept within an acceptable deviation from the preset target values. This control loop adjusts the raw mix composition in order to maintain the target values.
Re-Calculate current Recipe	In case of a new chemical analysis received from the laboratory for a material which is currently used in the active recipe, a new sinter calculation is performed.
Change sinter return fines	A harmonized sinter plant return fine bin level over a long term is the objective of the return fine control. The return fine consumption in the raw mix composition is adjusted to keep the return fine bin level within a range.
Change Coke for raw mix	In order to keep the FeO content of the sinter within an acceptable range, the coke control stabilizes the sinter return fine balance. Therefore, the control modifies the coke consumption in the raw mix composition in accordance with the sinter return fine balance.

Table 4. Expert System Rules.

3.3.5 Sinter plant productivity control

A very important part of the expert system represent the two controls described in section 3.3.4.3, namely the *Burn-Through-Point* and the *Transversal Burn-Through-Point Controller*. These controls improve the overall plant productivity.

There are many indicators for the strand-speed control with different precision. Some of these are available at an early stage in the process (e.g. the permeability), others only with long time delays after the process on the strand has been finished (e.g. the harmonic diameter). Generally, the information attained at a later stage is more precise than that attained early on.

The fundamental idea was therefore to use the early information to control the processes and to use the information that is attained at a later stage to self-tune the control system. With these two independent sources of information it is possible to achieve high control accuracy despite fast corrective actions. A general overview of the main parameters that affect productivity control is shown in Fig. 12. Since the availability and reliability of the listed data differs from plant to plant, the expert system can be based on individually selected entry data of the respective plant.

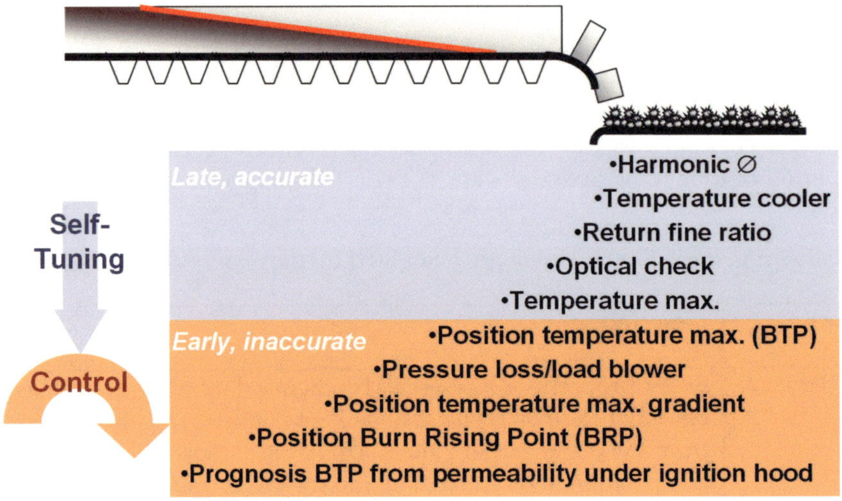

Fig. 12. Overview of Sinter Plant Productivity Control.

In the longitudinal BTP optimization described above, the objective is to obtain an average BTP position that is optimally distanced from the strand end. Full utilization of the surface of the sinter strand, however, can only be achieved when, at the same time, the flame front also reaches the lowest layer across the entire width of the strand (in a transversal direction). This is obtained through the transverse burn-through point control (Fig. 13). Here, feedback on the burn-through point is derived from the temperature conditions in a transverse direction from the last suction boxes and corrective measures are then executed online directly through proper adjustment of the angle of flaps near the drum feeder. In this manner, a uniform flame front can be achieved.

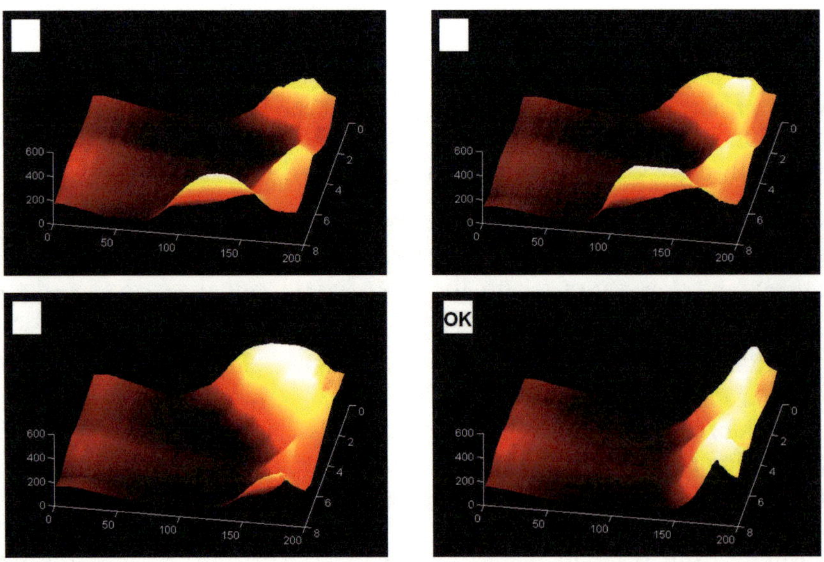

Fig. 13. Transverse Burn-Through Point Control

4. Savings and benefits

4.1 Advanced data management

From a broad spectrum of raw data sources, the data acquisition function pre-processes the plant data before storing it in the plant database. This database is of key importance to the advanced process optimization. The following data are collected from the process and connected systems:

- Continuously measured process data from the Level 1 system
- Amount of material charged
- Plant status data (runtime, shut-down, special process situations, etc.)
- Chemical and/or physical laboratory analyses data of all raw-materials and sinter
- Active raw mix recipes

Additionally, applications to visualize the above mentioned collected data are provided:

- Tag-Visualization program for graphical representation of any kind of time-based data in the database
- Lab-Browser application for material analyses visualization and evaluation
- Reporting system allowing for cyclical (e.g. daily) or on-demand report generation

The system can serve as a link between different automation levels in the customer's company: it is connected to the aggregate's Level 1 automation and it can be connected to the plant wide network. Therefore it can send production and consumption data as well as important process data to a Level 3 system.

4.2 Increase of operator know how

The expert system generates textual explanations for its diagnoses and suggestions. In combination with the graphical information provided by the system (see Section 3.3.2), the operator can understand the actual situation of the plant in detail. Using these facilities, the operator can permanently learn about the process and background of the knowledge system. In consequence, the system will improve the skills of the operational personnel.

4.3 Smooth plant operation

The expert system checks a number of process state indicators in a typical time cycle of five minutes:

- Several hundred measurement points from the Level-1 automation, and
- Related model calculations for internal process states which cannot be measured directly

Deviations from optimal process conditions can therefore be detected early. Small counter actions are sufficient to correct the process conditions at this early stage. Even experienced human operators are unable to cope with this flood of information and will detect such deviations later than the expert system.

In consequence, the main difference between manual operation and operation supported by the expert system is that the latter is characterized by more frequent, but smaller control actions. The resulting smooth operation of the sinter plant leads to:

- Higher availability of the sinter plant
- Longer overall lifetime of the sinter plant
- Reduced maintenance efforts and costs

4.4 Uniform operational philosophy

The VAiron expert system is customized for each individual plant where it is installed. In a first phase, the customer specific situation regarding raw materials, plant topology, equipment, etc. is analyzed. The specific rules are developed in cooperation between Siemens VAI specialists and experienced process engineers and operators of the customer. Specific customer operational philosophy is implemented instead of standard rules as a result of this cooperation during the engineering phase. During system commissioning the rules are fine-tuned together with the customer.

This approach has the following advantages:

- High acceptance of the system, because it reflects the internal operational philosophy
- The customer's operational philosophy of the most experienced personnel is followed 24 hours a day, 7 days a week
- Consistent sinter plant operation over all shifts, resulting again in smoother plant operation

4.5 Increased sinter plant productivity

The use of the Burn-Through-Point controller typically leads to an increase in the sinter plant productivity between 2% and 5%. An additional increase can be achieved if the flame front is uniformed along the full width of the sinter strand by means of the Transversal Burn-Through-Point Controller.

The usage of a higher percentage of sinter with stable quality in the blast furnace burden results in a further reduction of the blast furnace fuel consumption. Therefore, the increase of productivity of the sinter plant is a very important benefit of the expert system.

4.6 Reduced fuel consumption and stabilized sinter quality

The expert system ensures an economic fuel usage by keeping the return fines ratio at the optimal level. Whenever the average production of internal return fines deviates from the optimal value, the expert system corrects the process conditions (mainly the fuel addition) in order to compensate the deviation. Reduced fuel consumption and increased productivity are achieved by this control loop.

Stabilization of sinter quality is achieved by dedicated quality controllers considering incoming sinter analyses from laboratory and performing corrections of the raw mix recipe, if deviations in one of the following quality parameters from the target value are detected:

- Harmonic diameter of sinter
- Sinter basicity
- Sinter SiO_2 content
- Sinter MgO content
- Sinter total Fe content

Additionally, fluctuations in raw material analyses are detected and the new optimum raw mix composition is calculated immediately, downloaded to the Level 1 automation system, and executed there. Obviously this proactive compensation of raw material fluctuations is much faster than waiting to see effects in the produced sinter.

5. Summary and outlook

The sinter automation and optimization described in this chapter offers an integrated approach for ore preparation and sintering operations in one system, assuring optimal coordination of both plants. The application of the proven closed-loop expert system leads to transparent and reliable process control and shift-independent sinter quality at a high productivity level. The development of this system was an important step in the fulfillment of the vision of "fully automatic sinter plant operation".

Before the described VAiron Sinter automation system has been developed, an analogous automation package for blast furnaces was introduced by Siemens VAI in cooperation with voestalpine Stahl. These systems have in common that they optimize a single aggregate. If an iron making plant contains several blast furnaces and sinter plants, each system would optimize a single aggregate.

Optimal conditions for a group of aggregates differ in general from optimum conditions at each of the single aggregates. In consequence, the next step of development is the VAiron Productivity Control System, which consists of a superordinated expert system considering the whole iron making plant rather then single aggregates. The system considers the

- Coke oven plants
- Sinter plants
- Sinter stock yard
- Blast furnaces

of the iron making plant and coordinates all these aggregates. The system executes its suggestions by sending set-points to the individual Level 2 systems of the single aggregates. Up to now, a prototype of the VAiron Productivity Control System is installed at voestalpine Stahl with promising results.

6. References

Klinger, A.; Kronberger, T., Schaler, M., Schürz, B. & Stohl, K. (2010). Expert Systems: Chapt. 7 *Expert Systems Controlling the Iron Making Process in Closed Loop Operation*, InTech, ISBN 978-953-307-032-2, Vukovar, Croatia

Bettinger, D., Schürz, B., Stohl, K., Widi, M., Ehler, W. & Zwittag, E. (2008). Get More From Your Ore, metals & mining, 1/2008

Bettinger, D., Stohl, K., Schaler, M. & Matschullat, T. (2006). Automation Systems for Sustainable Energy Management, Proceedings of the Iron & Steelmaking Conference, Oct 9-10, 2006, Design Center Linz, Austria

Fan, X.H., Long, H.M., Wang, Y., Chen, X.L. & Jiang, T. (2006). Application of expert system for controlling sinter chemical composition, Ironmaking and Steelmaking, 3/2006

Sun Wendong, Bettinger, D., Straka, G. & Stohl, K. (2002). Sinter Plant Automation on a New Level!, *Proceedings of AISE Annual Convention*, Nashville, USA, Sep 30 - Oct 2, 2002

2

Modeling Sintering Process of Iron Ore

Jose Adilson de Castro
Graduate Program on Metallurgical Engineering -Federal Fluminense University
Brazil

1. Introduction

In this chapter, a methodology for simulating the sintering process of iron ore is presented. In order to study the process parameters and inner phenomena, a mathematical model based on transport equations of momentum energy and chemical species is proposed and the numerical implementations is discussed. The model is applied to simulate the impact of new technologies based on alternative sources of fuels, new operational design and the environmental concerns related to the new raw materials is estimated. The model is validated with pot test experiments and applied to industrial scale of sinter strand machine.

In the integrated steel industries the sintering process plays an important role furnishing raw material to the blast furnace. From the point of view of natural resources, the sintering process is key technology that allows recycling by products or dust produced within the steel plant and other facilities. However, the amount and quality of energy requirements have continuously changed and up to date is mainly based on nonrenewable energies resources such as coal, anthracite and oil. Several attempts to use new energy resources have been carried out and new technologies are continuously developed. The process is complex involving various physical and chemical phenomena. The raw materials used can vary to a wide extent, from iron ore to dust recycling and fluxing agents. The natural resources of iron ores varies widely depending on the mineral composition and mining technology applied to produce the sinter feed. Therefore, it is of special importance to developed comprehensive tools to draw good decision on what kind of available raw materials and their blending will meet the sinter quality requirements to use in the blast furnace or other reducing process(Castro et. al., 2000, 2011). The process is carried out on a moving strand, where a previously prepared mixture of iron ore (sinter feed), fine coke or anthracite(fuel), limestone, other additives and water is continuously charged together with returned sinter from the undersize of a sieving process to form a thick bed of approximately 800 millimeters. The strand width and length depends on the capacity of the machine and varies for each steel works. Along the first meters of the strand the charge is ignited by burners of natural gas or coke oven gas. The hot gas, generated by the combustion with air, is then sucked in through the packed bed from the wind boxes equipped with blowers placed below the grate. The strand can vary from small to large machines with the area and bed height compatible with the auxiliary equipments used for suction of the outlet gas. The area of the strand and the suction power together with the bed permeability determines the maximum speed and hence, the productivity of the process. However, depending on the selected operational parameters and raw materials the quality of the sinter produced can vary widely and can strongly effects the subsequent blast furnace process operation. The combustion of fines coke or other carbonaceous materials begins at the top of the layers, and as it moves, a relative narrow band of combustion front moves down through the bed. Several chemical reactions and phase transformations take place within the bed, part of the materials melt when the local temperature reaches the melting temperature and as it moves, the re-solidification phenomenon and phase transformations occur with considerable changes on phases composition and thermo physical properties (Jeon et al. 2010, Li et al. 2010, Nakano, 2005, 2010, Kasai, 2005, 2010, akiyama et al. 1993).

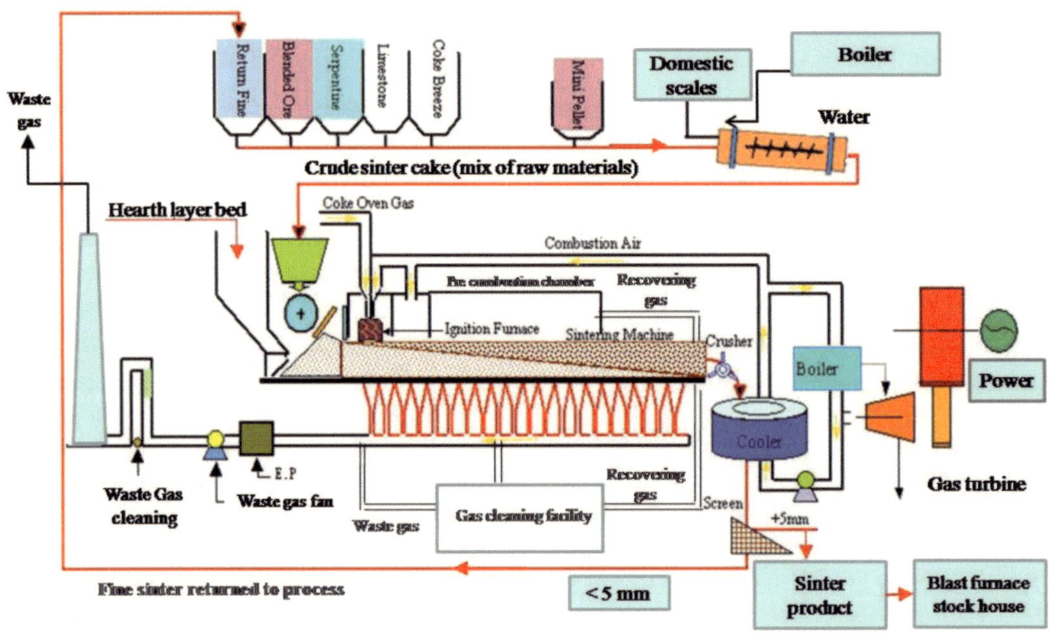

Fig. 1. Schematic overview of the sinter facilities of an integrated steel works with new facilities proposed for recycling waste gas.

The partial melting and diffusion within the materials causes the particle to agglomerate forming a continuous porous sinter cake. In general, the hot gas produced during sintering can also be re-circulated for better thermal efficiency. A schematic overview of the sinter machine with recycling gas concept is shown in **Fig. 1**. In addition, in this machine concept , part of the process gas is reutilized in a pre-combustion chamber with natural gas with ignition burners to keep heat supplying on the surface of the sinter strand and enlarge the heat affected zone. The physicochemical and thermal phenomena involved in this process are complex and numerous(Nakano et al, 2010, Kamijo et al. 2009, Kasai et al. 2008, 2005). Special mention is made to the phenomena of gas flow through the porous bed, gas-solid heat transfer, drying, vaporization and several chemical reactions and phase transformations (Umekage et al. 2009, Cieplik, 2003, Cores et al. 2010).

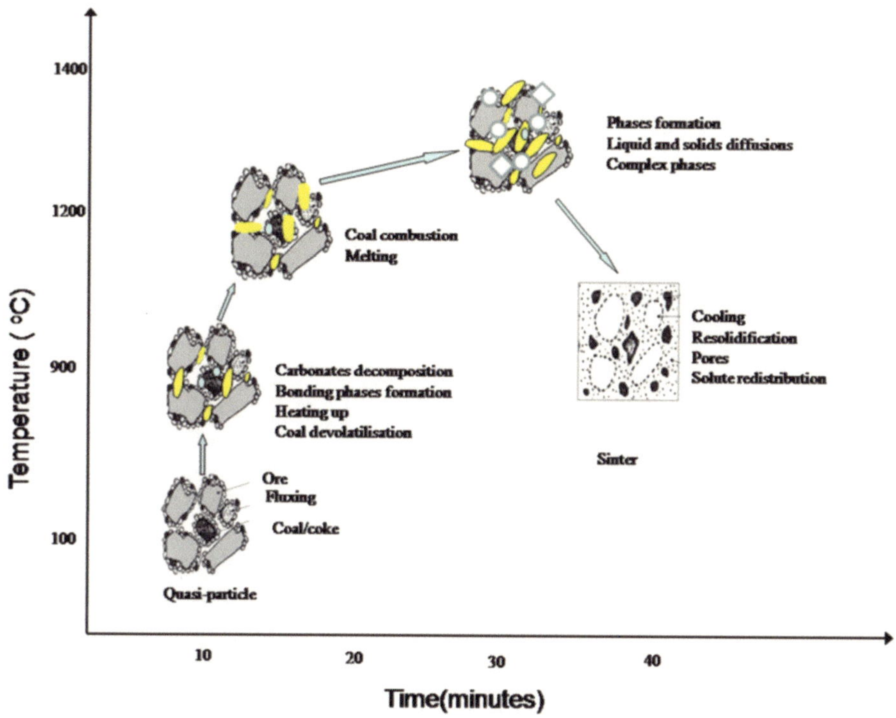

Fig. 2. Thermal cycle of the materials in the sinter strand.

Several attempts have been done to predict the final properties of the sinter product (Water et al. 1989, Kasai et al. 1989, 1991). The most important parameters are reducibility, degradability and the size distribution which influences strongly the sinter performance within the blast furnace. Waters and co-workers(Waters et al. 1989), developed a mathematical model to predict the final size distribution of the sinter product, however, as the authors pointed out, the model did not considered the kinetics of the sintering phenomena, which strongly affect the final size distribution, reducibility and degradation within the blast furnace. Kasai et al (Kasai et al. 1989, 1991) investigated the influence of the sinter structure into the macroscopic sinter properties. In their work a detailed explanation of the sintering mechanism and particles interaction is presented to clarify the bonding forces. The authors concluded that the void fraction and specific surface area are the main parameters influencing the sinter strength. They also concluded that the significant driving forces for structural changes in the sinter are compressive and capillary ones. Akiyama et al (Akiyama et al. 1992) investigated the heat transfer properties under the sinter bed conditions and established empirical correlations for the material conductivity. The mechanism of the oxidation and bonding phase formation in the sintering process conditions were studied by Yang et al (Yang et al. 1997). Yamaoka and Kawaguchi(Yamaoka & Kawaguchi, 2005) discussed 3D variations on sinter properties produced on a pot apparatus experimental facility and presented a mathematical model based on transport phenomena to simulate the experimental conditions and draw some correlations to predict sinter properties based on model variables and measurements. However, there are few comprehensive mathematical models describing the sintering process in an industrial machine such as the usual Dwight-Lloyd process. Mitterlehner et al (Mitterlehner et al.,2004), presented a 1-D mathematical model of the sinter strand focusing on the speed of the sintering front. Cumming and Thurnlby and Nath et al(Cumming & Thurnlby 1990 and Nath et al., 1996), developed a 2-D mathematical model based on transport equations, however, their analysis considered a few chemical reactions and the rate of phase transformations were simplified. Therefore, a comprehensive mathematical model able to describe the chemical reactions coupled with momentum, energy and species transport has yet to be considered. In the present work, a three dimensional mathematical model of the sinter strand is developed based on the multiphase multi-component concept and detailed interactions between the gas and solid phases are formulated: The main features of the model are as follows: a) dynamic interaction of the gas mixture with the solids; b) overall heat transfer of all phases which accounts for convection and radiation phenomena; c) kinetics of vaporization and condensation of water; d) decomposition of carbonates; e) reduction and oxidation of the iron bearing materials; f) coke combustion and gasification; g) volatile matter evolution; h) shrinkage of the packed bed; h) partial melt and re-solidification of the solids and i) phase changes to form alumina-calcium- silicates. **Figure 2** shows schematically a typical thermal cycle of the materials within the sinter strand and indicates main phenomena that occur along the sinter strand. In the present work, a comprehensive mathematical model to describe the phenomena within the sinter strand is presented. The present model differs significantly from the former ones due to the concept of multiple and coupled phenomena treatment, three -dimensional treatment of the sinter strand and detailed mechanism of chemical reactions involved in the process(Castro et al. 2005, Yamaoka et al 2005). The interphase interactions are considered via semi empirical sub-models for the momentum transfer, energy exchange due to chemical reactions, heat conduction, convection and radiation.

2. Model features

2.1 Conservation equations

In order to analyze the sintering process of an industrial strand machine, a multiphase, multi-component, three-dimensional mathematical model is proposed. The model considers the phases interacting simultaneously and the chemical species of each phase is calculated based on the chemical species conservation equations. The model concept and phase interactions are shown in **Fig. 3**. The model is based on conservation equations for mass, momentum energy(Austin et al. 1997, Castro et al. 2000, 2001, 2002, 2011) and mass fraction of chemical species of gas, solid phases: sinter feed, fine sinter(returned fine sinter), coke breeze(or other solid fuel), scales(fines of steel plant), fluxes and limestone. The liquid phase is composed of melted and formed components in the liquid phase. The re-solidified phase comprises the liquids re-solidified and phases formed during the re-solidification process. The final sinter cake will be formed by a mixture of these materials and its quality will depend upon the final compositions and volume fractions of each of these materials(Cores et al. 2010, Lv et al. 2009, Hayashi et al.2009).

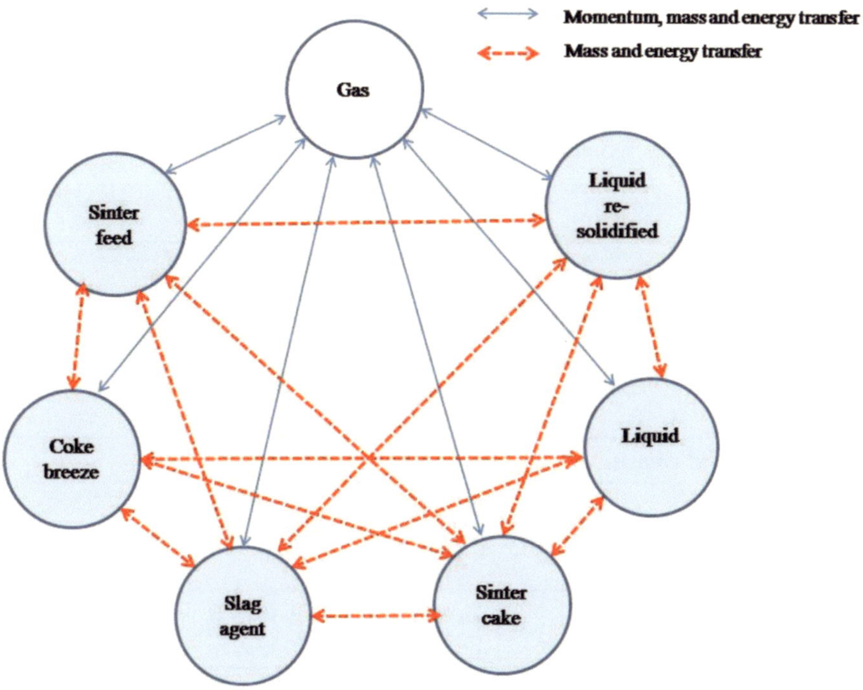

Fig. 3. Multiple phases considered in the present model.

In this model it is assumed that the liquid phase formed will move together with the remaining solid phase due to the viscosity, thus, equations for momentum transfer and enthalpy of the solids will account for this mixture of viscous liquid and solid materials. The equations for momentum, energy and chemical species are as follows:(Austin et al., 1997, Castro et al., 2000, 2001,2002, 2005).

Momentum:

$$\frac{\partial(\rho_i\varepsilon_i u_{i,j})}{\partial t}+\frac{\partial(\rho_i\varepsilon_i u_{i,k}u_{i,j})}{\partial x_k}=\frac{\partial}{\partial x_k}\left(\mu_i\frac{\partial u_{i,j}}{\partial x_k}\right)-\frac{\partial P_i}{\partial x_j}-F_j^{i-l} \tag{1}$$

Continuity:

$$\frac{\partial(\rho_i\xi_i)}{\partial t}+\frac{\partial(\rho_i\xi_i u_{i,k})}{\partial x_k}=\sum_{m=1}^{Nreacts} M_n r_m \tag{2}$$

Enthalpy balance:

$$\frac{\partial(\rho_i\varepsilon_i H_i)}{\partial t}+\frac{\partial(\rho_i\varepsilon_i u_{i,k}H_i)}{\partial x_k}=\frac{\partial}{\partial x_k}\left(\frac{k_i}{C_{pi}}\frac{\partial H_i}{\partial x_k}\right)+E^{i-l}+\sum_{m=1}^{Nreacts}\Delta H_m r_m \tag{3}$$

The chemical species are individually considered within the phase, for gas, or components of the solid or liquid phases as presented in Eq. 4.:

$$\frac{\partial(\rho_i\varepsilon_i\phi_n)}{\partial t}+\frac{\partial(\rho_i\varepsilon_i u_{i,k}\phi_n)}{\partial x_k}=\frac{\partial}{\partial x_k}\left(D_n^{eff}\frac{\partial \phi_n}{\partial x_k}\right)+\sum_{m=1}^{Nreacts} M_n r_m \tag{4}$$

Equations of the gas phase			
Gas	Momentum	$u_{1,g}$, $u_{2,g}$, $u_{3,g}$, P_g , ε_g	
	Energy	h_g	
	Chemical Species	N_2, O_2, CO, CO_2, H_2O, H_2, SiO, SO_2, CH_4, C_2H_6, C_3H_8, C_4H_{10}	
Equations of the solid phase			
Solid	Momentum	$u_{1,s}$, $u_{2,s}$, $u_{3,s}$, P_s , ε_s	
	Energy	h_s	
	Chemical Species	Fuels	C,Volatiles, H_2O, Al_2O_3, SiO_2, MnO, MgO, CaO, FeS, P_2O_5, K_2O, Na_2O, S_2
		Iron ore	Fe_2O_3, Fe_3O_4, FeO, Fe, H_2O, Al_2O_3, SiO_2, MnO, MgO, CaO, FeS, P_2O_5, K_2O, Na_2O
		Return Sinter (bed)	Fe_2O_3, Fe_3O_4, FeO, Fe, H_2O, Al_2O_3, SiO_2, MnO, MgO, CaO, FeS, P_2O_5, K_2O, Na_2O, $Ca_2Fe_3O_5$, Al_2MgO_4
		Scales	C, Volatiles, Fe_2O_3, Fe_3O_4, FeO, Fe, H_2O, Al_2O_3, SiO_2, MnO, MgO, CaO, FeS, P_2O_5, K_2O, Na_2O, $Fe_2Cl_6H_{12}O_6$, $Ca_2Fe_3O_5$, Al_2MgO_4
		Fused Materials	Fe_2O_3, Fe_3O_4, FeO, Fe, H_2O, Al_2O_3, SiO_2, MnO, MgO, CaO, FeS, P_2O_5, K_2O, Na_2O, $Ca_2Fe_3O_5$, Al_2MgO_4
		Fluxing agent	CaO, H_2O, Al_2O_3, SiO_2, MnO, MgO, TiO_2
		Sinter cake	Fe_2O_3, Fe_3O_4, FeO, Fe, H_2O, Al_2O_3, SiO_2, MnO, MgO, CaO, FeS, P_2O_5, K_2O, Na_2O, $Ca_2Fe_3O_5$, Al_2MgO_4

Table 1. Phases and chemical species considered in the model.

Where indexes i and l represent the phases, j and k are the indexes for coordinates component direction n is chemical species and m the indicator of the reactions, M is the molecular weight of the species, P is phase pressure, F is component of momentum interactions among the phases and r is the rate of chemical reactions. $\rho,\varepsilon,Cp,\ k\ and\ \Delta H$ are phase density, volume fractions, heat capacity, heat conductivity and heat due to chemical reactions, respectively. The quantity E^{i-l} is the heat transfer among the phases and accounts for convective and radiation heat transfer. The gas -solids momentum interactions are represented by F^{i-l} and detailed in the following section. The complete description and unit of these variable is presented in the list of variables and symbols. The chemical species for the solid and gas phases are presented in Table 1. Detailed chemical reactions and rate equations describing the in bed conditions of iron ore are found elsewhere(Austin et al., 1997, Castro et al., 2000, 2001,2002, 2005). Several authors have assessed particular phenomena and rate equations for interphase interactions and sintering process(Lv et. al., 2009, Kasai et. al, 2005, Jeon et al., 2010, Li et al.,2010, Nakano et al. 2005, 2009, 2010).

2.2 Momentum and energy transfer

The momentum transfer between the solid and gas are modeled based on the modified Ergun's equation as follows: (Castro et. al. 2005, Cumming et. al, 1990, Nath et al. 1997)

$$\vec{F}_g^s = \sum_m f_m \left[1.75\rho_g + \frac{150\mu_g}{\left|\vec{U}_g - \vec{U}_s\right|} \left(\frac{v_m}{(1-v_m) d_m \varphi_m} \right) \right] \left(\frac{v_m}{(1-v_m) d_m \varphi_m} \right) \left(\vec{U}_g - \vec{U}_s \right) \tag{5}$$

Where f is the phase component volume fraction and v is the phase component bulk void fractions, d is the average diameter of the phase component and φ is the shape factor (m=sinter feed, coke, limestone, mushy zone of liquid and solids, re-solidified and fines particles). $\vec{U}$ is the phase velocity vector, μ_g is gas dynamic viscosity and ρ_g the gas density given by the ideal gas state relationship. Each of these components has its own particle diameter, porosity, shape factor and density. The overall heat transfer coefficient between the gas and solid phases is given by the Ranz-Marshall equation modified by Akiyama et al for moving beds and incorporated together with the interfacial area to give the overall heat transfer of solid to gas phase and vice-versa(Akiyama et al. 1993, Castro et al. 2000, Austin et al. 1997, Castro et al. 2002, 2005, Yamaoka et al. 2005)

$$E^{g-s} = \sum_m f_m \frac{6(1-v_m)}{d_m \varphi_m} \frac{k_g}{\sum f_m (d_m \varphi_m)} \left[2 + 0.39 \left(\mathrm{Re}_{g-s} \right)^{1/2} \left(\mathrm{Pr}_g \right)^{1/3} \right] \left(T_g - T_s \right) \tag{6}$$

As shown in eq. 6, the gas-solid system inter-phase heat transfer is given by the product of the overall effective heat transfer coefficient, the interfacial area and the average temperature differences of the solid particles and gas phase. The average solids and gas temperatures used in Eq. 6 are calculated by solving **eq. (7)** for each phase, with the temperatures of each phase as incognita(Yamaoka et al. 2005, Castro et al. 2005, Nakano et al., 2010, Kamijo et al. 2009)

$$h_i = \int_n \varepsilon_i \left(C_p(T_i) \phi \right)_n dT_i \tag{7}$$

Where the quantities on the integral relation are averaged for each solid component and the component enthalpy is obtained by solving **eq. 3**, assuming that all solid and liquids components moves with the same velocity, although solid particles can assumes different temperatures depending on the heat exchanged with gas, chemical reactions and other particles.

2.3 Calculations of phase properties

The volume fractions occupied by each solid component are calculated based on empirical correlations solely dependent on the individual mean solid diameters as in eqs. 8 and 9.(Austin et al. 1997, Castro et al. 2005)

$$\varepsilon_{\text{coke}} = \left[0.153\log\left(d_{\text{coke}}\right) + 0.724\right] \tag{8}$$

$$\varepsilon_m = 0.403\left[100 d_m\right]^{0.14} \qquad (m = \text{sinter feed, sinter, fines, scales}) \tag{9}$$

The properties of the gas and solid phases are calculated considering the mixture rule based on the properties of the pure components. The density of the gas phase is calculated by using the ideal gas law.

$$\rho_{\text{g}} = \frac{P_{\text{g}}}{RT_{\text{g}}} \sum \phi_{j,\text{g}} M_j \tag{10}$$

Where j stands for gas species. Pure component viscosities are calculated from statistical mechanical theories as follows:(Reid et al. 1988, Bird et al., 1960)

$$\mu_{\text{g}} = 2.6693 \times 10^{-6} \frac{1}{\Omega_{\mu,j}} \sqrt{\frac{M_j T_{\text{g}}}{\sigma_j}} \tag{11}$$

With the parameters given by:

$$\Omega_{\mu,j} = \frac{1.16145}{\left(T^*\right)^{0.14874}} + \frac{0.52487}{\exp\left(0.77320 T^*\right)} + \frac{2.16178}{\exp\left(2.43787 T^*\right)} \tag{12}$$

$$T^* = \frac{k_{\text{Boltzmann}} T_{\text{g}}}{\epsilon_j} \tag{13}$$

Pure thermal conductivities are calculated using Eucken's polyatomic gas approximation (Wilke, 1950, Neufeld et al, 1972).

$$k_j = M_j \left(C_{P,j} + \frac{5R}{4M_j} \right) \tag{14}$$

The gas phase viscosity and thermal conductivity are calculated from pure components using Wilke's method (Wilke, 1950).

$$\lambda_g = \sum_{j \in g} \left[\frac{\gamma_{j,g} \lambda_j}{\sum_{jj \in g} \left(\gamma_{jj,g} \chi_{j,jj} \right)} \right] \quad (\lambda = \mu, k) \tag{15}$$

$$\gamma_{j,g} = \frac{\left(\phi_{j,g} / M_j \right)}{\sum_{jj \in g} \left(\phi_{jj,g} / M_j \right)} \tag{16}$$

$$\chi_{j,jj} = \left\{ 1 + \left(\lambda_j / \lambda_{jj} \right) \left(M_{jj} / M_j \right)^{1/4} \right\}^2 \tag{17}$$

The binary diffusivity of the gas species is calculated by:

$$D_{j,k}^{T} = 0.0018583 \times 10^{-4} \frac{T^{1/2} \sqrt{\left(\frac{1}{M_j} \right) + \left(\frac{1}{M_k} \right)}}{\left(P_g / 101325 \right) \sigma_{ave}^2 \Omega_{ave}} \tag{18}$$

And the parameters are calculated as follows:

$$\sigma_{ave} = 0.5 \left(\sigma_j + \sigma_k \right) \tag{19}$$

$$\Omega_{ave} = \frac{1.06036}{\left(T^* \right)^{0.15610}} + \frac{0.19300}{\exp\left(0.47635T^* \right)} + \frac{1.03587}{\exp\left(1.5299T^* \right)} + \frac{1.76474}{\exp\left(3.89411T^* \right)} \tag{20}$$

$$T^* = \frac{k_{Boltzmann} T_g}{\in_{ave}} \tag{21}$$

$$\in_{ave} = \sqrt{\in_j \in_k} \tag{22}$$

Where kboltzmann is the Boltzmann constant, σ and $\in$ are characteristics constants of the colliding gas species(Reid et al. 1988, Bird et al., 1960). The gas temperature is defined by Eq. 23 as a function of gas enthalpy and composition.

$$H_g = \sum_{j \in g} \phi_{j,g} \left[\Delta H_{j,g}^{298K} + \int_{298K}^{T_g} C_{p,j}(T) dT \right] \tag{23}$$

With

$$C_{P,j}(T) = a_j + b_j T + c_j T^{-2} \tag{24}$$

The solid properties are calculated based on the solid composition and the properties of the pure components. The pure component heat capacities are modeled by a polynomial function of the temperature as follows:

$$C_{P,k} = a_k + b_k T_s + \frac{c_k}{T_s^2} \tag{25}$$

$$C_{P,s} = \sum_{k \in s} C_{P,k} \phi_{k,s} \tag{26}$$

$$H_s = \sum_{k \in s} \phi_k \left[\Delta H_k^{298K} + \int_{298K}^{T_s} C_{P,k}(T) dT \right] \tag{27}$$

The solid conductivity is modified to take into account the intra-bed radiation and boundary layer convection(Akiyama et al. 1992, Reid et al. 1988, Bird et al., 1960).

$$k_{s,\text{eff}} = (1 - \varepsilon_s)(k_g + a) + \varepsilon_s \left[\frac{2}{3k_s} + \left(\frac{k_g}{0.274} + b \right)^{-1} \right]^{-1} \tag{28}$$

And the constants are given by:

$$a = \alpha \left[1 + \left(\frac{1 - \varepsilon_s}{\varepsilon_s} \right) \left(\frac{1 - e_s}{2e_s} \right) \right]^{-1} \tag{29}$$

Where es is the solid components emissivity, assumed 0.9 for iron bearing materials and 0.8 for coke breeze and anthracite throughout the calculations carried out in this study.

$$b = \alpha \left(\frac{e_s}{2 - e_s} \right) \tag{30}$$

$$\alpha = 0.1952 d_s \left(\frac{T_s}{100} \right)^3 \left(\frac{4.184}{3600} \right) \tag{31}$$

In the above equations the following nomenclature and Greek symbols are used:

A: surface area, ($m^2\ m^{-3}$)
C_p : heat capacity, ($J\ kg^{-1}\ K^{-1}$)

d_m : solid component diameter, (m)
d_s : solid phase mean diameter, (m)

F_g^s : interaction force on solid phase due to gas phase, ($Nm^{-3}\ s^{-1}$)

f_m : solid component volume fraction (m=sinter feed, sinter return, limestone, fines, coke breeze, mushy and bonding phases), ($m^3\ m^{-3}$)

H : enthalpy of the phase(kJ kg^{-1})

$\vec{U}_i$: phase velocity vector (i=gas and solid), (m s^{-1})

P : phase pressure (Pa)

$Pr_g = \dfrac{C_{p,g}\mu_g}{k_g}$: Prandtl number, (-)

$Re_{g-s} = \dfrac{\rho_g \left|\vec{U}_g - \vec{U}_s\right|}{\mu_g} d_s$: particle Reynolds number, (-)

R : gas constant, (J mol^{-1} K^{-1})

S_ϕ : source or sink terms for the ϕ variables, (various)

x_i: spatial coordinates,(m)

t: time, (s)

T: temperature, (K)

$\phi_{i,k}$: dependent variables in eq. 1, (calculated by the model), [various]

φ_m : solid diameter shape factor (m=sinter feed, sinter return, limestone, fines, coke, mushy and bonding), (-)

ν_m : porosity of solid component (m^3 m^{-3})

ρ_i : phase density (i= gas and solid), (kg m^{-3})

μ : phase effective viscosity (Pa s)

ε_i : phase volume fraction (i= gas and solid), (m^3 m^{-3})

2.4 Boundary and initial conditions

The system o f differential equations presented above are completed with their initial and boundary conditions. The computational domain is defined by the region of the sinter strand for the case of the industrial scale process simulation with the above equations solved for stationary conditions. When the pot test experiment is considered, the simulations are carried out for transient calculations with a cylindrical pot of 60 cm of diameter and the height is determined in order to account for the operational conditions which reproduce the similarity of permeability and gas flow within the packed bed with the industrial machine. Regarding to the boundary and initial conditions for the solid phases, the composition, initial and inlet particle diameters, charging particle and volume fractions distributions and moisture content are specified at the charging position of the strand. The outlet boundary condition for the solid phase is assumed fully developed flow and no sleep condition is assumed at the sinter strand. The other boundaries such as lateral and bed surface are assumed zero velocities components gradient. For the energy balance equations convective and radiation coefficients are assumed for each of these surfaces. The gas inlet and outlet flow rates are determined by the pressure drop specified for each wind box determined by the blower controls and it is calculated interactively by accounting simultaneously for the mass balance and pressure drop of each wind box. The gas inlet temperature is specified at the surface of the bed and the outlet temperature are calculate by using the integral relation given by **eq.(7)**.

2.5 Numerical features

The multiphase model is composed of a set of partial differential equations that can only be solved by numerical method due to their nonlinearities on the boundary, initial and source terms. In this work, the set of differential equations described above is discretised by using the finite volume method and the resulting algebraic equations are solved by the iterative procedure using the line by line method combined with the tri-diagonal matrix solver algorithm(Melaaen,1992, and Karki & Patankar, 1988). In this paper, the numerical grid used to simulate the pot test uses 32x8x16 = 4096 control volumes in cylindrical coordinate system while the industrial strand of the sinter machine was discretised based on the Cartesian coordinate system with 11x140x16 = 24640 control volumes. The numerical convergence was accepted for tolerance of the order of 10^{-6} for the velocity and temperature fields, meanwhile, for the chemical species the overall mass balance was accepted less than 1% for all chemical species calculated.

3. Results and discussions

3.1 Model validation

The model constructed was validated by using the pot test results for reproducing an actual sinter operation of average 37100 kg m^{-2} per day of productivity. The gas temperatures and compositions were recorded by using thermocouples and chromatography analysis. The temperature measurements carried out were averaged and plotted for intervals of 3 minutes within the sinter bed at 50, 350 and 750 mm, respectively, from the top to the bottom of the pot. The gas composition was measured at the outlet. The composition of the raw materials used for predicting both, pot experiments and industrial operations, is shown in **Table 2**. **Figure 4** shows the comparison of temperature measurements and model predictions of the pot experiment representing an actual operation (a-c) and for gas recycling of 10% on the first 10 wind boxes and pre-combustion with natural gas using pilot burners (d-f). As can be observed, the calculated results presented very good agreement with the measured values. Actually, these cases of calculations were used as base cases for the calculations of the industrial operations conditions. **Figure 5** shows vertical temperature profiles measured and predicted by the model at 5, 15 and 30 minutes, respectively. As can be seen in **Fig 5** a-c the calculated results are also very close to the measured ones at the top, meddle and bottom of the pot reactor.

	C	VM	Fe_2O_3	Fe	H_2O	SiO_2	Al_2O_3	MgO	CaO
Iron ore (%)	-	-	88.2	-	5.2	3.5	1.4	0.4	0.8
Fuel (%)	86.9	1.4	-	-	4.3	4.1	3.0	0.1	0.2
Limestone (Flux) (%)	-	-	-	-	0.8	2.2	1.6	13 ($MgCO_3$)	82 ($CaCO_3$)
Scale (%)	31.9	3.1	10.8	35.5	5.8	6.6	4.1	0.4	1.7

Table 2. Chemical compositions of the raw materials used.

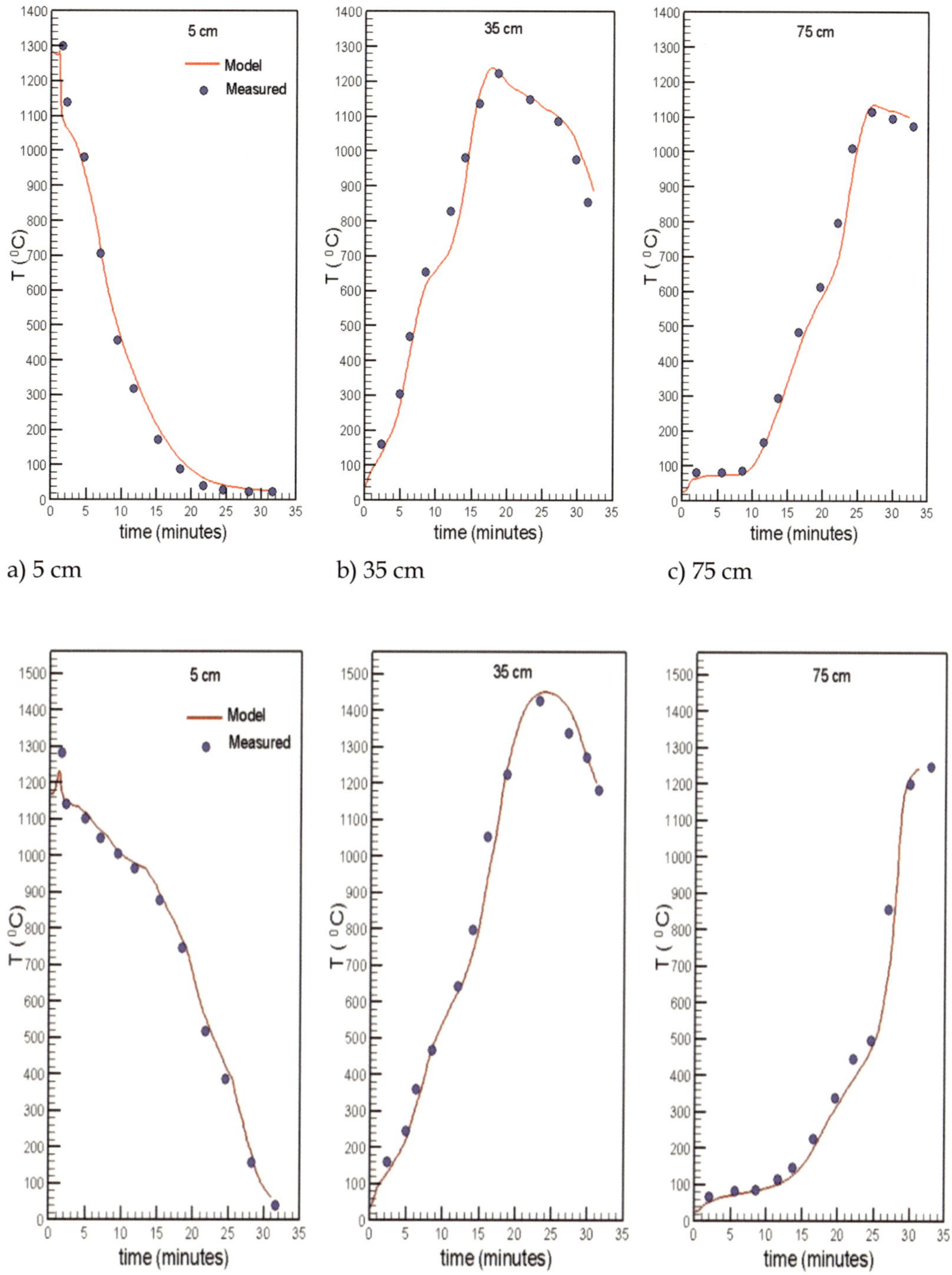

a) 5 cm b) 35 cm c) 75 cm

d) 5 cm e) 35 cm f) 75 cm

Fig. 4. Comparison of temperature predictions and measurements for the pot test experiment (a-c: actual operation technique and d-f: gas recycling and pre combustion of natural gas).

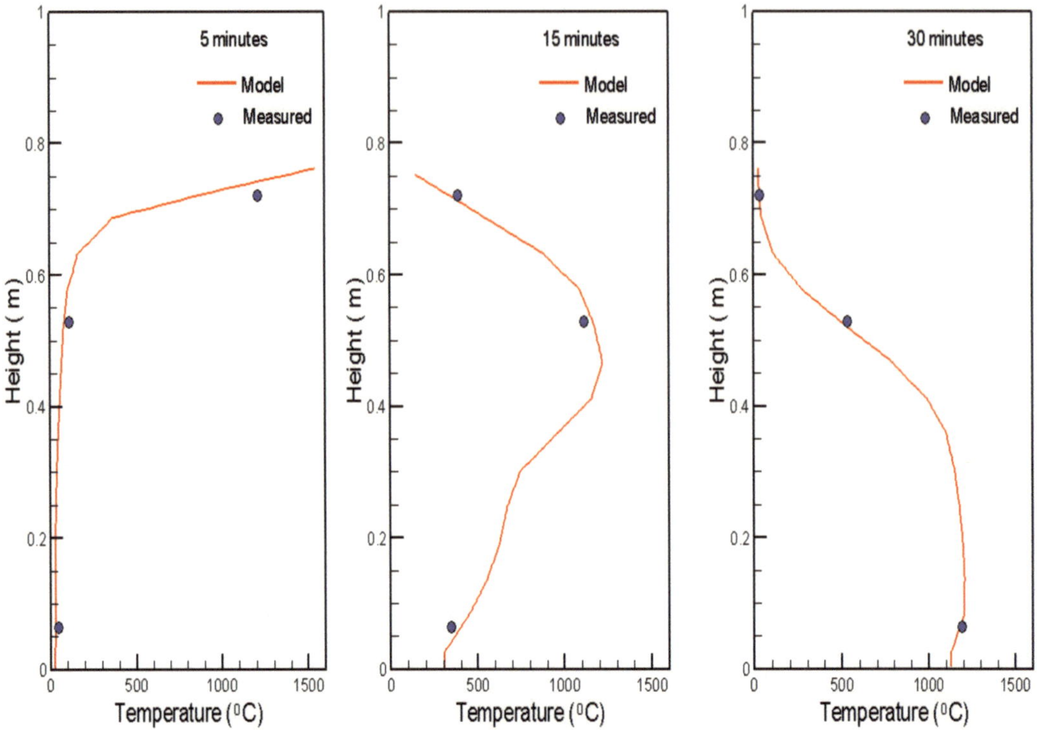

Fig. 5. Comparison of vertical temperature profiles within the pot experiment.

The measured and calculated temperatures used in Figure 5 for the sake of comparison were taken as the averaged solid and gas temperature predicted by the model and compared with the thermocouples measurements positioned in the interior of the packed bed. The reason for this approximation is due to the thermocouples are in direct contact with the gas and solid packed bed. In this comparisons it is assumed the local arithmetic average values. These results confirms the validity of the model to predict the thermal conditions within the sintering process. Therefore, in this investigation, it is assumed that the main sub-models to predict the heat transfer and chemical reactions are suitable to model the inner phenomena. Next step of the study was to apply the model to predict industrial sintering operation scenarios. The starting point was to apply the model to predict the actual operation of a typical machine and confirm the validation of the model into a industrial sinter machine. Thus, it was carried out a campaign of measurements in bed temperatures by inserting thermocouples within the sintering bed and recorded along the sinter strand until the sinter discharging position to the cooling system. **Figure 6** shows industrial data predictions and measurements of temperature distributions along the sinter strand for a conventional operation of a large sinter machine. The measured values are obtained by averaging the recorded values passing through the control volume which represents the numerical values.

The model predictions for the temperature is also in excellent agreement with the measured ones, similarly as observed in previous cases of pot experiment validations. Therefore, it is assumed that the model is able to predict new operations techniques and draw newly scenarios using recycling outlet gas and increasing amount of inner residues such as scales and fines. In the following section these possibilities are considered.

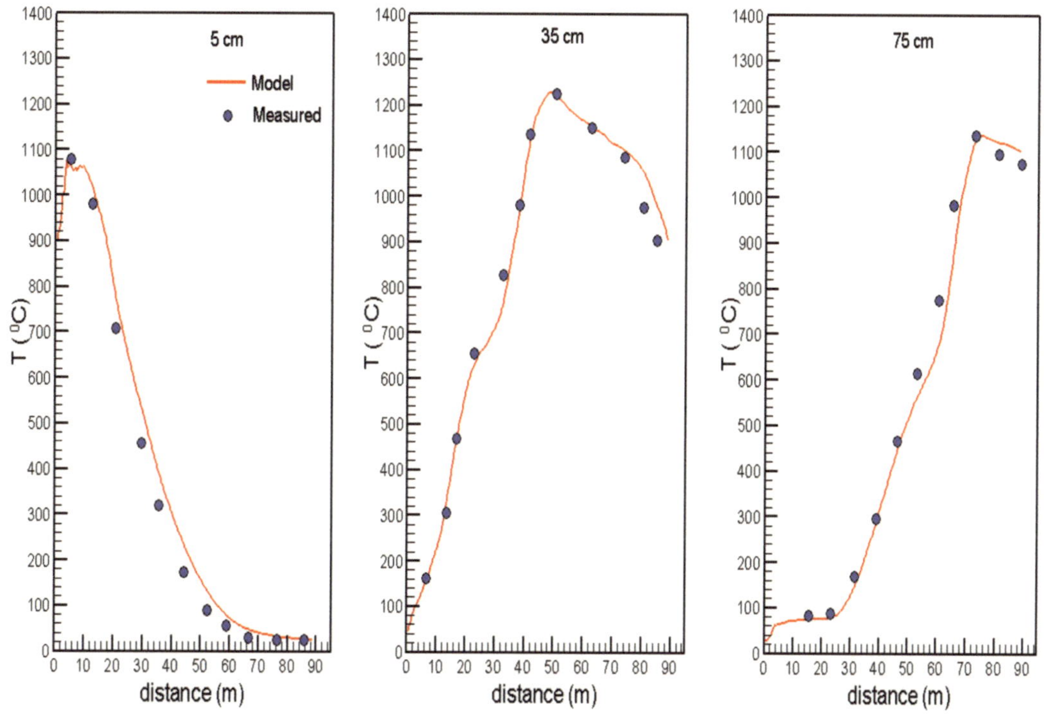

a) 5 cm from surface   b) 35 cm from surface   c) 75 cm from surface

Fig. 6. Industrial data predictions and measurements of temperature distributions along the sinter strand- conventional operation.

3.2 Simulations of advanced operations techniques

In this section, 8 operation scenarios for the iron ore sinter process are simulated and compared with the actual practice. A set of 4 cases of outlet gas recycling with pre-combustion of natural gas and another set of 4 cases of combined scenarios of increasing recycling of scales and outlet gas recycling are selected. **Table 3** summarizes the operational results predicted by the model and the base case (actual operation). As can be seen, from cases 1-4 increasing amount of outlet gas recirculation is proposed and additional natural gas is used in a pre-combustion chamber to promote post combustion and increase the temperature. From cases 5-8, increasing amount of recycling scale is proposed and combined with gas recycling aiming at searching better combination of the both practices. In this study is concluded that case 6 showed better combination of high gas and scale re-utilization with low fuel consumption. The di-calcium ferrite($Ca_2Fe_3O_5$) formation in the final sinter product is of special interest for the technological application due to its strong effect on the reducibility and mechanical strength.

	Conventional	Recycling gas only				Recycling gas and scale			
	Base	**Case 1**	**Case 2**	**Case 3**	**Case 4**	**Case 5**	**Case 6**	**Case 7**	**Case 8**
Productivity ($t\ m^{-2}\ day^{-1}$)	37.1	40.5	41.3	42.7	46.3	47.4	46.3	45.9	46.5
Bed height (mm)	770	820	840	860	900	900	900	900	900
Basicity $\left(\frac{CaO}{SiO_2}\right)$	1.87	1.72	1.81	1.82	1.84	1.54	1.75	1.54	1.38
$\left(\frac{CaO + MgO}{SiO_2 + Al_2O_3}\right)$	1.41	1.32	1.37	1.38	1.41	1.12	1.35	1.22	1.04
Fuel rate ($kg\ t^{-1}$) (%)	49.6 5	49.3 5	40.4 4	39.3 3.9	39.3 4	32.7 3.8	37.8 3.8	34.5 3.5	34.5 3.5
Iron ore (%)	56	56	55.2	55.3	55.4	54.1	53.5	51.8	55.9
Sinter return (%)	28.8	28.7	31	31.1	30.9	30.3	28.8	29.3	23.1
Scale (%)	0.0	0.0	0.0	0.0	0.0	2.1	4.3	6.1	8.2
Limestone (%)	10.2	10.3	9.8	9.7	9.7	9.7	9.6	9.3	9.3
$Ca_2Fe_3O_5$ (%)	28.4	24.8	26.2	26.2	26.1	22.6	25.3	23.1	20.4
Exhaust gas ($Nm^3\ t^{-1}$)	706	683.9	634.9	613.5	639.3	660.5	666.8	691.2	694.8
Recycling gas (%)	0.0	5.6	6.5	10.4	13.7	13.2	13.3	12.9	12.7
Additional Fuel (CH_4) ($kg\ t^{-1}$)	0.0	4.5	6.8	7.3	7.8	4.2	4.7	5.3	5.8
PCDD/F Emissions ($ng\ Nm^{-3}$)	1.9	2.3	2.6	2.7	2.5	2.2	1.9	2.0	2.1

Table 3. Operational parameters for simulated scenarios.

Next, the inner temperature fields are compared for selected cases. It is worth to mention that feasible distributions profiles were obtained for all scenarios, however only cases 6 and 8 are shown due to their attractive operational parameters, better performance and higher re-utilization methods. The temperature distributions within the sinter strand are shown in **Fig. 7** for actual sintering machine operation and the case of higher scale recycling combined with gas and pre-combustion(case 8). As can be seen in **Fig. 7(b)** the sintering temperature was not uniformly obtained in the sinter strand, and therefore, this operation could lead to produce sinter of lower mechanical strength.

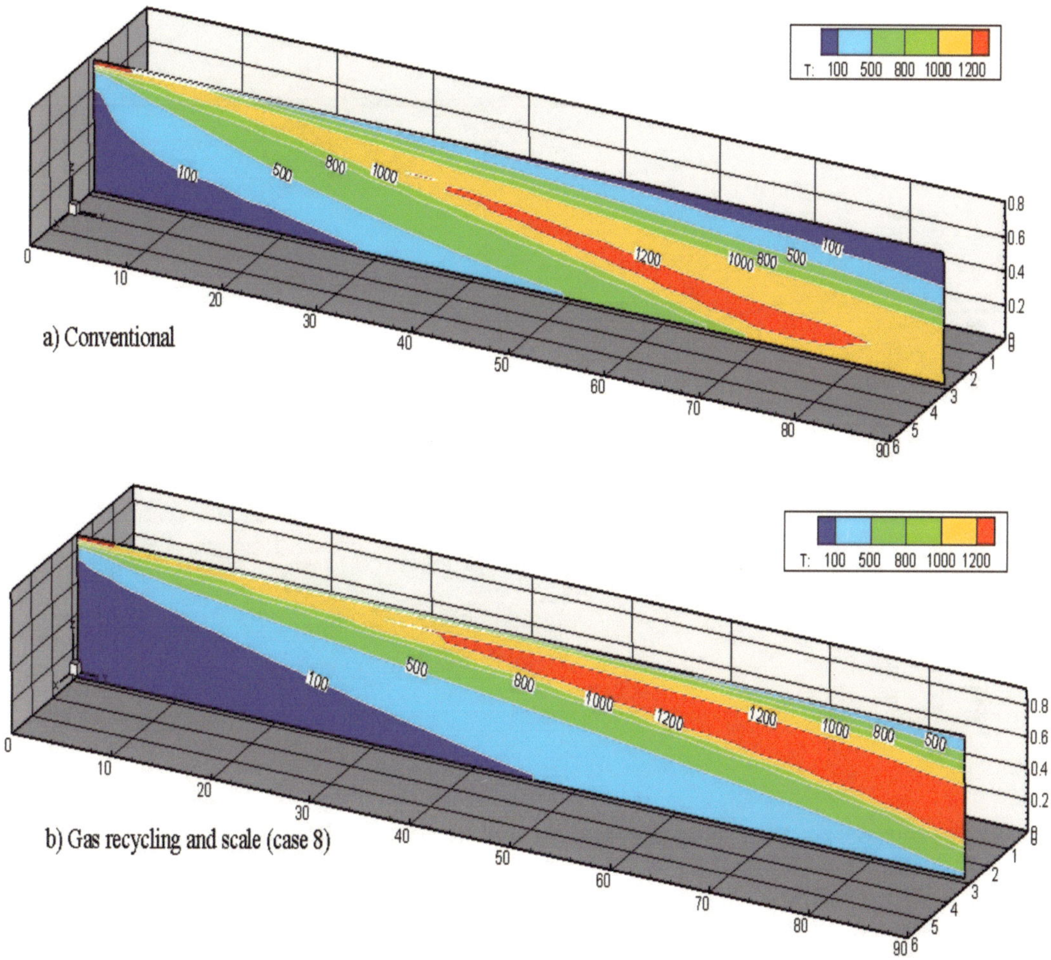

Fig. 7. Comparison between the sinter bed temperature distributions for conventional and high gas and scale recycling methods.

Figure 8 shows a comparison of the inner temperature predictions of the actual operation with the optimum scenario for gas and scale recycling. As observed, the high temperature region is enlarged and the residence time within the sintering zone is higher. Although all the scenarios presented in **Table 2** could be considered operational, case 6 was selected due to good temperature distribution combined with overall operational parameters.

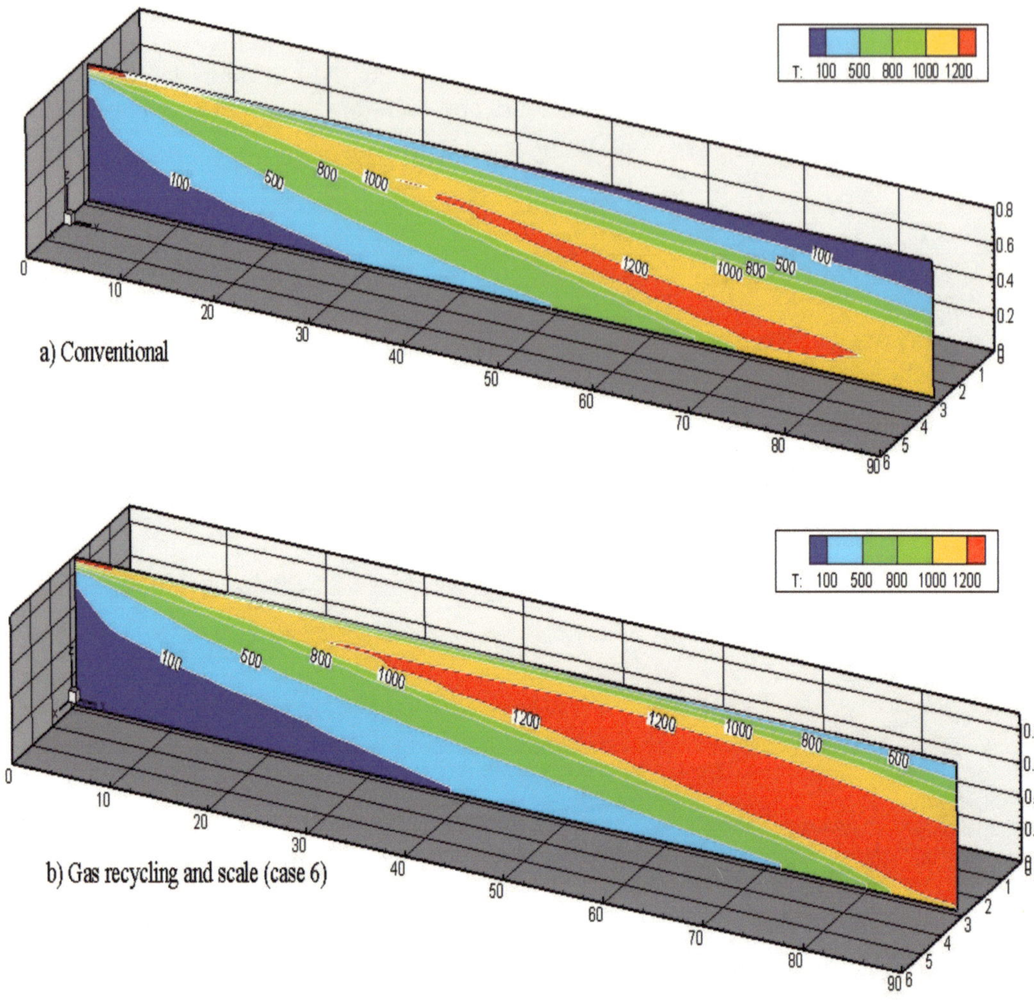

Fig. 8. Optimized sinter bed temperature distributions for recycling methods in comparison with conventional one.

Therefore, the calculation results was used to indicate promising operational conditions able to give higher productivity combined with low fuel consumption and high recycling amount of outlet gas and powder residues such as scale and others. Finally, this scenario was compared with the actual operation with regard to the emissions of dioxin (PCDD- Poly Chlorinated Di-benzene Dioxin) and furans(PCDF - Poly Chlorinated Di-benzene Furan). The sintering process have been strongly requested to reduce polychlorinated substances emissions, therefore, new technologies need to be compared not only from the point of view of economics but also from the environmental load (Castro et al. 2005, Cieplik et al. 2003, Kasai et al. 2008, Nakano et al. 2005, 2009). **Figure 9** and **10** show the comparisons of PCDD and PCDF distributions within the sinter strand for actual operation and case 6. Although the concentration distributions are slightly higher when the recirculation methods of scale and outlet gas is considered, when outlet gas re-utilization is adopted the PCDD and PCDF are also recycled into the higher temperature zone and partially decomposed and depending on the gas recirculation method the total amount could be reduced. As observed in **Table 2**, when the recycling amount is considered the specific emissions are reduced.

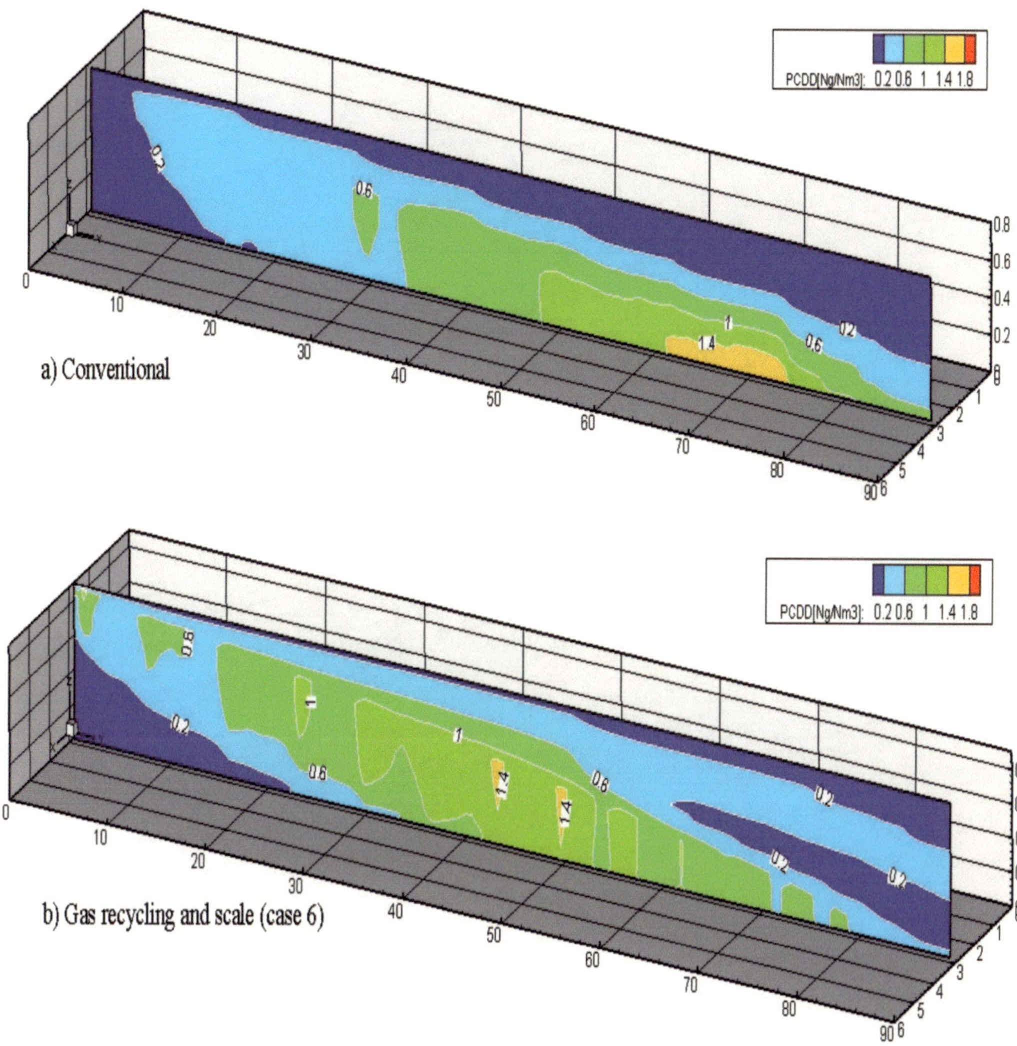

Fig. 9. Predictions of PCDD distributions within the sinter bed.

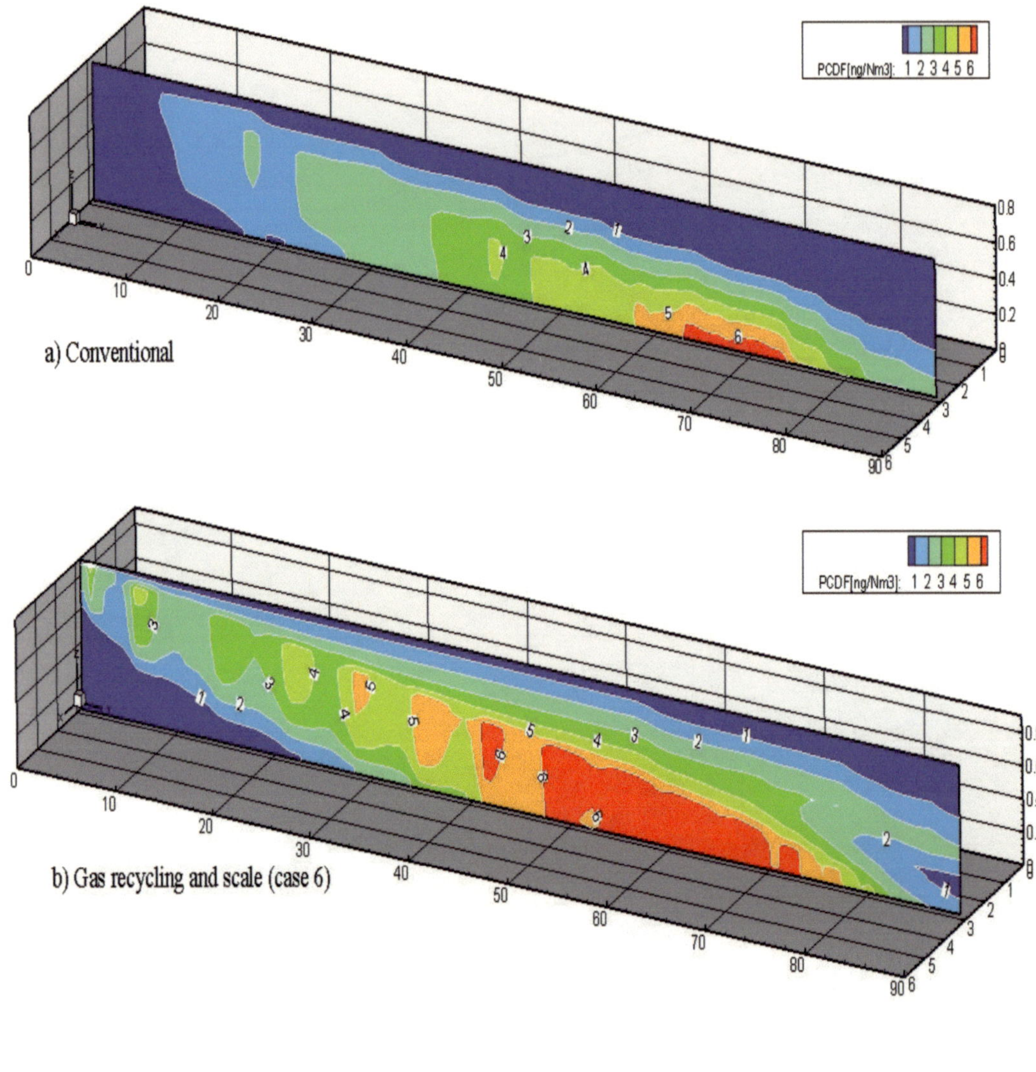

Fig. 10. Predictions of PCDF distributions within the sinter bed.

4. Conclusions

In this chapter a mathematical model of the sintering process of a steel works has been derived in order to predict optimum scenarios of fines and gas recycling. The model is based on transport equations of momentum, energy and chemical species coupled with chemical reactions and phase transformations. The model predicted the temperature distributions inside the sintering bed together with the final sinter composition. The model was validated against the measured values for pot experiments and industrial monitoring temperature distributions. For all validation cases considered in this study the model predictions closely agreed with the measured values. Finally, the model was used to predict advanced operations such as high amount of scale recycling combined with outlet gas recirculation and pre-combustion of auxiliary fuels. The model developed in this study has shown potential applications to search for new technologies and point out advantages when compared with actual operation techniques. The model showed potential applications in order to monitoring industrial sintering machines and develop high efficiency operational techniques from the point of view of environment load and economical aspects.

5. Acknowledgment

This work was partially supported by CNPq - Conselho Nacional de Desenvolvimento Científico e Tecnológico and Faperj - Fundação Carlos Chagas Filho de Amparo a Pesquisa do Estado do Rio de Janeiro - Brazil.

6. References

Akiyama, T. Hohta, H., Takahashi, R., Waseda, Y. and Yagi, J.,(1992), Measurement and modeling of thermal conductivity for dense iron oxide and porous iron ore agglomerates in stepwise reduction. ISIJ International, vol. 32, No. 7, (July, 1992), pp. 829-837. ISSN 0915-1559

Aizawa, T. and Suwa, Y..,(2005), Meso-porous modeling for theoretical analysis of sinter ores by the phase-field, unit cell method. ISIJ International, vol. 45, No. 4, (April 2005), pp. 587-593. ISSN 0915-1559

Austin, P.R., Nogami, H. and Yagi, J. (1997), A mathematical model for blast furnace reaction analysis based on the four fluid model. ISIJ International, vol. 37, No. 8, (August, 1997), pp. 748-755. ISSN 0915-1559

Castro, J. A , Nogami, H. and Yagi, J. ,(2000), Transient Mathematical Model of Blast Furnace Based on Multi-fluid Concept, with Application to High PCI Operation. ISIJ International, vol. 40, No. 7, (July, 2000), pp. 637-646. ISSN 0915-1559

Castro, J. A , Nogami, H. and Yagi, J.,(2001), Numerical Analysis of Multiple Injection of Pulverized Coal, Prereduced Iron Ore and Flux with Oxygen Enrichment to the Blast Furnace ISIJ International, vol. 41, No.1, (January, 2001), pp. 18-24. ISSN 0915-1559

Castro, J.A. Nogami, H. and Yagi, J., (2002), Three-dimensional Multiphase Mathematical Modeling of the Blast Furnace Based on the Multifluid Model. ISIJ International, vol. 42, No. 1, (January, 2002), pp. 44-52. ISSN 0915-1559

Castro, J. A., Silva, A.J., Nogami, H. and Yagi, J., (2004), Simulação computacional da injeção de carvão pulverizado nas ventaneiras de mini altos-fornos Tecnologia em Metalurgia e Materiais, Vol. 1, No. 2, (October, 2004), pp.59-62, ISSN 2176-1515

Castro, J. A., Silva, A.J., Nogami, H. and Yagi, J., (2005), modelo matemático tridimensional multi-fásico da geração de dioxinas no leito de sinterização. Tecnologia em Metalurgia e Materiais, Vol. 2, No. 1, (July, 2005), pp.45-49, ISSN 2176-1515

Castro, J. A., Silva, A.J., Sasaki, Y. and Yagi, J., (2011), A Six-phases 3-D Model to Study Simultaneous Injection of High Rates of Pulverized Coal and Charcoal into the Blast Furnace with Oxygen Enrichment. ISIJ International, vol. 51, No. 5, (May,2011), pp. 748-758. ISSN 0915-1559

Cieplik, K. M., Carbonell, J. P., Munoz, C., Baker, S., Kruger, S., Liljelind, P., Marklund, S. and Louw, R.,(2003), On dioxin formation in iron ore sintering. Environment Science Technology, vol. 37, No 15, (June, 2003), pp. 3323-3331,ISSN 101021

Cores, A., Babich, A. Muniz, M., Ferreira, S. and Mochon, J. (2010), The influence of different iron ores mixtures composition on the quality of sinter. ISIJ International, vol. 50, No. 8 , (August, 2010), pp.1089-1098. ISSN 0915-1559

Cumming, M.J. and Thurlby, J. A: (1990), Developments in modeling and simulation of iron ore sintering, Ironmaking and Steelmaking, vol. 17, No. 4, (April, 1990), pp. 245-254, ISSN 0301-9233

Jeon, Ji-Won, Jung, Sung-Mo and SASAKI, Y., (2010), Formation of Calcium Ferrites under Controlled Oxygen Potentials at 1 273 K. ISIJ International, vol. 50, No 8, (August 2010), pp. 1064-1070, ISSN 0915-1559

Kamijo, C., Matsumura, M. and Kawaguchi, T.(2009), Production of carbon included sinter and evaluation of its reactivity in blast furnace environment. ISIJ International, vol. 49, No. 10, (October, 2009), pp. 1498-1504. ISSN 0915-1559

Karki, K. C. and S. V. Patankar, S. V. (1988), Calculation Procedure for Viscous Incompressible Flows in. Complex Geometries. Numer. Heat Transfer, B, vol. 14, pp. 295-307, ISSN: 1040-7790

Kasai, E., Rankin, W.J., . Lovel, R.R and Omori,Y.,(1989), An analysis of the structure of iron ore sinter cake, ISIJ International, vol. 29, No. 8, (August, 1989), pp. 635-641. ISSN 0915-1559

Kasai, E., Batcaihan, B., Omori, Y., Sakamoto, N. and Kumasaka, A.,(1991), Permeation characteristics and void structure of iron ore sinter cake. ISIJ International, vol. 31, No. 11, (November, 1991), pp. 1286-1291. ISSN 0915-1559

Kasai, E, Komarov, S., Nushiro, K. and Nakano, M., (2005), Design of bed structure aiming the control of void structure formed in the sinter cake. ISIJ International, vol. 45, No. 4,(April 2005), pp.538-543, ISSN 0915-1559

Kasai, E. Kuzuhara, S., Goto, H. and Murakami, T., (2008), Reduction in dioxin emission by the addition of urea as aqueous solution to high-temperature combustion gas. ISIJ International, vol. 48, No. 9 (September, 2008), pp. 1305-1310. ISSN 0915-1559

Hayashi, N. Komarov, S. V. and Kasai, E. (2009), Heat transfer analysis of mosaic embedding iron ore sinter (MEBIOS) process. ISIJ International, vol. 49, No. 5, (May 2009), pp.681-686, ISSN 0915-1559

Li, L., Liu, J., Wu, X., Ren, X., BING, W. and Wu, L. (2010). Influence of Al2O3 on Equilibrium Sinter Phase in N2 Atmosphere. ISIJ International, vol. 50, No 2, (February, 2010), pp. 327-329. ISSN 0915-1559

Lv. X., Bai, C., Qiu, G., Zhang, S. and Shi, R, (2009), A novel method for quantifying the composition of mineralogical phase in iron ore sinter, ISIJ International, vol. 49, No. 5, (May, 2009), pp. 703-708. ISSN 0915-1559

Melaaen. M.C. (1992), Calculations of fluid flows with staggered and nonstaggered curvilinear nonorthogonal grids - The theory. Numerical Heat Transfer, B, vol. 21, No.1. (January, 1992), pp. 1-19, ISSN: 1040-7790

Mitterlehner, J. , Loeffler, G., Winter, F. Hofbauer, H., Smid, H., Zwittag, E. , Buergler, T.H., Palmer, O. and Stiasni, H.,(2004), Modeling and Simulation of Heat Front Propagation in the Iron Ore Sintering Process ISIJ International, vol. 44, No. 1, (January, 2004), pp. 11-20. ISSN 0915-1559

Nakano, M., Morii, K. and Sato, T. (2009), Factors accelerating dioxin emission from iron ore sintering machines. ISIJ International, vol. 49, No. 5, (May, 2009) , pp.729-734. ISSN 0915-1559

Nakano, M., Hosotani, Y. and Kasai, E., (2005), observation of behavior of dioxins and some reating elements in iron ore sintering bed by quenching pot test. ISIJ International, 2005, vol. 45, No.4, (April, 2005), pp. 609-617. ISSN 0915-1559

Nakano, M., Katayama, K. and Kasama, S. (2010), Theoretical characterization of steady-state heat wave propagating in iron ore sintering bed. ISIJ International, vol. 50, No. 7, pp.1054-1058. ISSN 0915-1559

Nath, N. K. Silva, A.J. and Chakraborti, N., (1997), Dynamic Process Modeling of Iron Ore Sintering. Steel Research, vol. 68, No.7, (July, 1997), pp. 285-292, ISSN 1611-3683

Neufeld, P.D., Janzen, A.R. and Aziz, R.A. (1972), Empirical equations to calculate the transport collision integrals for Lennard-Jones potentials. Journal Chemical Physics, vol. 57, No. 3, pp.1100-1102, ISSN 0021-9606

Umekage, T. and Yuu, S., (2009), Numerical simulation of Particle agglomeration and bed shrink in sintering process. ISIJ International, vol. 49, No. 5, (May, 2009) , pp. 693-702. ISSN 0915-1559

Waters, A.G., Lister, J.D. and Nicol, S.K, (1989), A Mathematical Model for the Prediction of Granule Size Distribution for Multicomponent Sinter Feed. ISIJ International, vol.29, No. 4, (April, 1989), pp. 274-283. ISSN 0915-1559

Yamaoka, H and Kawaguchi, T, (2005). Development of a 3-D sinter process mathematical simulation model. ISIJ International, vol. 45, No.4,(April 2005), pp. 522-531. ISSN 0915-1559

Yang, L. X. and Matthews, E., (1997), Sintering reactions of magnetite concentrates under various atmospheres. ISIJ International, vol. 37, No. 11, (November, 1997), pp. 1057-1065. ISSN 0915-1559

Bird, R.B., Stewart, W.E. and Linghtfoot ,E.N, (1960), *Transport Phenomena*, Wiley Int. New York.

R.C.Reid, J.M.Prausnitz and B.E.Poling: (1988), *The properties of Gases and Liquids*, 4th Ed., McGraw-Hill, New York.

Wilke, C.R.. (1950), A viscosity equation of gas mixtures. Journal Chemical Physics, vol.18, pp. 517-519, ISSN 0021-9606

3

Development of Sintered MCrAlY Alloys for Aeronautical Applications

Fernando Juárez López and Ricardo Cuenca Alvarez
Instituto Politécnico Nacional-CIITEC
México

1. Introduction

Thermal barrier coatings (TBCs) are widely used in turbines for propulsion and power generation (Bose & DeMasi-Marcin ,1995; Choi et al., 1998; Cruse et al. 1988; DeMasi-Marcin & Gupta, 1994; DeMasi-Marcin et al., 1990; Eaton & Novak, 1987; Golightly et al., 1976; Hillery, 1996; Lee & Sisson, 1994; Mariochocchi et al., 1995; Meier et al. ,1991 & Gupta, 1994; Miller, 1984; Kingery et al., 1976; Rigney et al., 1995; Strangman, 1985; Stiger et al., 199;, 1999 & Evans, 1999). They comprise thermally insulating materials having sufficient thickness and durability that they can sustain an appreciable temperature difference between the load bearing alloy and the coating surface. The benefit of these coatings results from their ability to sustain high thermal gradients in the presence of adequate back-side cooling. Lowering the temperature of the metal substrate prologs the life of the component: whether from environmental attack, creep rupture, or fatigue. In addition, by reducing the thermal gradients in the metal, the coating diminishes the driving force for thermal fatigue. Both of these benefits can be traded off in design for greater component durability, or for reduced cooling air or for higher gas temperature/improved system efficiency. As a result, TBCs have been increasingly used in turbine engines. Successful implementation has required comprehensive testing protocols, facilitated by engineering models (Cruse et al. 1988; Eaton & Novak, 1987; Meier et al. ,1991; Wright, 1998). Expanded application to more demanding scenarios (Fig. 1) requires that their basic thermo-mechanical characteristic be understood and quantified. This need provides the opportunities and challenges discussed in this article.

There are four primary constituents in a thermal protection system (Fig. 2). They comprise (i) the TBC itself, (ii) the superalloy substrate, (iii) an aluminum containing bond coat (BC) between the substrate and the TBC, and (iv) a thermal grown oxide (TGO), predominantly alumina that forms between TBC and the BC. The TBC is the insulator, the TGO on the BC provides the oxidation protection and the alloy sustains the structural loads. The TGO is a reaction product. Each of these elements is dynamic and all interact to control the performance and durability.

The thermal barrier coating is a thermally insulating, "strain tolerant" oxide. Zirconia has emerged as the preferred material, stabilized into its cubic/tetragonal forms by the addition of Yttria in solid solution. This material has low thermal conductivity (~1 W/m^2) with

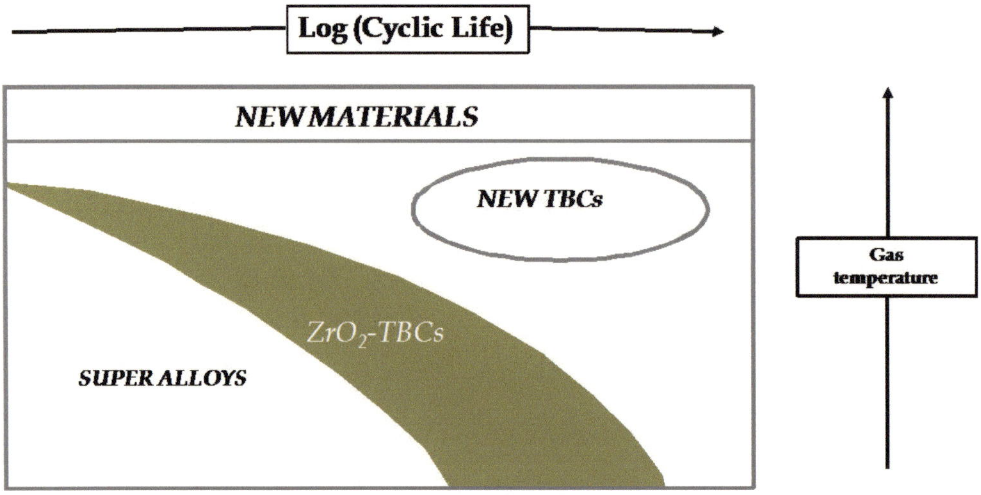

Fig. 1. Schematic indicating the operating domain for TBCs and the challenge for a new generation of materials.

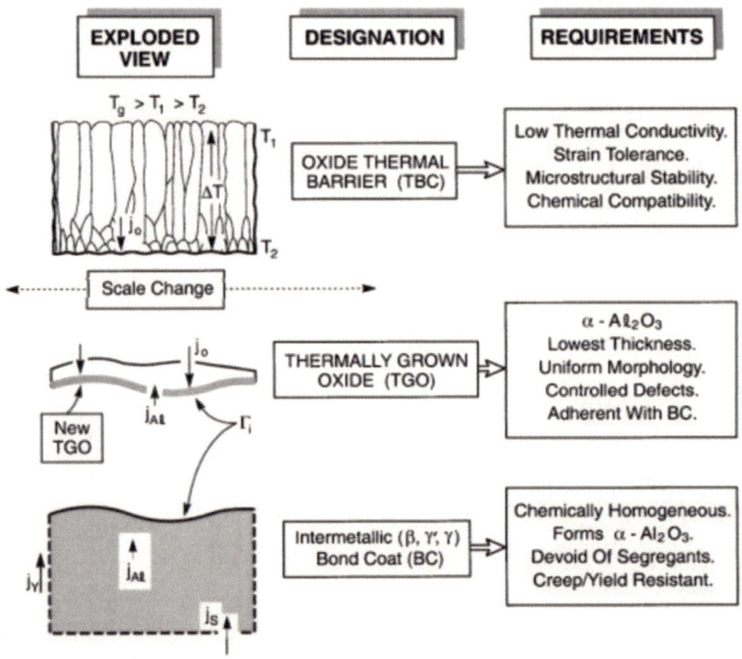

Fig. 2. The four major elements of a thermal barrier system: each element changes with exposure / cycling.

minimal temperature sensitivity (Fig. 3) (Kingery, 1976). The thermal resistance at lower temperatures corresponds to a phonon mean free path governed by structural vacancy scattering. Complex oxides having even lower conduction are being investigated, but there is not affirmation of their viability as TBCs. Strain tolerance is design into the material to avoid instantaneous delamination from thermal expansion misfit. Two methods are used to deposit strain tolerant TBCs. Electron beam physical vapor deposition (EB-PVD) evaporates the oxide from an ingot and directs the vapor onto the preheated component (Hillery, 1996; Mariochocchi et al., 1995; Rigney et al. 1995). The deposition conditions are designed to create a columnar grain structure with multiscale porosity (Fig. 2) that provides the strain tolerance and also reduces the thermal conductivity (to about 0.5 W/m K, Fig. 3). Air plasma spray (APS) deposition is a lower cost alternative (DeMasi-Marcin et al. 1990; Lee & Sisson, 1994; Choi et al., 1998). The deposition is designed to incorporate intersplat porosity and a network of crack-like voids that again provides some strain tolerance, while lowering the thermal conductivity.

The thermally grown oxide has a major influence on TBC durability (Cruse et al. 1988; Eaton & Novak, 1987; Golightly et al., 1976, Meier et al. ,1991; Stiger et al., 1999; Quadakkers et al., 1999; Wright, 1998, 1999). The bond coat alloy is design as a local Al reservoir (Fig. 2), enabling α- alumina to form in preference to other oxides, as oxygen ingresses through the TBC (which is transparent to oxygen).

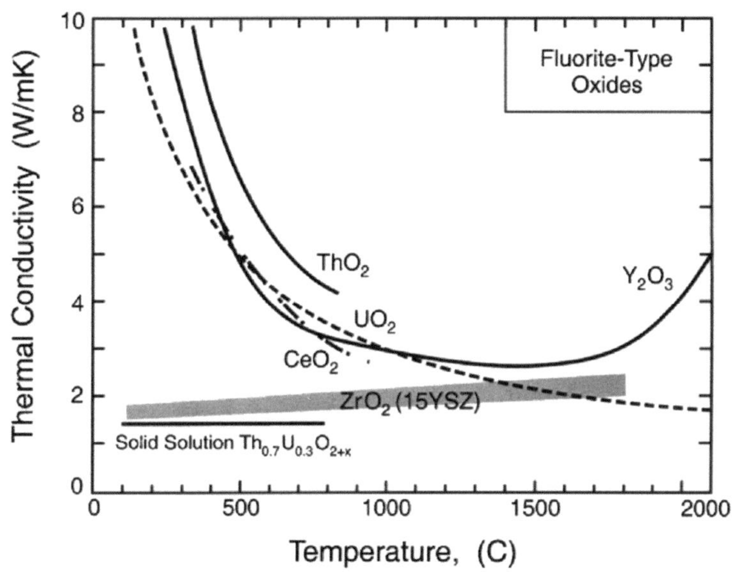

Fig. 3. The thermal conductivity of several insulating oxides illustrating the major role of solid solutions in affecting phonon transport (Evans et al., 2001).

Alumina is the preferred oxide because of its low oxygen diffusivity and superior adherence. This layer develops extremely large residual compressions (3-6 GPa, Fig. 4), as the system cools to ambient, primarily because of its thermal expansion misfit with the substrate (Fig. 5, Table 1) (Christensen et al., 1997; Lipkin & Clarke, 1996; Mennicke et al., 1997; Sarioglu et al. 1997; Sergo & Clarke, 1998; Tolpygo & Clarke, 1998, 1998). Stresses also arise during TGO growth (Lipkin & Clarke, 1996; Quadakkers et al., 1999; Stiger et al., 1999). They are much smaller (generally less than 1 GPa), but still important. Though thin (3-10 um), the high energy density in the TGO motives the failure mechanisms.

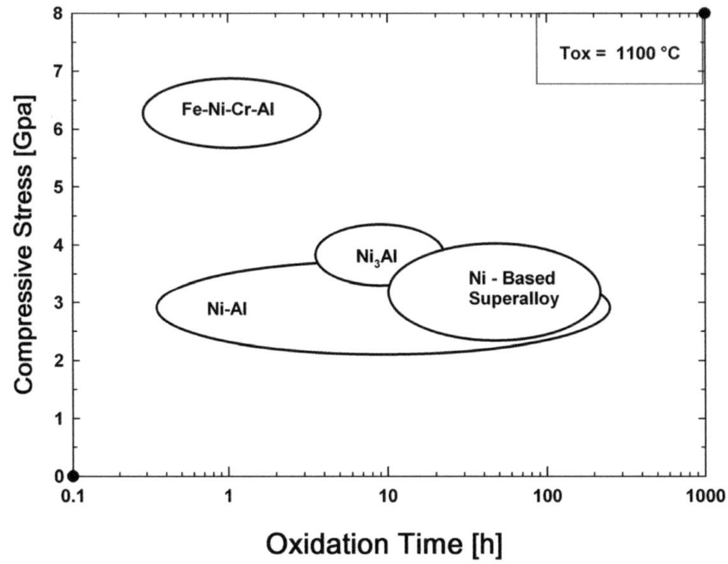

Fig. 4. Ambient residual compressions measured in the TGO developed on several alloy system (after (Lipkin & Clarke, 1996).

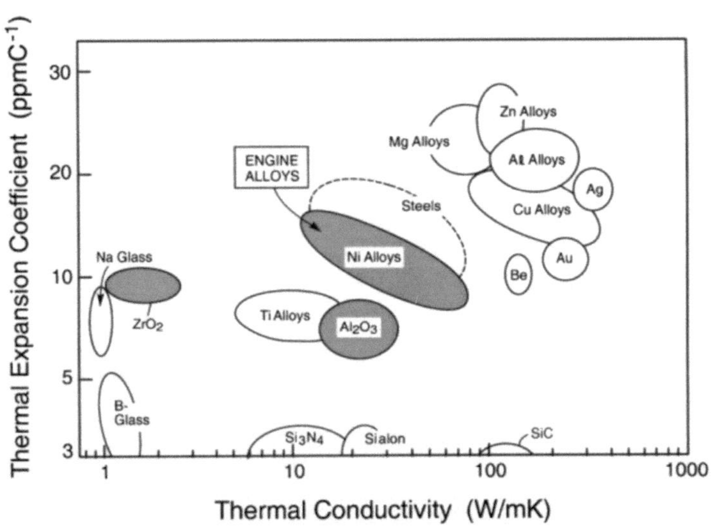

Fig. 5. Cross plot of the thermal expansion coefficient and thermal conductivity of the major materials constituents in the TBCs system (Evans et al., 2001).

The **bond coat** is arguably the most crucial component of the TBC system. Its chemistry and microstructure influence durability through the structure and morphology of the TGO created as it oxidizes (Stiger et al., 1999). Moreover, system performance is linked to its creep and yield characteristics. Bond coats are in two categories. One is based on the **NiCoCrAlY (MCrAlY)** system, typically deposited by low-pressure plasma spraying (LPPS). It is usually two-phase (β-NiAl and either γ - Ni solid solution or γ'-Ni_3Al). The γ/γ' phases have various other elements in solution. The Y is added at low concentrations to improve the adhesion of the

TGO, primarily by acting as a solid state gettering site for S (Haynes, 1999; Meier & Pettit, 1999; Smegil, 1987; Smialek et al., 1994), which diffuses up from the substrate. In some cases, small amounts of a Ni-Y intermetallic may also be present. The second category consists of a Pt-modified diffusion aluminide, fabricated by electroplating a thin layer of Pt onto the superalloy and then aluminizing by either pack cementation or chemical vapor deposition. These coatings are typically single-phase-β, with Pt in solid solution (Stiger et al., 1999). Their composition evolves during manufacture and in-service. Diffusion of Al into the substrate results in the formation of γ' at β grain boundaries (Stiger et al., 1999).

TGO (α-Al_2O_3)	
Young´s Modulus, E_o (GPa)	350-400
Growth stress, σ^g_{xx} (GPa)	0 - 1
Misfit compression, σ_0 (GPa)	3 - 4
Mode I fracture toughness, Γ_0 (J m^{-2})	20
Thermal expansion coefficient, α_o (C^{-1} ppm)	8 - 9
Bond coat	
Young´s modulus, E_s (GPa)	200
Yield strength (ambient temperature) σ_Y (MPa)	300-900
Thermal expansion coefficient, α_s (C^{-1} ppm)	13-16
Interface (α - Al_2O_3/bond coat)	
Mode 1 adhesion energy, Γ^0_1 (J m^{-2})	
Segregated	5-20
Clean	>100
TBC (ZrO_2/Y_2O_3)	
Thermal expansion coefficient, α_{tbc} (C^{-1} ppm)	11-13
Young´s modulus, E_{tbc} (GPa)	0-100
Delamination toughness T_{tbc} (J m^{-2})	1-100

Table 1. Summary of material properties.

The interface between the TGO and Bond coat is another critical element. It can be embrittled by segregation, particularly of S (Haynes, 1999; Meier & Pettit, 1999; Smegil, 1987; Smialek et al., 1994). During thermal exposure, S from the alloy migrates to the interface. Dopant elements present in the BC getter much of this S and suppress (but not eliminate) the embrittled. As already noted, bond coat based on NiCoCrAl contain Y for this purpose. The Pt-aluminide BCs do not contain elements which getter S. Nevertheless, they are durable and can have longer lives in cyclic oxidation than NiCoCrAlY systems Meier & Pettit, 1999). While it has been proposed that the Pt mitigates the effects of S [30], there is no fundamental reason to expect this. A number of effects of the Pt on the behavior Pt-modified aluminides have been documented (Schaeffer, 1988). But a complete understanding of the "Pt effect" is an important goal for future research.

A system approach to TBC design and performance requires that several basic bifurcations be recognized and characterized. Three of the most important are addressed.

i. The NiCoCrAlY and Pt-aluminide bond coats result in distinct TGO characteristics as well as differing tendencies for plastic deformation. Accordingly, the failure mechanisms are often different.

ii. TBCs made by APS and EB-PVD are so disparate in their microstructure, morphology and thermo-physical properties that different failure mechanisms apply.
iii. The failure mechanisms may differ for the two predominant areas of application (propulsion and power generation), because of vastly different thermal histories. Systems used for propulsion and for power peaking experience multiple thermal cycles, whereas most power systems operate largely in an isothermal mode with few cycles. The frequency affects coating durability.

Then NiCoCrAlY (MCrAlY) alloys are subjected to extensive research efforts to develop applications in gas turbine due to their high specific young's modulus and strength, and to their good oxidation and corrosion resistances. However, such alloys suffer from limited ductility at room temperature and creep resistance at service temperature (950- 1100°C) (Czech et al., 1994; Nickel et al., 1999; Monceau et al., 2000; Van de Voorde & Meetham, 2000). From a technological point of view, the current limitations are due to a large scattering in mechanical properties resulting from correlated chemical and structural heterogeneities, to manufacturing difficulties and high costs. In this context, the present work aimed to produce MCrAlY alloys with refined and homogenous microstructure by using the spark plasma sintering process (SPS) and hot pressing.

Indeed the need for improved oxidation and hot corrosion resistance of the protective oxide scale led to doping of the different bond coats by various metals like HF, Ir, Pd, Pt, Re, Ru, Ta, Zr (Alperine et al. 1989; Czech et al., 1994, 1995; Taylor & Bettridge, 1996) These, as well as other reports that have been published in the open and in the patent literature, (Taylor et al., 1995) conclude that doping of the bond coats by such elements was globally beneficial. However, for plasma-sprayed MCrAlY bond coats, the difficulty of performing such an investigation may be due to the cost of the dopants associated with the relatively large quality of powder required for the coating operation, even if it is performed on the laboratory scale, or to the mixing process involving either alloying high melting temperature elements or mechanical mixing with possible contamination from the atmosphere and the mixing apparatus.

A solution to this problem could be an economic, versatile, and time saving process allowing the doping of quantities of commercially available material with well established performance. The reference and doped powders could then be used for the preparation and compaction samples to produce the bond coat. These samples could, in turn, be subjected to thermal and cyclic oxidation and corrosion test, followed by adhesion test of the superficial scale that is produced.

In this sense a collaborative program was currently underway to satisfy this need. The process being investigated is superficial doping of commercially available MCrAlY powders which are actually used in industry, with Ruthenium as a series of platinum group metals, using the SB-MOCVD (Spouted Bed Metal-Organic Chemical Vapour Deposition) technique. In contrast to the direct mixing of the MCrAlY powder with the additive, this process ensures homogeneous distribution of the metallic additives on the surfaces of each particle and, consequently homogeneous distribution of the doping elements throughout the volume of the bond coats (Juarez et al., 2003). Initial results introducing the SB-MOCVD doped process have presented in (Caussat et al., 2006; Juarez et al., 2001; Vahlas et al., 2002).

The choice of Ruthenium (Ru) and Rhenium (Re) for this study is based on recent results, following which doping of monocrystalline nickel superalloys with this element reduces their high-temperature creep (Feng et al., 2003; Fleischer, 1991; Lu & Pollock, 1999, 1999; Noebe et al., 1993; Pollock et al., 2001; Tryon, 2006; Tryon et al., 2004; Wang et al. 2011; Wolff, 1997; Wolff, & Sauthoff, 1996).

In this context, the present study aimed to produce MCrAlY alloys with refined and homogenous microstructure by using the uniaxial hot pressing and spark plasma sintering process (SPS).

Then, firstly MCrAlY powders were doped by Ru and Re. The doping level, purity, microstructure and, Ru and Re distribution of the powders were established. TEM samples of Ru and Re nanometric coatings on the surfaces MCrAlY particles are presented. They allowed the investigations of morphology, the microstructure and the composition of the Ru-coatings. Second, at moment of this manuscript Ru-doped MCrAlY and undoped MCrAlY powders were sintered at 1473 K by only uniaxial hot pressing. Finally, MCrAlY undoped powders were sintered by SPS.

2. MCrAlY powder characteristics

The commercial MCrAlY powder is a pre-alloyed material, which is mainly composed of Ni with additions of Co (21 wt.%), Cr (19 wt.%), Al (8 wt.%), Y (1 wt. %) and Ta (5 wt.%). Fig. 6a shows a SEM micrograph of the as-received powder featuring spherical shape and agglomerates. Their skeleton-like shape report a theoretical density of 7700 kg/m^3, while the apparent tap density is 4300 kg/m^3. Specific surface area was computed from the N_2 adsorption isotherms (recorded at -196 ^{o}C with a Micrometrics Flowsorb II2300), using the BET method and was found to be 0.83 m^2/g. This low value is characteristic of a non porous material. Powder size distribution was determined with a Malvern Mastersizer laser diffractometer. It was found that mean size distribution of the particles is 23 μm, with minimum 0.05 μm and maximum 556 μm. Fig. 6b presents the particle size distribution measurement of the MCrAlY superalloy.

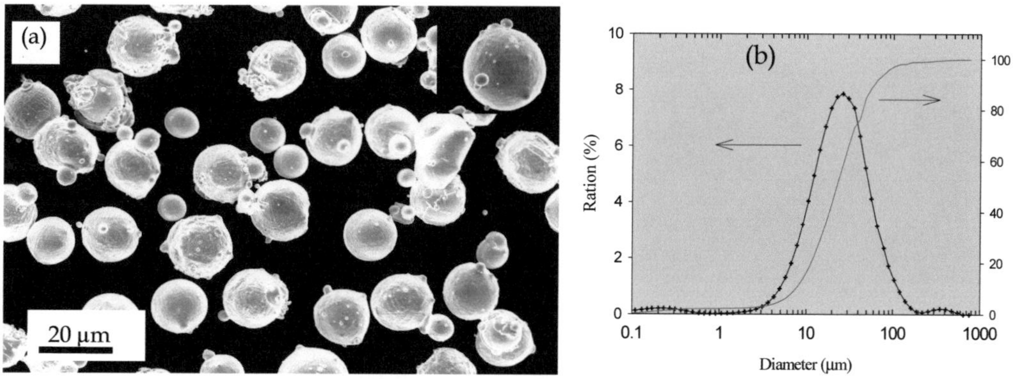

Fig. 6. Scanning electron micrograph of the as-received powders illustrating the spherical morphology and small agglomerated (a). Particle size distribution measured of the MCrAlY powder (b).

3. MCrAlY doped

We used a spouted bed reactor in a research program aiming the superficial doping with ruthenium and rhenium of as received commercial powders of MCrAlY alloys by MOCVD (Caussat et al., 2006; Juarez et al., 2001, 2005; Vahlas et al., 2002). Our results revealed that; since the uniform modification of the composition of commercial raw material is possible by SBMOCVD, the end user could dispose of a valuable tool to adjust the properties of use as a function of the aimed application. However, to end the validation of this process for the wanted application, the hydrodynamic behavior of such peculiar powders in a lab-scale SB contactor remained to be studied.

3.1 Characteristics MCrAlY powder doped

3.1.1 SEM analysis

The evolution of materials science towards nanometric scales requires appropriate microstructure characterization techniques and the corresponding specimen preparation. Correlating the processing conditions, in terms of both the quality of spouting (coat and MOCVD) and the efficiency of the MOCVD, with the oxidation resistance required an insight into the nature and the morphology of typically 50 nm thick films deposited on the surface of particles with a mean diameter of 25 um.

The positive results validated the use of a SB for the superficial doping by Ru and Re of MCrAlY powders by MOCVD. It allowed homogenous deposition of the two metals on the surface of the MCrAlY powder.

The morphology of the Re and Ru films, on the surface of MCrAlY powders is shown in the SEM micrographs of Fig. 7. The Ru films were deposited in the presence of SB-MOCVD.

EDS maps of the corresponding elements are also presented. Re and Ru films are uniforms and their morphology is smooth. Deposition of Ru is efficient on the entire available surface of the particles in contrast to Re which covers only parts of the particle.

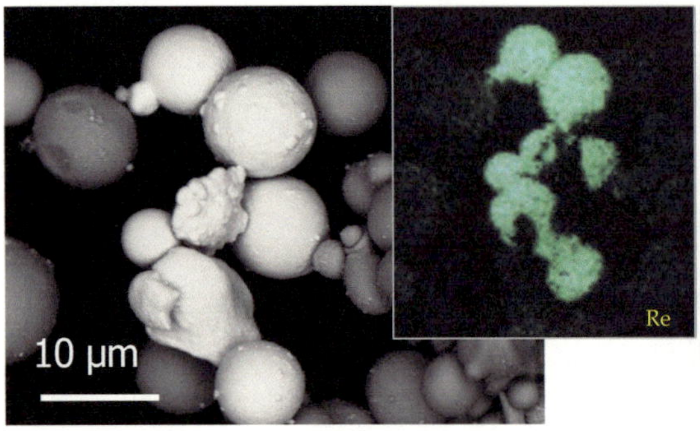

(a)

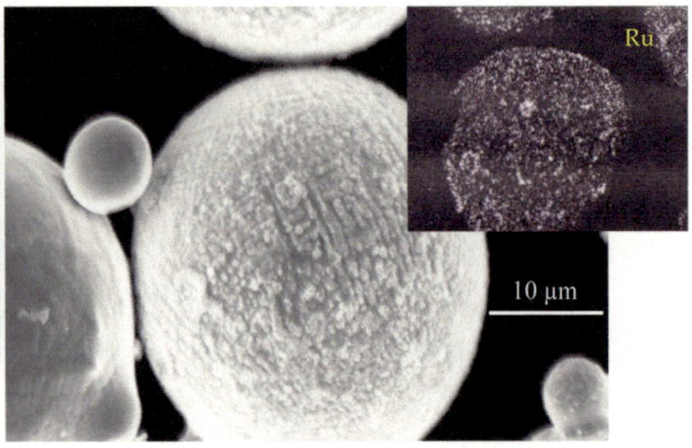

(b)

Fig. 7. Secondary electrons SEM micrograph of MCrAlY powders with Re (a) and Ru (b) deposited on their surface, and corresponding EDS maps.

The SEM images in the Fig. 8 show the Ru deposit on a particle. The secondary electron micrograph on the left reveals a wrinkled surface morphology, compared to the smooth one of the as-received powders, illustrated in the Fig. 8. This morphology is due to the Ru being deposited on the whole surface of the powders as shown in the EDS Ru map of the same particle, on the right. The doping level of the MCrAlY powders was, in this case, 0.8 wt. %.

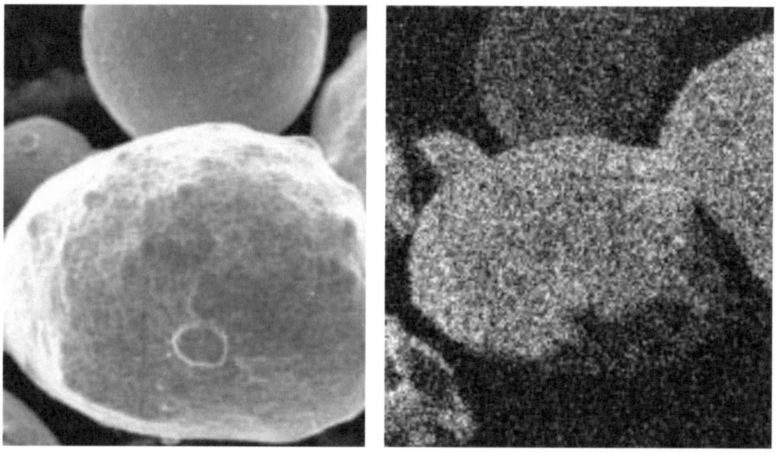

Fig. 8. SEM images of Ru deposited on MCrAlY powder. The micrographs are secondary electron (left) and Ru EDS mapping (Right) of a particle covered by Ru under adequate SB-MOCVD conditions.

3.1.2 TEM analysis

It is particularly important in transmission electron microscopy (TEM) observations of either thin areas in bulk materials such as interfaces in compositions, fine powder particles (Shiojiri et al., 1999; Yoshioka et al., 1996) or other complex in-shape samples (Yoshioka et al., 1995). The need to dispose of an appropriate TEM specimen preparation protocol was reported in our work for which 0.70 wt. % Rhenium (Re) and 0.90 wt. % Ruthenium (Ru) were added to the surface of commercial pre-alloyed MCrAlY powders (Caussat et al., 2006; Juarez et al., 2001, 2005; Vahlas et al., 2002).

The preparation of thin sections of Ru-doped powders was based on a method proposed by Yoshioka et al. (Shiojiri et al., 1999; Yoshioka et al., 1995, 1996). The sample was mixed with Gatan G1 epoxy resin in a Teflon cup and a drop of the suspension was placed on electron microscopy grid (100 mesh) that was positioned on a potassium bromide (KBr) crystal. It was aimed to obtain a sample that contained enough particles for convenient observation but not too much, to present satisfactory cohesion with enough resin. After polymerization of the epoxy at 373 K, the grid was wet-stripped from the KBr and mechanically polished with a South Bay Tech Tripod R to reduce the overall thickness down to 70 um. The sample was finally ion-milled during 5-6 h in Gatan Precision Ion Polishing Systems (PIPS) equipment until a hole was detected. PIPS operating conditions were; acceleration voltage of the ion gun 5 keV, rotation frequency 3rpm, and incidence angle of the two ion beams on both sides of the sample 8-10 deg.

The above presented way to prepare cross section for TEM is rapid and relatively easy. However, it can only be used in the case of particles with a homogenously distributed deposit on their surface. As shown below, it was more difficult to localize the Re-containing zones of the particle surfaces due to the less homogeneous distribution of Re than of Ru of the sample. In this case, the sample and G1 epoxy resin mixture was transferred to a brass tube whose internal diameter and wall thickness were 2.4 and 0.3 mm, respectively. After curing at 373 K the tube was sliced into a series of 300 um thick discs using a wire diamond saw. Finally, the discs were mechanically ground to a thickness of 100 um, and dimpled to about 40 um prior to ion thinning to electron microscopy transparency in the PIPS. At that time, the sample was very brittle and it was necessary to stick it to a specimen support grid with silver paste before the observation.

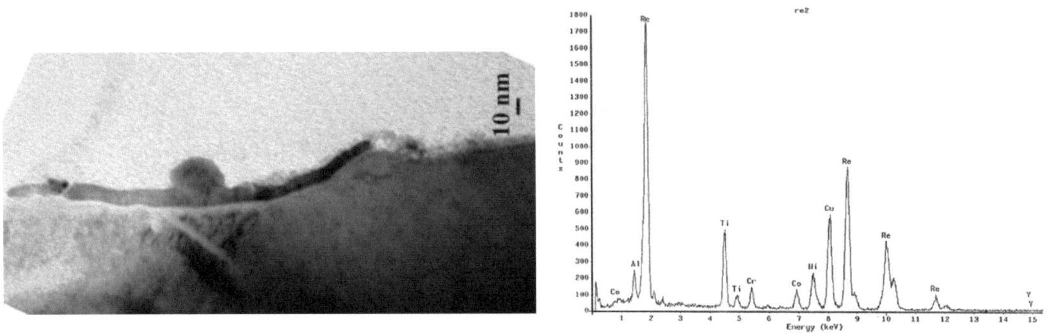

Fig. 9. Bright field TEM micrographs of the nanocrystallite Re coating on MCrAlY powder. Re cluster on the right of the micrograph reveals homogeneous side- nucleation.

Fig. 9 presents bright field TEM images of the nanocrystallite Re coating at two regions on MCrAlY powder. The film is continuous and its thickness varies between 10 and 40 nm. It is composed of particles typically sized around 10 nm. A larger cluster composed of Re grains is also shown in the micrograph. EDS revealed pure Re for both the film and the cluster.

Fig. 10 shows a dark field TEM micrograph of coating deposited from SB-MOCVD. The micrograph was obtained from an electron beam centered on the D1 and D3 spots of the corresponding EDP shown in the insert. The measured interplanar distances on this sample are also reported in Table 2. They reveal that the film is composed of pure crystalline Ru in agreement with X-Ray diffraction analysis.

	D (cm)	d_{exp} (nm)	Plan	$d_{JCPDS\ \#6\text{-}663}$
1	0,85	0,235	100	0,2343
2	1,0	0,211	002	0,2142
3	1,28	0,205	101	0,2056
4	1,68	0,157	102	0,1581
5	2,10	0,119	103	0,1219

Table 2. Interplanar distances of pure crystalline Ru.

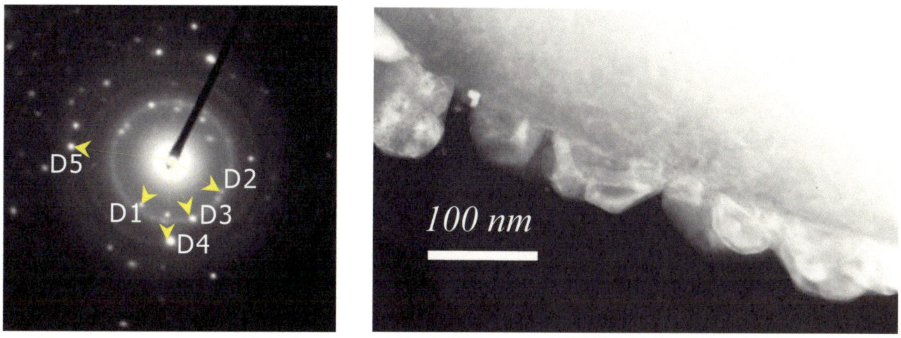

Fig. 10. Bright field TEM micrograph of Ru deposit and corresponding electron diffraction pattern.

Continuous Ru films deposited on the surface of MCrAlY particles are composed of the grains whose diameter approaches 100 nm. These grains are in turn composed of smaller crystallites. Fig. 11 illustrates this microstructure. In the Table 2 the interplanar spacing measured from the diffraction spots of Fig. 10 are presented and compared with the corresponding JCPDS values. Similar data from Ru samples, as presented in the following paragraphs are also included in this table.

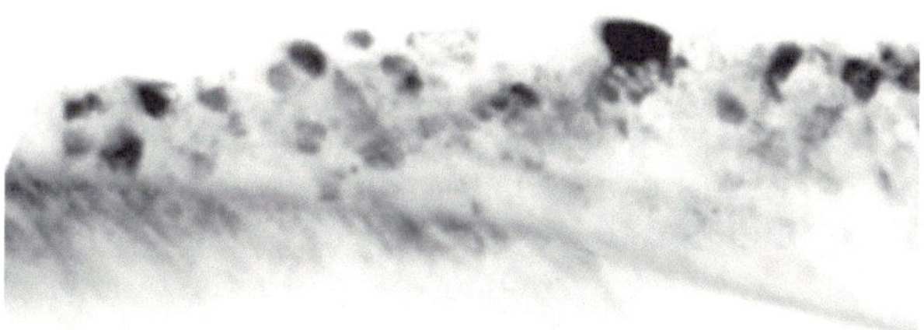

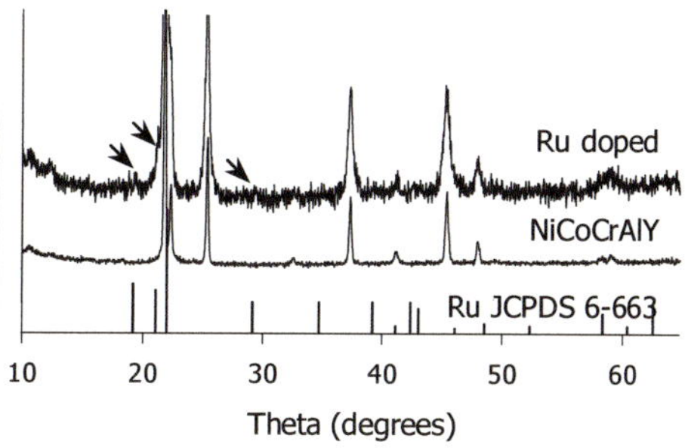

Fig. 11. Bright filed TEM image of Ru.

This parameter, together with the simplicity of this process should be considered for the final selection of the deposition route on the surface of MCrAlY powders of these, as well as other, metals.

4. Sintering of MCrAlY powders

4.1 Hot-pressing MCrAlY powders undoped and Ru-doped powders

Sintering of MCrAlY powder was carried out by uniaxial hot pressing at temperatures ranging between 1173 and 1473 K and during periods 0 to 60 minutes. The powders were poured into graphite dies coated with boron nitride and were sintered in a graphite furnace by uniaxial hot pressing. As shown below, this temperature is high enough: (i) to ensure densification of the samples; and (ii) to stabilize the microstructure in view of the subsequent heat treatments. A load pressure of 10 MPa is initially applied and the system operated under primary vacuum up to 523 K. Then, Ar flow was established and the pressure was gradually increased to achieve 40 MPa at 1273 K. During cooling, the pressure was gradually decreased. Heating and cooling rates were 20 °C min^{-1}. Densification is practically up to 94%.

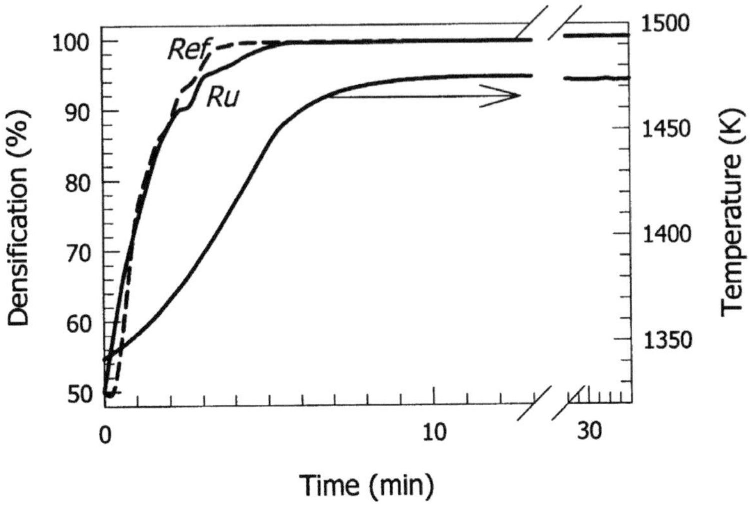

Fig. 12. Densification of the MCrAlY powders versus the duration of the uniaxial hot pressing performed at 1473 K. 100% densification corresponds to a pore-free sample.

Fig. 12 presents densification of reference and Ru-doped powders as a function of time for a temperature of 1473 K and during 30 min. Densification of both is practically identical up to 94%. During this period under the operating conditions used, grain boundaries were deformed and reorganized, especially due to the plastic deformation of the Al-rich b phase. Contact among surfaces is favored and the resulting product is exempt from open porosity. Elimination of closed porosity involves diffusion through grain boundaries and is somewhat slower in the doped sample. In view of this behavior, the same hot-pressing duration of 1 h was selected for both samples, ensuring complete densification and a similar microstructure.

The Fig. 13 (a) corresponds to the sintered sample at 1323 K, this micrograph shows characteristics of the powders in charge, where the powders present certain plasticity. Indeed, there is a pressure contact between faces of the powder due to the action mechanical exerted by the piston, consequently some powder boundaries begin to disappear. Intergranular diffusion starts but plasticity or diffusion rates do not contribute to the complete elimination of porosity. In practical the process is even slower in the periphery of the sample due to the frictional forces against the die walls and the slight contamination by boron nitride. An increase in the sintering temperature of 150 °C (1473 K Fig. 13 (c) micrograph) allowed an improvement of the densification but only after 60 min.

Fig. 14 presents a backscattered scanning electron micrograph of a polished surface of the Ru-doped sintered sample. Two phases, shown in dark and light gray, are present. They correspond to β-NiAl and to γ-Ni, respectively. White dots also appear in this micrograph; they correspond to tantalum carbide. The microstructure revealed is comparable to that of the corresponding bond coat applied by plasma torch in terms of the nature and distribution of the phases present. Consequently, the different properties of both materials, particularly their oxidation behavior, are also expected to be comparable.

Fig. 13. Scanning electron micrograph of sintered sample by hot pressing;
(a) 1323 K- 30 minutes, (b) 1473 K-0 minutes and (c) 1473 K- 60 minutes respectively.

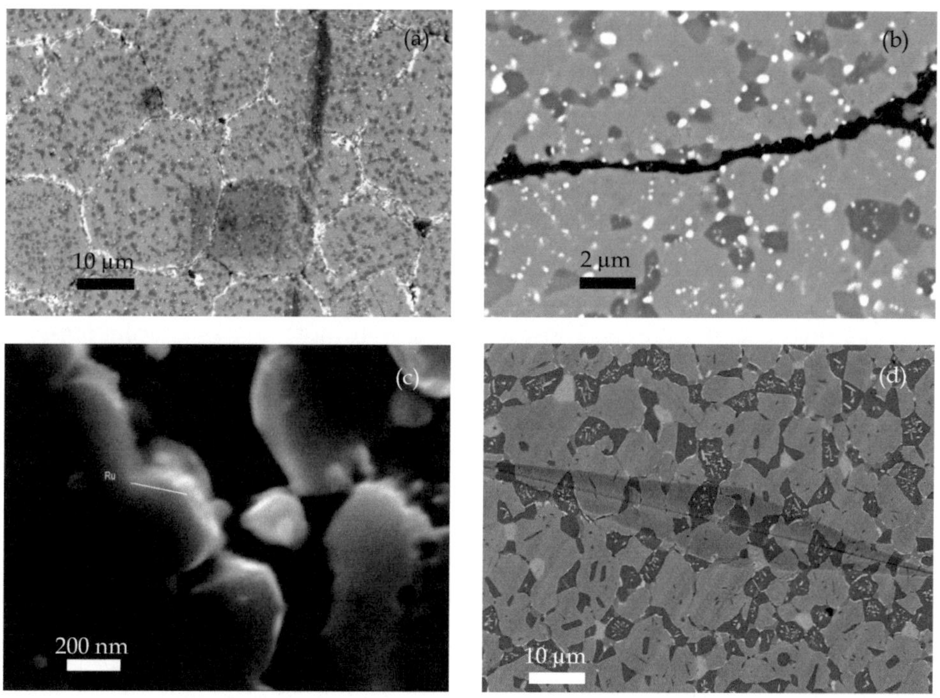

Fig. 14. Scanning electron micrograph of sintered sample Ru-doped by hot pressing;
(a) 1323 K- 30 minutes, (b) 1473 K-0 minutes and (c) 1473 K- 60 minutes respectively.

SIMS maps for 27Al and 99Ru of the same 50=50 mm2 area of the sintered doped sample are shown in Fig. 15. Bright zones in each map correspond to element rich phases, namely β-NiAl for the Al-rich zones. From a comparison of the two maps, it appears that Ru is found in β-NiAl. This re-organization is expected from the enhanced plasticity of β-NiAl in the sintering conditions, and is in agreement with literature information reporting that platinum group metals present a remarkable affinity for this phase β. Consequently, Fig. 15 reveals that sintering conditions allow Ru to diffuse completely from the outer surface to the core of the particles, towards the same position as it is expected to occupy in the bond coat. In view of this behavior, the same hot-pressing temperature 1473 K and duration of 1 h was selected for ensuring a complete densification and a reproducible microstructure.

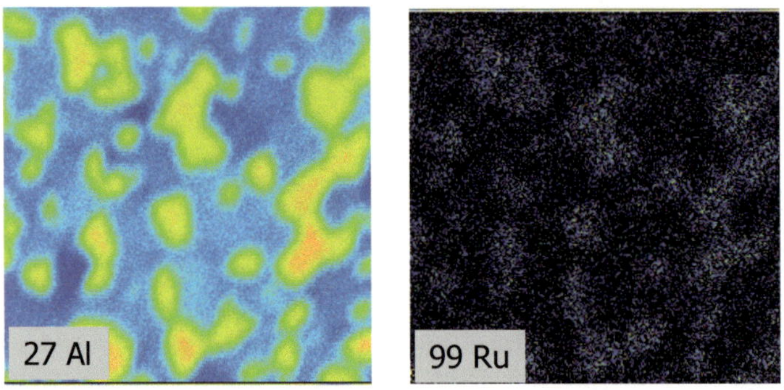

Fig. 15. ^{27}Al and ^{99}Ru SIMS maps of the same 50=50-mm2 area of the sintered MCrAlY doped sample.

4.2 Spark plasma sintering of MCrAlY powders undoped

SPS is found to be an effective technique to consolidate powder through the simultaneous application of direct pulsed current and uniaxial pressure (Munir et al., 2006). Assisted by an applied pressure, the electric current density induces a temperature elevation within the sample through the Joule´s effect, thus leading to powder sintering.

The emerging SPS theme from the large majority of investigations of current activated sintering is oriented to the advantages over conventional methods including pressureless sintering, hot-pressing, and others. These advantages include: lower sintering temperature, shorter holding time, and markedly improvements on the properties of materials consolidated by this method (Courat et al., 2008; Oquab et al., 2006).

The use of the hot-pressing techniques to optimize microstructure of MCrAlY alloys for high temperature application has been scarcely documented (Jeandin et al., 1988; Menzies et al., 1982; Somani et al., 1998; Prakash et al., 1988). The present study shows, the results of MCrAlY powder sintered by spark plasma sintering.

The SPS experiments were carried out on a commercial Dr Sinter Sumitomo 1050 apparatus (Sumitomo Coal Ming Co., Japan). This equipment can supply a direct current intense of 5000 Amp under a maximum voltage of 5 V. The powder was poured into a graphite die set of 20 mm wall thickness, placed between two graphite punches of 20 mm diameter (Fig. 16). Elements of graphite play both the role of electrodes and plates imparting the pressure. The sinter chamber is kept under vacuum (10^{-2} Pa) along the experiments.

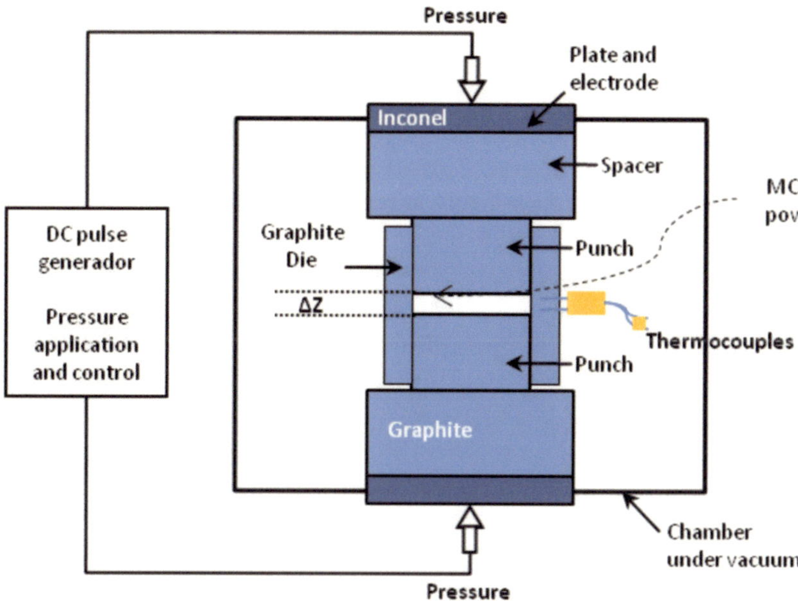

Fig. 16. Schematic illustration of the SPS technique.

MCrAlY alloy powders were consolidated at temperatures ranging between 1173 and 1323 K, range 15 KN in-load and processing time 0 and 30 minutes. To avoid grain coarsening of the microstructure of sintered specimen, SPS was conducted below the γ prime temperature, which is about 1140 °C (1413 K) (Prakash et al., 1988). The Fig. 17(a) shows an experimental record of the SPS-processing parameters, i.e., temperature, applied load and relative displacement of the punches, as a function of time. The relative displacement of the punches is expressed in percentage of the maximum displacement attainable. The temperature curve

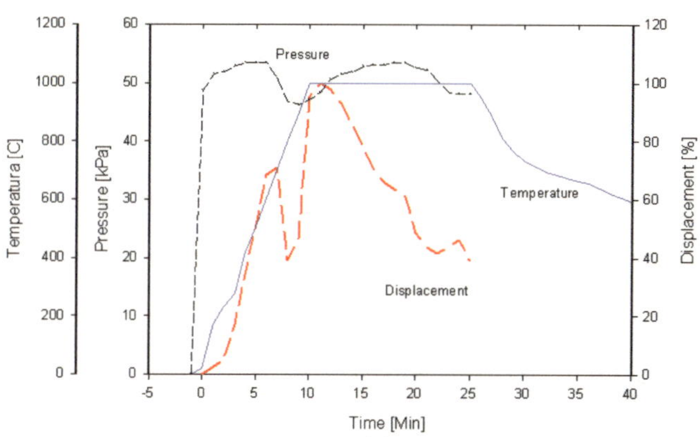

(a)

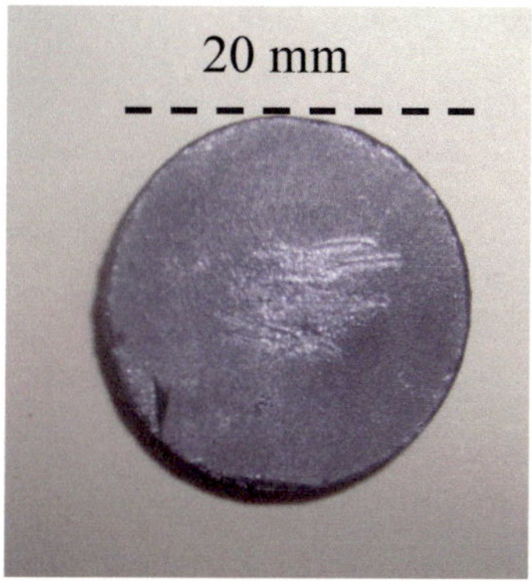

(b)

Fig. 17. Experimental record of a number of SPS-parameters such as; (a) Temperature, applied pressure and relative displacement of the punches as a function of time for an experiment MCrAlY performed at 1273 K. (b) Pastille obtained by SPS 20 mm diameter.

corresponds to the variation measured by the internal pyrometer. For this illustration, the selected holding temperature was 1273 K. Meanwhile pressure and current pulses are applied simultaneously. The set pressure was 50 KPa and was applied for 3 min. A heating rate of 150 °C/min was programmed, in such a way that the sintering temperature was reached in 9 minutes. After 15 minutes of holding time at the maximum temperature, the pressure and vacuum are then removed. This results in a pressure which falls quickly, whereas the temperature reduction takes 7 min to reach 600 °C. The cooling stage drop from 1000 to 600 °C occurs at a rate of 90 °C/min.

With above experimental parameters set the sintering of a tablet-like specimen of 20 mm diameter and 3 mm thick was accomplished in less than 15 minutes. No subsequent thermal treatment was applied to the tablets, in such a way that the final microstructure was obtained by one single step (Fig. 17(b)).

The Fig. 18 shows the microstructures of the MCrAlY alloys sintered at temperature of 1173, 1223, 1273 and 1323 K. SEM analysis of polished surfaces does not reveal presence porosity for the sintered samples at 1223 K, which confirms the major densification. Fig. 18(a) shows a surface of a sintered sample at 1173 K. At this sintering temperature, the surface still shows characteristics from original powder structure and is only observed one plastic deformation of the powders. For all temperatures, the microstructure is mainly composed of two phases, identified in a dark gray and light gray. EDS analysis on the sintered samples indicates that; dark gray area with high aluminum content would correspond to a β phase ($NiAl_3$), and the light gray area rich in nickel and chromium which would correspond to a γ phase nickel and, finally some rich precipitates in Tantalum would correspond to tantalum carbide TaC. The latter is shown in the analysis by energy dispersion spectroscopy.

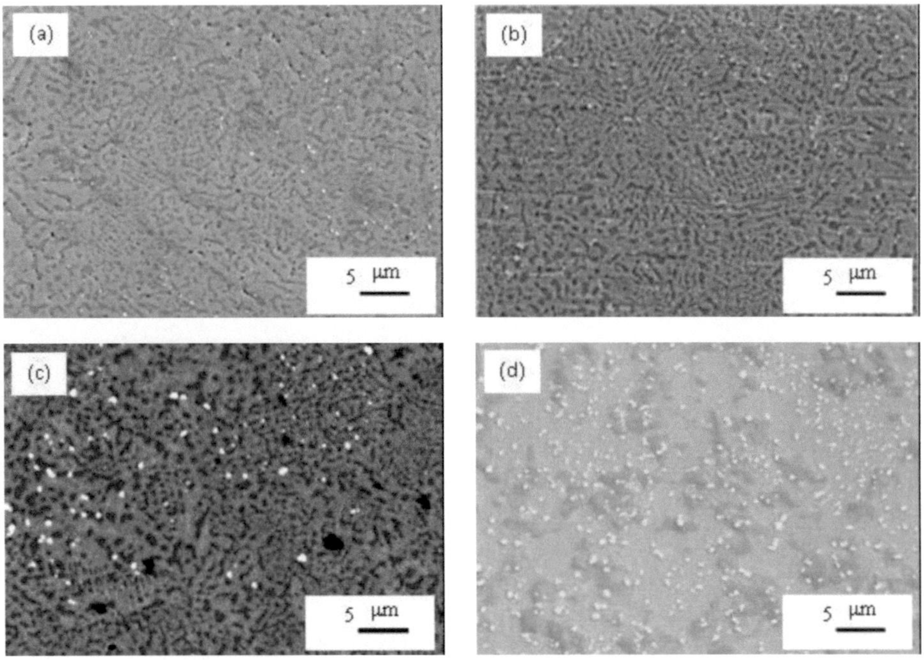

Fig. 18. Scanning electron micrographs of MCrAlY powder sintered by setting SPS at (a) 1173 K, (b) 1223 K, (c) 1273 K and (d) 1323 K.

Fig. 19 shows the XRD patterns evolution of the sintered MCrAlY. Two phases were identified: β $NiAl_3$ and γ Nickel. The set of conditions of temperature and pressure in SPS allowed to keep the same microstructure and grain coarsening is not appreciated at 1273 K.

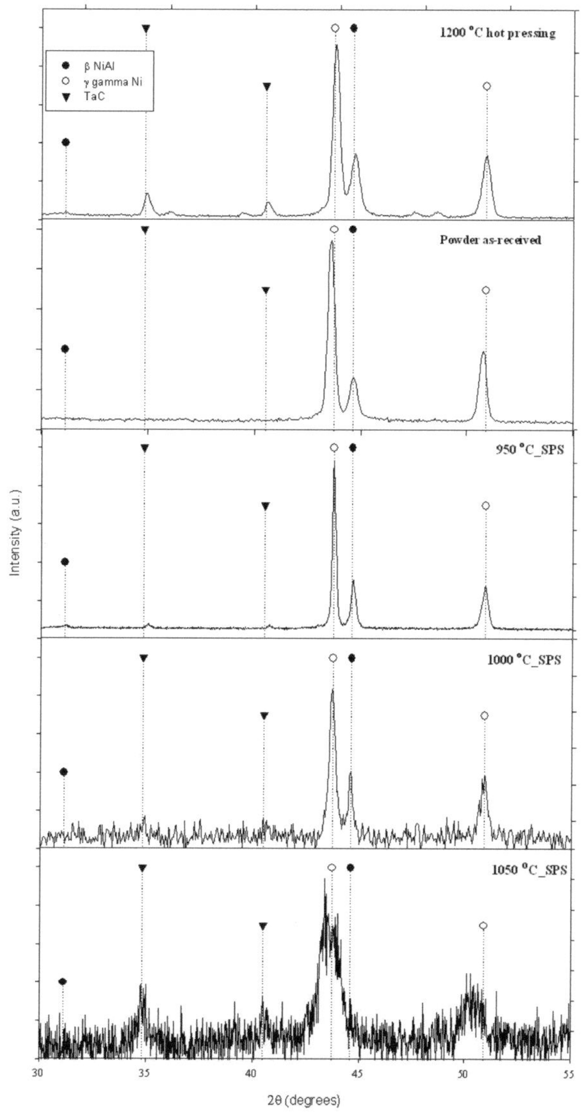

Fig. 19. XRD patterns for corresponding MCrAlY specimens sintered by SPS and hot pressing.

Fig. 20(a) shows the variation in porosity of sintered samples at different SPS temperatures between 1073 and 1323 K. A significant reduction in porosity is observed when an increasing sintering temperature is applied. Also this Fig. 20(b) shows an increasing Vickers hardness in function of SPS sintering temperature how effect of microstructural evolution. However, at temperature of 1323 K it is observed a diminution of hardness with a major porosity that could be associated with the start of the melting of components.

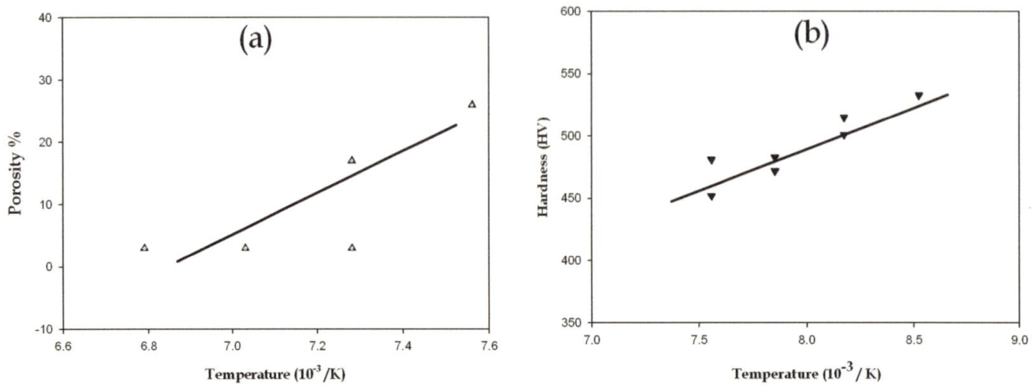

Fig. 20. Analysis of MCrAlY samples sintered by SPS for temperatures range between 1073 and 1273 K: (a) porosity (b) Vickers hardness.

To study the advantages of spark plasma sintering against hot pressing, Fig. 18(b) shows a secondary electron micrograph of surface of MCrAlY powders sintered by SPS. The present work clearly demonstrates that MCrAlY alloys can be rapidly sintered by SPS (in less than 30 min). As illustrated by the presence of little number of porosities, a full densification can be achieved by using SPS. From Fig. 6(b), both larger powder particles and a relative proportion of surface area/per volume unit limits the interstitial hardening, and tend to favor elastic deformation of powders. Moreover, the current density at contacts among larger particles should be higher because of a smaller number of connections and on other case of smaller particles the current density should be lower in where the sintering of powders started early, as it see at the Fig. 21. This shows a surface of MCrAlY sample sintered at 1173 K after a fracture procedure. Thus, a faster consolidation was achieved mainly due to the presence of a wide size distribution of powders allowing to fill the interstices present between the larger powders per the smaller ones.

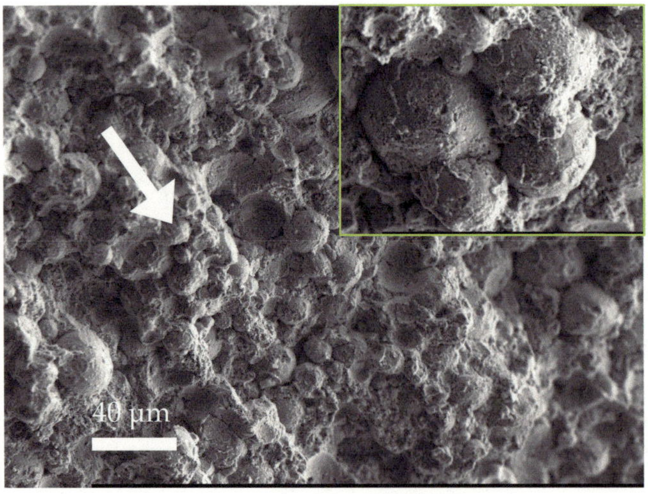

Fig. 21. Fracture surface of MCrAlY sample sintered at 1173 K.

Fig. 17(a) shows experimental record of a number of SPS parameters such as the densification of MCrAlY powders as function of time. During this period, grain boundaries were deformed and reorganized, especially due to the plastic deformation of the β phase rich in aluminum. In this Fig. 17(a), three main steps can be identified for the displacement variation, which could be interpreted as follows; in a beginning (5 min), when pressure increasing, the powder is compressed as a green body. This is followed by a flat stage. Sintering occurs in the final stage as the temperature reaches about 750 °C under 50 KN. For this condition, the pressure (applied load) exceeds the yield stress at a temperature below the brittle-ductile transition (950 °C). Then sintering begins taking place in less than 12 min. which means that full densification is achieved at a temperature of 950 °C, namely before the holding temperature has been reached. Finally the decrease of the relative displacement is interpreted as a result of the system dilation.

This latest β phase contains a major amount of aluminum, thus plastic deformation during the SPS process, is expected to be most important with respect to the presence of deformation in the γ phase that contains a major amount of nickel (Fig. 22). The measurement of the relative displacement of the punches (Fig. 17(a)) indicates that the compaction can be completed at 950 °C. Thus, as long as a transus temperature is not reached, a quite similar microstructure is generated. A short hold-time also helps to avoid grain ripening due to diffusion controlled by phase transformations. MCrAlY alloys might be satisfactorily described by a Ni-Al binary diagram with only a slight effect of Tantalum on the related transus temperature. The transus temperature for the Ni-Al alloys have been measured at 1100 °C, cf. (Noebe et al., 1993). Fig. 22 presents a backscattered scanning electron micrograph of a polished surface of sintered sample. Two phases, shown in dark gray and light gray, are present corresponding to β-NiAl and to γ-Ni, respectively. Tantalum carbide is detected for positions in white dots that also appear in this micrograph. For SPS temperatures ranging between 1173 and 1273 K two mainly phases are formed in the microstructure: β and γ. The SPS temperatures sintering are lower than the α transus temperature and no grain coalescence is observed due to the short duration of the experiments.

Contact among surfaces is favoring during SPS and the resulting product is exempt of open porosity. Fig. 18 shows the microstructures of the MCrAlY alloy sintered at SPS temperatures of 1173, 1223, 1273 and 1323 K. SEM analyses of polished surfaces do not reveal a porosity, which confirms a major compaction. Munir et al. (Munir et al., 2006) have not reported contribution of the time in the process consolidation of materials treated by SPS, from here for time range sintering between 0 min and 15 min, the MCrAlY alloy shows a similar microstructure of individual grains.

XRD diffraction patterns (Fig. 19) reveal the presence of an γ phase and an β phase for all temperatures, and there is neither difference in the positions of the peaks nor new peaks indicating the formation of a new phase. Besides, according with the peak shape, it is suggested that there was not coarsening of grain in the temperature range of 1223-1273 K for all sintered samples.

Courat et al. have mentioned that the current density conditions prevailing during the SPS process (Courat et al., 2008) do not allow mass transport. Thus the Tantalum carbide located on the surface of atomized powder before sintering process, it remains during sintering within the limits of powders, and is only transported by the powder limit, as is observed in Fig. 22 for all sintering temperatures. Indeed the Tantalum carbide has a high melting point.

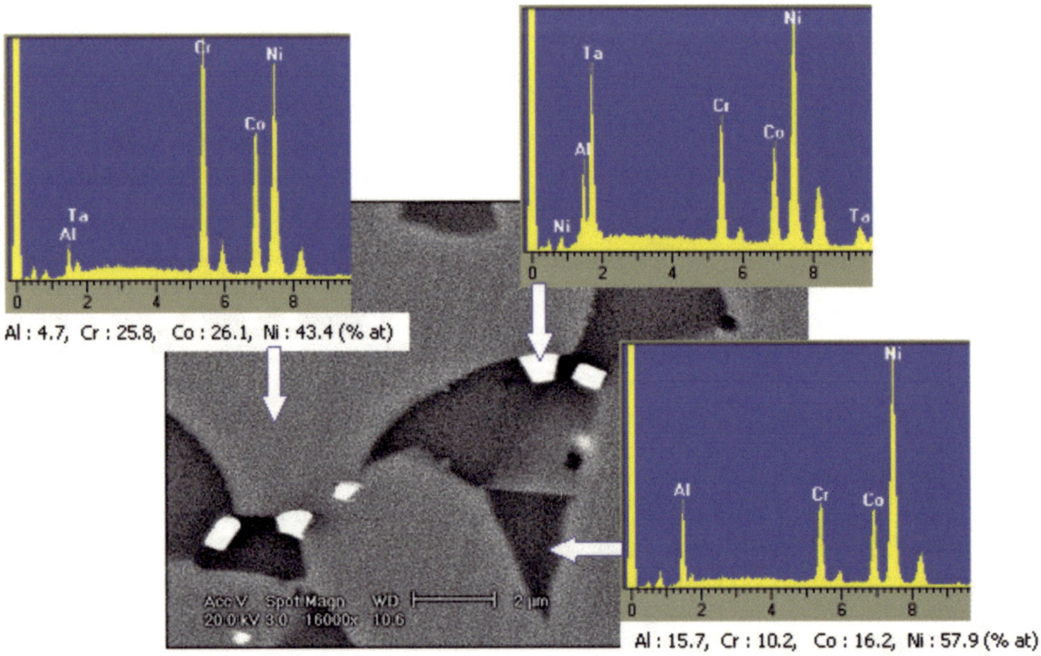

Fig. 22. Backscattered scanning electron micrograph of a polished surface of MCrAlY sintered sample at 1273 K.

5. Conclusions

Film characteristics show the SB-MOCVD process to be compatible with doping of MCrAlY powders prior to their use in the preparation of bond coats by sintering on turbine blades and vanes for improved mechanical and oxidation resistance during high temperature operation of gas turbines.

MCrAlY alloys were sintered by using hot-pressing and spark plasma sintering. For a temperature 1273 K and duration 30 min, specimens disclose a good compacting and a microstructure homogenous. Two phases in the microstructure are mainly formed under all sintering conditions: β Beta and γ gamma. Such, microstructures represents a real advantage by using spark plasma sintering with respect to hot-pressing conventional processing by a decrease of 200 °C degree in the sintering temperature. Indeed without any significant change in the structural and mechanical property.

However, the mechanical properties are still under investigation for SPS samples and it should be reported in the future.

Finally, spark plasma sintering process appears to be a promising route to produce MCrAlY alloys doped with Ruthenium and/or others materials.

6. Acknowledgments

The authors are grateful for the support to conduct this work to COFAA, EDI-IPN and CONACYT-SNI. And the authors are also indebted to Red de Nanotecnología of IPN, for the SPS facilities.

7. References

Alperine, S.; Steinmetz, F.; Josso, P. & Constantini, A. (1989). High temperature-resistant palladium-modified aluminide coatings for nickel-base superalloys. *Materials Science Engineering*. Vol. A121, pp 367-372, ISSN: 0921-5093.

Bose, S. & DeMasi-Marcin J. T. (1995). Thermal Barrier coatings experience in gas turbine engines at Pratt & Whitney, Workshop; *Thermal Barrier Coatings*, Cleveland Ohio, NASA CP 3312. pp. 63-77.

Caussat, B.; Juarez, F. & Vahlas C. (2006). Hydrodynamic study of fine metallic powders in an original spouted bed contactor in view of chemical deposition vapour treatments. *Powder Technology*. 135, pp. 63-70. ISSN 0032-5910.

Choi, S.R.; Zhu, D. & Miller, R.A. (1998). High-temperature slow crack growth, fracture toughness and room-temperature deformation behavior of plasma-sprayed ZrO2-8 wt % Y2O3. *Ceramic Engineering and Science proceedings*. Vol. 19, No 4, 293-301, ISSN 01966219.

Christensen, R.J.; Tolpygo, VK. & Clarke, DR. (1997). The influence of the reactive element yttrium on the stress in alumina scales formed by oxidation. *Acta Materialia*. Vol. 45, No. 4, pp. 1761-1766, ISSN: 1359-6454

Courat, A.; Molenat, G. & Galy, J. (2008). Microstructures and mechanical properties of TiAl alloys consolidated by spark plasma sintering. *Intermetallics*. Vol. 16, pp. 1134-1141, ISSN 0966-9795.

Cruse, TA.; Stewart, SE. & Ortiz, M. (1988): Thermal Barrier Coating Life Prediction Model Development. *Journal of Engineering for Gas Turbines and Power*. Vol. 110, pp. 610-616, ISSN 0742-4795.

Czech, N.; Schmitz, F. & Stamm, W. (1995). Microstructural Analysis of the Role Rhenium in Advanced MCrAlY Coatings. *Surface and Coatings Technology*. Vol. 76 - 77, pp. 28 - 33. ISSN 0257-8972.

Czech, N.; Schmitz, F. & Stamm, W. (1994). Improvement of MCrAlY coatings by addition of rhenium. *Surfaces and Coatings Technology*. Vol. 68-69, pp. 17-21, ISSN 0257-8972.

DeMasi-Marcin, J.T. & Gupta D. K. (1994). Protective coatings in the gas turbine engine. *Surfaces and Coatings Technology*. Vol. 68/69, pp. 1-9, ISSN 0257-8972.

DeMasi-Marcin, J.T.; Sheffler, KD. & Bose, S. (1990). Mechanisms of Degradation and Failure in a Plasma-Deposited Thermal Barrier Coating. *Journal of Engineering for Gas Turbines and Power*. October 1990 --Volume 112, Issue 4, pp. 521- 526, ISSN 0742-4795.

Eaton, H. E. & Novak, R. C. (1987). Sintering studies of plasma-sprayed zirconia. *Surface and Coatings Technology*. Vol. 32, pp. 227-236, ISSN 0257-8972.

Feng, Q. Nandy, T.K. Tin, S. & Pollock, T. (2003). Solidification of high-refractory ruthenium -containing superalloys. *Acta Materialia* 51 (1) (2003), pp. 269-284, ISSN 1359-6454.

Fleischer, R.L.; Field, R.D. & Briant, C.L. (1991). Mechanical properties of high-temperature alloys of AlRu. *Metallurgical Transactions A*. Vol. 22A, pp. 403-414, ISSN 1073-5623.

Golightly, F.A.; Sttot, F.H. & Wood, G.C. (1976). The influence of yttrium additions on the oxide-scale adhesion to an iron-chromium-aluminum alloy. *Oxidation of Metals*. Vol. 10, No. 3 pp. 163-187, ISSN 0030- 770X.

Haynes, J.A.; Zhang, Y.; Lee, W.Y.; Pint, B.A.; Wright, I.G. & Cooley, K.M. (1999). Effects of Pt Additions and S Impurities on the Microstructure and Scale Adhesion Behavior of Single-Phase CVD Aluminide Bond Coatings Hampikian JM, Dahotre NB, editors. *Elevated temperature coatings: science and technology III, Symposium Proceeding Warrendale* (PA) TMS, pp. 185-196, ISBN 0873394216.

Hillery R. editor. (1996) NRC report. Coatings for high temperature structural materials: Trends and opportunities. *National Academy Press*, pp 43-45, ISBN-10: 0-309-08683-3.

Jeandin, M.C.M. Koutny, J.-L. & Bienvenu, Y.C. (1988). Procédé d'Assemblage de Piéces en Superalliages à base de Nickel par Frittage en Phase Liquide et Compaction Isostatique à Chaud, *Institut National de la Propiété Industrielle*, France, 2610856 Patent. (2610856).

Juarez, F.; Castillo, A.; Pieraggi, B. & Vahlas, C. (2001) Spouted bed metallorganic chemical vapor deposition of ruthenium on MCrAl-Y powders. *Journal de Physique IV*. Vol. 11, pp. 1117-1123. ISSN 1155-4339.

Juarez, F.; Lafont, M-C.; Senocq, F. & Vahlas, C. (2005). Spouted bed of MCrAlY powders in applications CVD. *Electrochemical Society Proceeding*, No. 08, pp. 501-508, ISBN 978-1-56677-793-3

Juarez, F.; Monceau, D.; Tetard, D.; Pieraggi, B. & Vahlas, C. (2003), Chemical vapor deposition of ruthenium on NiCoCrAlYTa powders followed by thermal oxidation of the sintered coupons. *Surfaces and Coatings Technology*, Vol. 163-164, pp. 44- 49. ISSN 0257-8972.

Kingery, W.D.; Bowen H.K. & Uhlmann, D.R. (1976) Introduction to ceramics. New York: *Wiley and Sons*, pp. 1056. ISBN0471478601.

Lee, EY. & Sisson, R.D. (1994). The effect of bond coat oxidation on the failure of thermal barrier coating: thermal spray industrial applications In: Berndt CC, Sampath S. editors. Proc. *7th National Spray Conference, Boston* MA, 20-24 June. Materials Park, OH: ASM International, pp. 55-59. ISBN 0871705095.

Lipkin, D.M. & Clarke, D.R. (1996), Measurement of the stress in oxide scales formed by oxidation of alumina-forming alloys. *Oxidation of Metals*. Vol. 45, No. 3-4, pp. 267-280. ISSN 0030-770X.

Lu D.C. & Pollock, T.M. (1999). Low temperature deformation and dislocation substructure of ruthenium aluminide polycrystals. *Acta Materialia*. Vol. 47 No. 3, pp. 1035-1042, ISSN 1359-6454.

Lu, D.C. & Pollock, T.M. (1999). Low temperature deformation kinetics of ruthenium aluminide alloys. *Material Research Society Symposium Proceedings*. Vol. 552, pp. KK7.11.1-KK7.11.5, ISSN 02729172. Symposium KK - High-Temperature Ordered Intermetallic Alloys VIII , MRS Fall Meeting, Boston MS, USA, 1998.

Mariochocchi, A.; Bartz, A. & Wortman, D. (1995). PVD TBC Experience on GE Aircraft Engines. In *proceedings 1995 Thermal Barrier coatings workshop* W. J. Brindley, Ed., NASA CP 3312. pp. 79-89.

Meier, G.H. & Pettit, F.S. (1999). Interaction of steam/air mixtures with turbine airfoils alloys and coatings. Report on AFOSR Contract F49620-981-0221. Univ. of Pittsburgh, 1 September, pp. 1-9.

Meier, S.M.; Nisseley, D.M. & Sheffer, K.D. (1991) Thermal barrier coating life prediction model development phase II. NASA CR-18911, July.

Meier, SM. & Gupta, D.K. (1994). The evolution of thermal barrier coatings in gas turbine engine applications. *Journal of Engineering for Gas Turbines and Power, Trans ASME*, vol. 116, No. 1, pp. 250-257, ISSN 0742-4795, January 1994.

Mennicke, C.; Schumann, E.; Ulrich, C. & Ruehle, M. (1997). The Effect of Yttrium and Sulfur on the Oxidation of FeCrAl. *Materials Science Forum*. Vol. 251-254, pp. 389-396, ISSN: 1662-9752.

Menzies, R.G.; Davies, G.J. & Edington. J.W. (1982). Effect of the Treatment on Superplastic Response of Powder - Consolidated nickel - Base Superalloy IN100. *Metal Science*. Vol. 16, No. 7, pp. 356-362, ISSN 0306-3453.

Miller, R.A. (1984). Oxidation-Based Model for Thermal Barrier Coating Life. *Journal American Ceramic Society*. Vol. 67, No. 8, C-154-C 170 pp. 517-521, ISSN 1551-2916.

Monceau, D.; Boudot-Miquet, A.; Bouhanek, K.; Peraldi, R.; Malie, A.; Crabos, F. & Pieraggi, B. (2000), Oxydation et protection des matériaux pour sous -couches (NiAlPd, NiAlPt, NiCoCrAlTa, CoNiCrAlY) de barrières thermiques, *Journal Physical IV* France 10, pp. 167-171, ISSN 1155-4339.

Munir, Z. A.; Anselmi -Tamburini U. & Ohyanagi, M. (2006). The effect of electric field and pressure on the synthesis and consolidation of materials: A Review of the spark plasma sintering method. *Journal of Materials Science*. Vol. 41, pp. 763-777, ISSN 0022-2461.

Nickel, H.; Clemens, D.; Quadakkers, W.J. & Singheiser, L. (1999). Development of NiCrAlY Alloys for Corrosion - Resistant Coatings of Gas Turbine Components. *Journal of Pressure Vessel Technology*. Vol. 121, pp. 384-387, ISSN 0094-9930.

Noebe, R.D.; Bowman, R.R. & Nathal, M.V. (1993). Physical and mechanical properties of the B2 compound NiAl. *International Materials Reviews*. Vol. 38 No. 4, pp. 193-232. ISSN 0950-6608.

Oquab, D.; Estournes, C. & Monceau, D. (2006). Oxidation resistant aluminized MCrAlY coating prepared by Spark Plasma Sintering (SPS). *Advanced Engineering Materials*. Vol. 9, No. 5, pp. 413-417, ISSN 1527-2648.

Pollock, T.M.; Lu, D.C.; Shi, X. & Eow, K. (2001). A comparative analysis of low temperature deformation in B2 aluminides. *Materials Science and Engineering A.* Vol. 317 No. 1-2, pp. 241-248, ISSN 0921-5093.

Prakash, T. L.; Somani M.C. & Bhagiradha Rao, E.S. (1988). Structure property correlation of as- HIPped and HIP + forged P/M alloy nimonic AP-1, *Powder Metallurgy Related High Temperature Materials.* Vol. 29-31, pp. 179-198, ISBN 978-0-87849-577-1

Quadakkers, WJ.; Tyagi, AK.; Clemens, D.; Anton, R. & Singhesir, L. (1999). The Significance of Bond Coat Oxidation for the Life of TBC Coatings. Hampikian JM, Dahotre NB, editors. *Elevated temperature coatings: science and technology III, Symposium Proc., Warrendale* (PA) TMS, pp. 119-130, ISBN 0873394216.

Rigney, D.V.; Viguie, R.; Wortman, D.J. & Skelly, W.W. (1995). PVD thermal barrier coatings applications and process development for aircraft engines, *in Proc of the workshop on Thermal Barrier Coatings, NASA-CP-3312.* NASA Lewis Research Center, pp. 135-150.

Sarioglu, C. ; Blachere, JR. ; Petit FS, & Meier, GH. (1997). Room temperature and in situ high temperature strain (stress) measurements by XRD techniques. *Proceedings of the Third International Conference held at Trinity Hall,* Cambridge 1996In: Newcomb S.B. Little J.A. Editors. London: Microscopy of oxidation Vol. 3. The institute of materials, pp. 41-51, ISBN: 9781861250346.

Schaeffer, J.; Kim, G.M.; Meier, G.H. & Pettit, F.S. In: Lang E. editors. (1988). *Proceeding of the European Colloquium;* The role of the active elements in the oxidation behavior of the high temperature metals and alloys: The effects of precious metals on the oxidation and hot corrosion of coatings. Elsevier Applied Science pp. 231-270. ISBN: 1-85- 166-420-3.

Sergo, V. & Clarke, D.R. (1998). Observation of Subcritical Spall Propagation of a Thermal Barrier Coating. *Journal American Ceramic Society.* Vol. 81, No. 12, pp. 3237-3242, ISSN 1551-2916.

Shiojiri, M.; Kawasaki, M.; Fujii, M.; Wakayama, M. & Yoshioka, T. (1999). High-resolution transmission electron microscopy of Fe-Al powder particles. *Journal Electron Microscopy.* Vol. 48, No. 4, pp. 367- 373, ISSN 0022-0744.

Smegil, J.G. (1987). Some comments on the role of yttrium in protective oxide scale adherence. *Materials Science Engineering.* Vol. 87, pp. 261-265, ISSN: 0921-5093.

Smialek, J.L.; Jayne, D.T.; Schaeffer, JC. & Murphy, WH. (1994). Effects of hydrogen annealing, sulfur segregation and diffusion on the cyclic oxidation resistance of superalloys: a review. *Thin Solid Films,* Vol. 253, No. 1-2, pp. 285-292, ISSN 0040-6090.

Somani, M.C.; Muraleedharan, K.; Prasad, Y.V. & Sigh, V. (1998). Mechanical Processing and Microstructural Control in hot Working of hot Isostatically Pressed P/M IN-100 Superalloy. *Materials Science and Engineering* A. Vol. 245, pp. 88-99, ISSN 0921-5093.

Stiger, M.J.; Yanar, N.M.; Topping, MG.; Pettit, F.S. & Meier, GH. (1999), Thermal Barrier coatings for the 21st century, Z. *Metallkd.* Vol. 90, No. 12, pp. 1069-1078. ISSN 0044-3093.

Strangman T.E. (1985), Thermal barrier coatings for turbine airfoils. *Thin Solid Films.* Vol. 127, pp. 93-106, ISSN 0040-6090.

Taylor, T.A.; Bettridge, D.F.; Tucker, Jr. & Robert, C. (1995). Coating composition having good corrosion and oxidation resistance Praxair S.T. Technology, Inc. Rolls-Royce PLC, *USA Patent* 5455119, October 1993.

Taylor, T.A. & Bettridge, D.F., (1996). Development of Alloyed and Dispersion-Strengthened MCrAlY Coatings. *Surface and Coatings Technology,* Vol. 86- 87, pp. 9-14, ISSN 0257-8972.

Tolpygo, V.K. & Clarke, D.R. (1998). Competition Between Stress Generation and Relaxation During Oxidation of a Fe-Cr-Al-Y Alloy. *Oxidation of Metals.* Vol. 49, No 1-2, pp. 187-211. ISSN 0030-770X.

Tolpygo, VK. & Clarke, DR. (1998) Wrinkling of α-alumina films grown by thermal oxidation-I. Quantitative studies on single crystals of Fe-Cr-Al alloy. *Acta Materialia.* Vol. 46, No. 14, pp. 5153-5166, ISSN 1359-6454

Tryon, B.; Cao, F.; Murphy, K.S.; Levi, C.G. & Pollock, T.M. (2006). Ruthenium-containing bond coats for thermal barrier coating systems. Journal of Metals. Vol. 58 No. 1, pp. 53-59, ISSN 0148-6608.

Tryon, B.; Pollock, T.M.; Gigliotti, M.F. & Hemker, K. (2004) Thermal expansion behavior of ruthenium aluminides. *Scripta Materialia.* Vol. 50, No. 6, pp. 845-848, ISSN 1359-6462

Vahlas, C.; Juarez, F.; Feurer, R.; Serp, P. & Caussat, B. (2002), Chemical vapor deposition on fluidized particles; Application to the metalorganic CVD of platinum group metals. *Adv. Mater. - Chemical Vapour Deposition,* Vol. 8, No. 4, pp. 127-143, ISSN 0948-1907.

Van de Voorde, M. H. & Meetham, W. G. (2000). Materials for High Temperature *Engineering Applications, Springer-Verlag,* Berlin, pp. 0-173, ISBN: 3540668616, ISBN-13: 97835406686192000.

Wang, Y.; Guo, H.; Peng, H.; Peng, L. & Gong, S, (2011), Diffusion barrier behaviors of (Ru,Ni)Al/NiAl coatings on Ni-based superalloy substrate. Intermetallics, *Proceedings of the 7th International Workshop on Advanced Intermetallics and Metallic Materials* 2008, Volume 19, Issue 2, February 2011, pp 191-195, ISSN 0966-9795.

Wolff, I.M. & Sauthoff, G. (1996), Mechanical properties of Ru-Ni-Al alloys. *Metallurgical and Materials Transactions A: Physical Metallurgy and Materials Science.* Vol. 27A, No. 5, pp. 1395-1400, ISSN 1073-5623.

Wolff, I.M. (1997). Toward a better understanding of ruthenium aluminide. *Journal of Metals.* 49 (1) (1997), pp. 34-39, ISSN 0148-6608.

Wright PK, (1998), Influence of cyclic strain on life of a PVD TBC. *Materials Science and Engineering A.* Vol. 245, pp. 191-200, ISSN 0921-5093.

Wright, P.K. & Evans, A.G. (1999). Mechanisms governing the performance of thermal barrier coatings. *Current Opinion in Solid State and Materials Science.* Vol. 4, No. 3, pp. 255-265, ISSN 1359-0286.

Yoshioka, T.; Kawasaki, M.; Kitano, M.; Nishio, K. & Shiojiri, M. (1995), Cross-Sectional Transmission Electron Microscopy of ZnO Tetrapod-Like Particles, *Journal Electron Microscopy,* Vol. 44, No. 6, pp- 488-492, ISSN 0022-0744.

Yoshioka, T.; Kawasaki, M.; Yamamatsu, J.; Nomura, T.; Isshiki, T. & Shiojiri, M. (1997). A preparation method of sections of fine particles and cross-sectional transmission electron microscopy of Ni powder. *Journal Electron Microscopy*. Vol. 46, No. 4, pp. 293-301, ISSN 0022-0744.

4

Metal Laser Sintering for Rapid Tooling in Application to Tyre Tread Pattern Mould

Jelena Milovanovic, Milos Stojkovic and Miroslav Trajanovic
University of Nis, Faculty of Mechanical Engineering in Nis
Republic of Serbia

1. Introduction

SLS[1], DMLS[2] and SLM[3] belong to the family of additive manufacturing technologies (we will use term Metal Laser Sintering technologies or MLS abbreviation further in text due to simplicity) that build the geometry of the part by solidification of metal powders using laser power (Kruth et al., 2005; Khaing, 2001).What particularly distinguishes them from other additive technologies is the possibility to produce fully functional metal parts. This feature as well as the ability to create highly complex geometrical shapes, which are often not possible, or at least very difficult to make by conventional manufacturing processes, promote these technologies as perfect candidates for moulding and rapid tooling (RT) (Simchi et al., 2003; Pessard et al. 2008). This is why MLS technologies attract a great attention of mouldmakers for more than a decade. On the other side, the whole range of features of the parts that are manufactured by MLS technologies such as high price of metal powder, porosity, chemical reactivity, then the limitations regard to geometric accuracy, available materials, size of building chambers and necessity for additional post-processing create a barrier for the application of these technologies in manufacturing of moulds.

Especially, production of large parts and moulds rich in small and complex geometric details, such as tyre tread pattern mould, still remains a great challenge for MLS. Nevertheless, it is very interesting to find out whether and how MLS technologies could be employed for manufacturing of these kind of moulds. In these cases, there is a clue that usage of MLS could be made worthwhile, but it could require specific tooling approaches to be considered. In this chapter, an application study is presented which concerns application of metal laser sintering technologies for rapid tooling in application to tyre tread pattern mould.

2. Review of applications of DMLS/SLS/SLM in RT

Over the years there are more and more examples of application of MLS technologies in rapid tooling. What makes them a particularly attractive in the mould industry is the ability to create and integrate so called conformal channels into injection molds or other tooling,

---

[1] Selective Laser Sintering
[2] Direct Metal Laser Sintering
[3] Selective Laser Melting

which can reduce injection cycle times between 30 and 60 percent over conventional tools and even increase parts quality (Xu et al., 2001).

Some industrial case studies of laser sintered injection moulds using DMLS and SLS have been reported in (Voet et al., 2005). These cases show that it is possible to manufacture moulds with MLS technologies, where cavity depth and complexity do not limit of the process. However, finishing of laser sintered parts using traditional manufacturing technologies is proved to be necessary.

There are also reports on successful use of combination of indirect SLS and machining processes to create injection mould tools which have been evaluated in industrials trials. (Ilyas et al., 2010; King & Tansey, 2003).

One of many examples that confirm the benefits and importance of making conformal channels using MLS technologies (in this case SLM) is presented in (Campanelli et al., 2010) where SLM is used for creation of jig for welding of constituent parts of titanium alloy intramedullary nail.

MLS technologies are also used for die-casting applications. In (Ferreira, 2004) author presented good results in application of DMLS for manufacturing of die inserts for shoot squeeze moulding under full production conditions (for 3750 sand moulds). DMLS shows benefits in reducing manufacturing time and achieving acceptable geometrical accuracy of die, both for low and medium production.

Mainly, all of these examples are related to application of these technologies for manufacturing of relatively small parts and small size moulds. One of the main features that have been observed regarding the use of MLS technologies in tools manufacturing is that the quality of tools significantly depends on the composition and grain sizes of the powder as well as sintering conditions. Small grain sizes of powder provide better overall geometrical accuracy and surface quality. Yet, some practical problems as powder removal and porosity should be further solved.

3. Fabrication of tyre vulcanization mould

Tyre vulcanization mould is a very specific and complex kind of mould that can be considered as a large tool (Fig.1).

Fig. 1. Tyre vulcanization mould.

It is characterized by large number of very complex geometrical features, which are often very difficult to machine in a conventional way. The most complex part of the mould is tread pattern ring (further in text *tread ring*) (Fig.2) which is characterized by inverse tyre tread geometry. The complexity of tread ring is manifested primarily by its toroidal shape containing different kind of ribs (circumferential, lateral, esthetic and sipes or lamellas) (Stojkovic & Trajanovic, 2001; Stojkovic et al. 2003; Stojkovic et al. 2005a, 2005b; Chu et al 2006; Lee, 2008) .

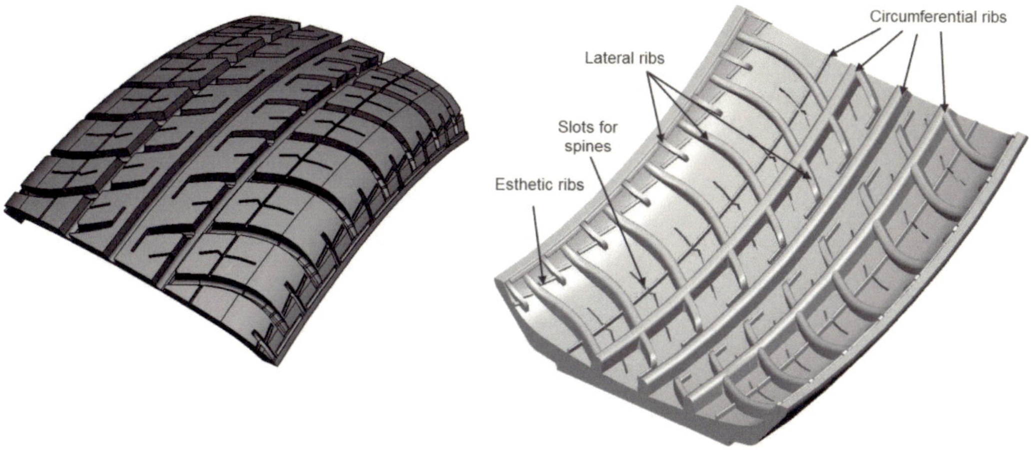

a) b)

Fig. 2. CAD models of: a) tyre tread segment and b) segment of tread ring of tyre mould.

In addition, the mould is characterized by considerable difference in dimensions between the smallest and biggest geometrical features (Fig.3).

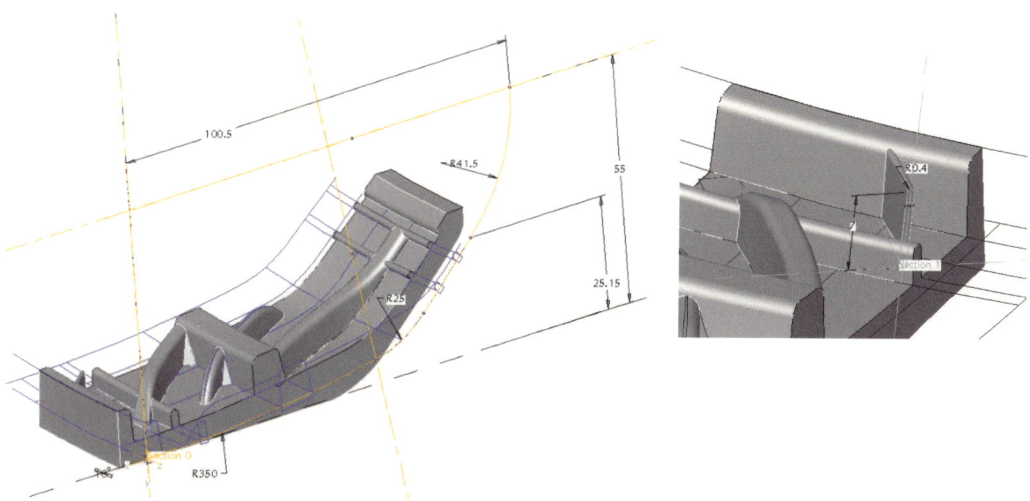

Fig. 3. Dimensional range of geometrical features of the tread ring segment.

3.1 Conventional tooling

Conventional manufacturing process of tread ring involves complex (4- or 5-axis) machining and precise die-casting, which make tooling process very intricate, time consuming and expensive (Fleming, 1995). Therefore, development of new tyre models is limited in great extent by the manufacturability of tread ring and its geometrical features. There are two main conventional approaches in manufacturing of the tread ring as it is explained by (Salaorni & Pizzini, 2000; Knedla, 2000).

1. Direct manufacturing of mould segments is used for getting the so-called *engraved* moulds, which can be produced by two different manufacturing procedures:
   a. Engraving of mould segments by 5- axis CNC milling machines (Fig.4).
   b. Electro-discharge machining of tread ring by sinking tread-like copper or brass electrode.

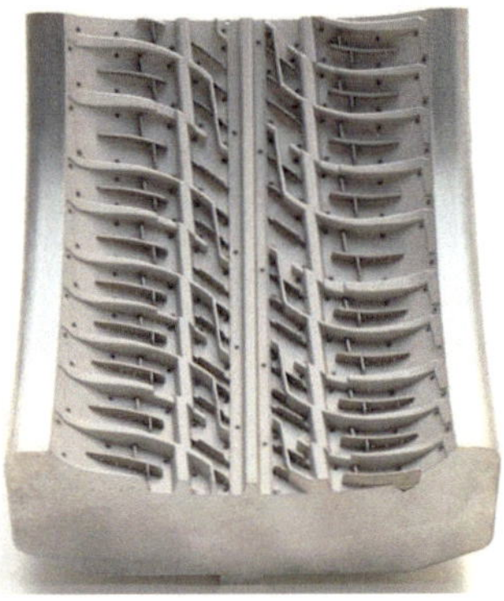

Fig. 4. Tread ring segment made by engraving on CNC.

2. Indirect manufacturing is used for getting the so-called casted moulds. These moulds are usually produced by standard three-step casting procedure (Chu et al., 2006; Fleming, 1995). This is the most often used manufacturing procedure for this purpose and it is appropriate for manufacturing of all kind of segmented moulds.

The moulds which are manufactured by this procedure are shown in Fig. 5_a and Fig. 5_b.

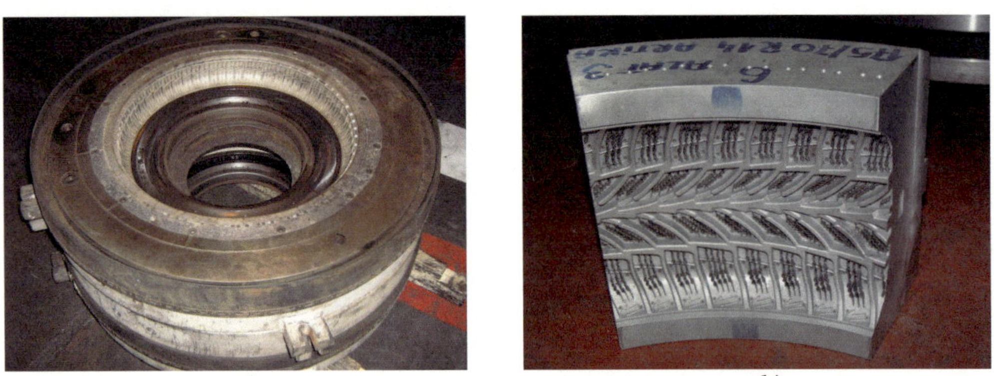

a) b)

Fig. 5. a) *Monoblock* tyre vulcanization mould system b) 1/8 of tread ring segment.

3.1.1 Utilization features of the conventionally manufactured moulds

The tread rings that are produced by conventional manufacturing procedures should meet certain specifications which are shown in Table 1.

Feature	Value
Nominal dimension tolerances	±0.2mm on overall diameter of tyre (app. up to 600mm)
Surface roughness	Ra=3.2 - 6.30 (μm) [4]
Working temperature and pressure	180 °C / 21 bar
Hardness	70HRB
Thermal conductivity	122 - 134 W/(mK) 25 °C - 200 °C [4]
Corrosion and wear resistance	2500 to 3000 vulcanization cycles before regular maintenance procedure (sandblasting, laser cleaning, etc.)

Table 1. Specifications (utilization features) of tread ring (data source: Rubber product company Tigar Tyres, now incorporated in Michelin).

3.2 MLS technologies for tyre tread ring mould fabrication

First of all, one should notice that all kind of layer manufacturing technologies (the whole family of so-called rapid prototyping (RP) technologies) attract a great attention of tyre mould makers because of their remarkable capabilities to produce almost any kind of shape in a relatively simple way. Although, there is a possibility to use all kind of RP technologies, MLS technologies appear as the most suitable for alternative production process of tread rings of tyre moulds (Milovanovic et al., 2005; Milovanovic, 2006; Milovanovic et al., 2009).

Compared to other RP technologies the greatest strength of MLS technologies is certainly the possibility to create fully functional parts i.e. parts of the mould with any shape that can be found on the tread ring. In addition, the simplicity of digital model preprocessing and fabrication process planning usually takes 10% of time which has to be invested in CAPP/CAM[5] activities that precede CNC machining[6]. The time savings for the model preprocessing cause a secondary, but very important advantage. It is manifested in the extraordinary flexibility of tyre development. The simplicity of building the mould prototype contributes to the easier and faster development and testing of new tyres.

One of the most important limitations of MLS technology applications in the production of tyre mould is the size of chambers (Table 2). Concerning the application of MLS in tyre moulding, the largest piece of tread ring for the passenger tyre 205/60 R15 is 1/8 segment (Table 3).

---

[4] This values are given for silumin, which is commonly used alloy for tyre mould manufacturing (eutectic alloy AlSi , with 12 % Si).

[5] CAPP - Computer Aided Process Planning, CAM - Computer Aided Manufacturing

[6] Summary report on project No. 0231 „Computer aided tire development"(2002-2005) that was was conducted at University of Nis, Faculty of Mechanical Engineering in Nis under the sponsorship of Ministry of Science, Technology and Development of Serbia for Rubber Products Company TIGAR from Pirot, Serbia. (in Serbian)

Method	*SLM (MCP Realizer^{SLM})*	*SLS (Vanguard SLS)*	*DMLS (EOSINT 270)*
Building area	250 x 250 x 240 (mm)	370 x 320 x 445 (mm)	250 x 250 x 215 (mm)

Table 2. Maximal part building area of SLM/SLS/DMLS chambers.

Tyre size	*1/12 segment (mm)*	*1/9 segment (mm)*	*1/8 segment (mm)*
205/60 R15 OD = 630 (mm) (*overall diameter*)	L<164 W<200	L<216 W<200	L<242 W<200

Table 3. Segment size for particular tyre dimension 205/60 R15.

In order to identify whether these technologies meet specific values of utilization features, an application study was performed.

4. Application study

The application study was consisted of two parts. The first one is focused on question whether the MLS technologies could be used for manufacturing the tread ring finding out what are the utilization features of samples produced by MLS. In addition, this part of application study had to show what MLS technology is the most suitable one for tread ring manufacturing.

The second part of application study was devoted to answer following question: if there is any kind of MLS technology that fulfills requirements in regard to utilization features then, how that or these MLS technology(ies) should be employed to manufacture tread ring in a worthwhile manner.

4.1 Application study of SLM/SLS/DMLS – Technology issues

Application study of SLM, SLS and DMLS technology is performed with one-pitch-segments (1/128 of the tread ring), which are shown on Fig. 6.

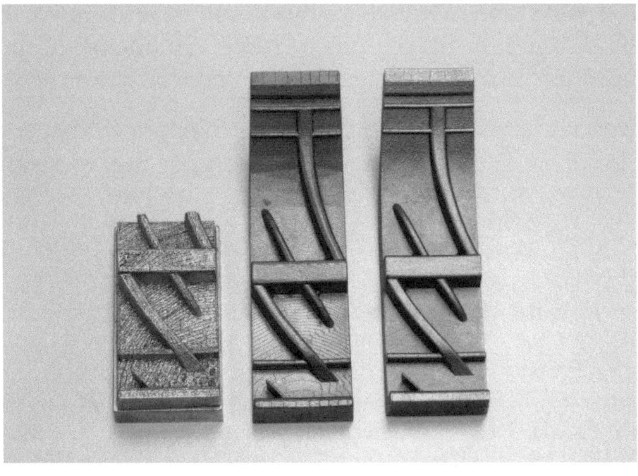

Fig. 6. The one-pitch-segments made by SLM, SLS and DMLS technology.

Materials

Materials used for building the tread ring segment are the latest member to the family of metals for use with the SLM, SLS and DMLS systems and have the best properties among the available metal materials for these technologies. Selected materials are 1.4404 (316L) stainless steel metal powder (SLM), Laserform A6 (SLS) and Direct Steel H20 (DMLS).

- 316 L is stainless steel metal powder of single composition. No heat treatment or infiltration of other material is required. The variety of possible used powder materials which are commercially available is one of the very important advantages of SLM process. The material powder pallet started from aluminum, zinc, bronze, over high grade steel powders, titanium, chromium-cobalt, silicon carbide up to the tool steel powder.
- Laserform A6 is polymer (binder) coated steel powder. During the build process the binder is sintered. The resulting part is exposed to a 24h furnace cycle, where the binder is burned off and bronze is infiltrated into the part. As a result, the metal prototype with 80 % stainless and 20% bronze is obtained.
- Direct Steel H20 is a very fine grained steel-based metal powder with properties similar to conventional steel tool. As a result, after sintering we obtain alloy steel prototype containing Cr, Ni, Mo, Si, V, and C.

Important information for the application study is the material costs. Here is material price ratio based on data that has come down from 2006: 1(SLM): 1.96(SLS): 2.31(DMLS).

Machines

Machines used for making these segments were:

- SLM - MCP Realizer^{SLM} with Nd: YAG laser 100W.
- SLS - Vanguard HS with CO_2 laser 100W.
- DMLS - EOSINT M270 with Yb fiber laser 200W.

4.1.1 Utilization features of mould segments made by SLM/SLS/DMLS

Density

Density is measured using a test cubes. Results of density are shown in Table 4. It should be noted that the porosity of the segments is not necessarily disadvantage. To some extent, porosity can be useful if it provides a better, i.e. thorough ventilation of the mould in the process of vulcanization.

Surface roughness

Surface roughness directly affects on tyre appearance and wear resistance. In addition, surface roughness of tread ring segments affects on mould durability and its maintenance costs. Surface roughness was measured by Mitutoyo surftest SJ-301 (Fig. 7).

The obtained values of surface roughness for the one-pitch segments are shown in Table 4.

The results of measurement show that SLM segment has poor surface quality:

- The values of roughness on vertical surfaces are in range of 20μm to 50μm.
- The roughness of tread ring bottom surface, which is almost horizontal, is out of the measuring range.

- The slanted lateral sides of the mould ridges show staircase structure and inappropriate surface roughness of 75µm.

Considering that mould segment requires a good surface quality and a very good accuracy SLM segment needs more than five hours of post processing.

a) b) c)

Fig. 7. Surface roughness measurement; a) SLM segment - measurement in slanted lateral side of central ridge region; b) SLS segment - measurement in circular ridge region; c) DMLS segment - measurement in tread ring bottom surface region.

Hardness

Hardness is measured using Rockwell and Brinell device. Results of hardness are shown in Table 4.

Results show that the segment made by DMLS technology has significantly higher values than SLS and SLM segments. All segments can be heat treated and increase those values if necessary.

Geometrical accuracy

One of the most important issues is the geometrical i.e. dimensional accuracy of the segment. This is especially significant according to the fact that accuracy here is maintained on large difference in dimensions between the smallest and biggest geometrical features.

Considering the importance of this parameter, the accuracy of the model is diagnosed in two ways. The first approach is based on geometry comparison between the native CAD model, which was the input file for the RP machines, and the healed CAD models (Fig. 8). These healed models were reversely designed from scanned metal segments (SLM, SLS and DMLS).

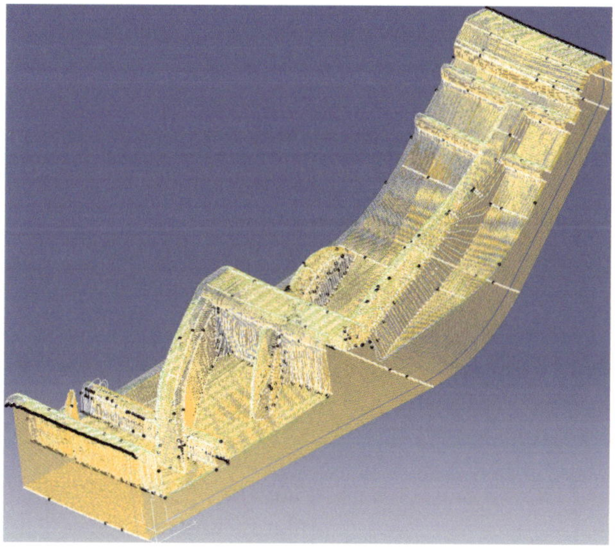

Fig. 8. Comparison between the native CAD model versus scanned metal segments.

The second approach of geometrical accuracy analysis employs Steinbichler optical measurement system. The Fig. 9 shows result window of dimensional accuracy for DMLS segment measured by Steinbichler optical measurement system.

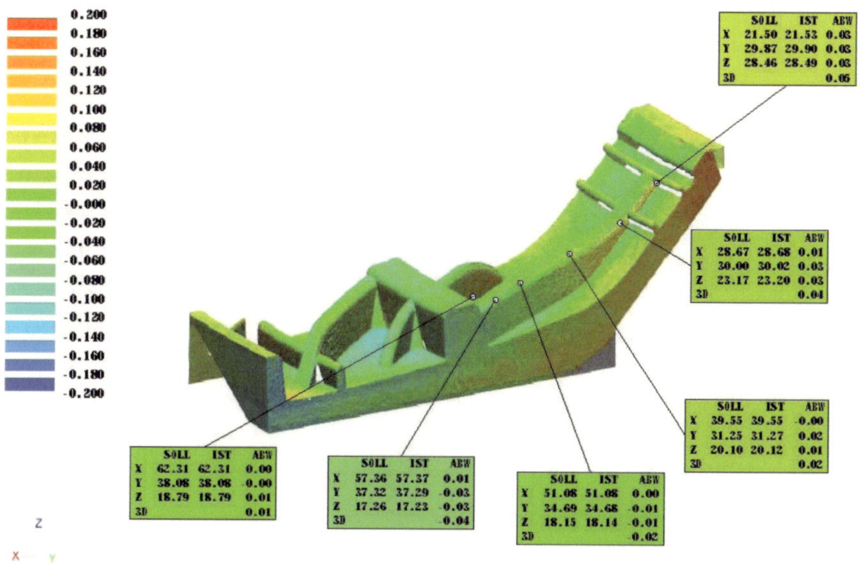

Fig. 9. Geometrical accuracy of DMLS segment.

The results of accuracy for SLM, SLS and DMLS segments (≈ 88 × 33 × 25 mm) are shown in Table 4.

Manufacturing time and post processing

Manufacturing time and post processing are parameters which have a very important role in determining whether these technologies are competitive to conventional technologies or not. The shortest manufacturing time (including post-processing) is needed for the DMLS segment (Table 4).

Method	*SLM*	*SLS*	*DMLS*
Density	97%	99.5%	98%
Surface roughness	Rz = 20 - 50 μm without any post treatment	Ra ≈ 5 - 9 μm Rz ≈16.60 μm - 35 μm without any post processing Ra ≈ 4 μm after polishing	Ra ≈ 3.78 – 5.60 μm Rz ≈ 16-25 μm after shootpeening. Ra ≈ 1 μm after polishing
Hardness	72 HRB 130 HB	89 HRB 180 HB. It can be significantly increase by heat treatment	34 HRC 319 HB
Accuracy	0.10 mm	0.120 mm	0.05 mm
Manufacturing time	15h	3-4h + 24h	14h
Post processing	Because of very poor surface finish at least few hours of polishing.	Necessary – at least one hour of polishing	Sawing from building platform, support removal, shotpeening (30 min)

Table 4. Utilization features of mould segments made by *SLM, SLS and DMLS.*

The general comparison between technologies which were the subject of this experimental study is shown in Table 5.

RP technology	*Geometrical accuracy*	*Hardness*	*Surface roughness*	*Density*	*Chemical reactivity* [7]
SLS	Appropriate	Acceptable	Acceptable	Acceptable	Reactive
DMLS	Appropriate	Appropriate	Appropriate	Appropriate	Non-reactive
SLM	Appropriate	Appropriate	Coarse	Appropriate	Non-reactive

Table 5. Final comparison table.

[7] In the specific case of tyre vulcanization environment.

According to the results of measured parameter, it can be concluded that DMLS technology shows better results than SLS and SLM in all relevant areas (hardness, accuracy, surface finish, manufacturing time, post processing time). Considering this DMLS appears to be the most appropriate alternative to the conventional manufacturing methods.

In order to obtain results for other features of DMLS segment (temperature and pressure endurance, thermal conductivity, wear resistance and chemical reactivity) a test tread ring was assembled in which one of its segments was DMLS segment. The test mould was exposed to the real conditions in standard vulcanization process cycle. The results of the test demonstrated that the DMLS one-pitch-segment fulfills this set of utilization features, too.

5. Application study – Tooling issues

Even though DMLS segments that are sintered by DMLS fulfill all the utilization features, still tooling strategy and related economic issues remain to be considered. Generally, there are two different tooling strategies, which can utilize DMLS for tyre tread ring fabrication: fully-direct strategy and semi-direct strategy (Milovanovic et al., 2009).

5.1 Fully-direct rapid tooling strategy

This production strategy anticipates direct laser sintering of large segments or one-pitch-segments of the tread ring. In addition, this RT strategy appears to be the simplest and the most flexible. Low speed of volume sintering (about 1650 mm^3/h) makes this RT strategy economically unacceptable for the particular case of tread ring volume. Results from case study showed that DMLS system can make 16 one-pitch-segments for 150 working hours. Considering that one tread ring usually includes 128 one-pitch-segments fully-direct RT strategy needs 1200hr of DMLS.

In order to be competitive to CNC direct engraving, fully direct RT should be much faster process (no more than 240 hours), and costs should not exceed the costs of the production mould (12000$).

Another important shortcoming of fully-direct RT strategy is the high price of H20 steel powder. Material costs of H20 increase total production costs cumulatively. In the case of rapid tooling of prototype mould by DMLS, economic indexes are better.

5.2 Semi-direct strategy

The so-called semi-direct tooling strategy does not aim to utilize DMLS for direct fabrication of tread ring or its segments. This strategy attempts to optimize DMLS utilization in order to get the very best of the technology (easiness to plan the production process and manufacture complex geometry, fulfillment of utilization features) and, at the same time, to reduce the costs by reducing volume to sinter (reducing the time and usage of expensive material consequently).

Semi-direct RT strategy employs DMLS technology for sintering the form of tyre tread segments, so called *master* models (Fig. 10).

Fig. 10. Sintered master model of the tyre tread pitch.

After the post processing (removing from platform, cleaning and shotpeening as well as additional machining), the master model is used as an insert for die-casting (Panjan et al., 2005) of the AlSi tread ring segments. The next step is die-casting of the aluminum one-pitch-segments of the mould as it is shown in Fig.11 and Fig. 12.

After casting of the one-pitch-segment of the mould (Fig. 13), it is necessary to post-process it by removing the gates, runners and burrs.

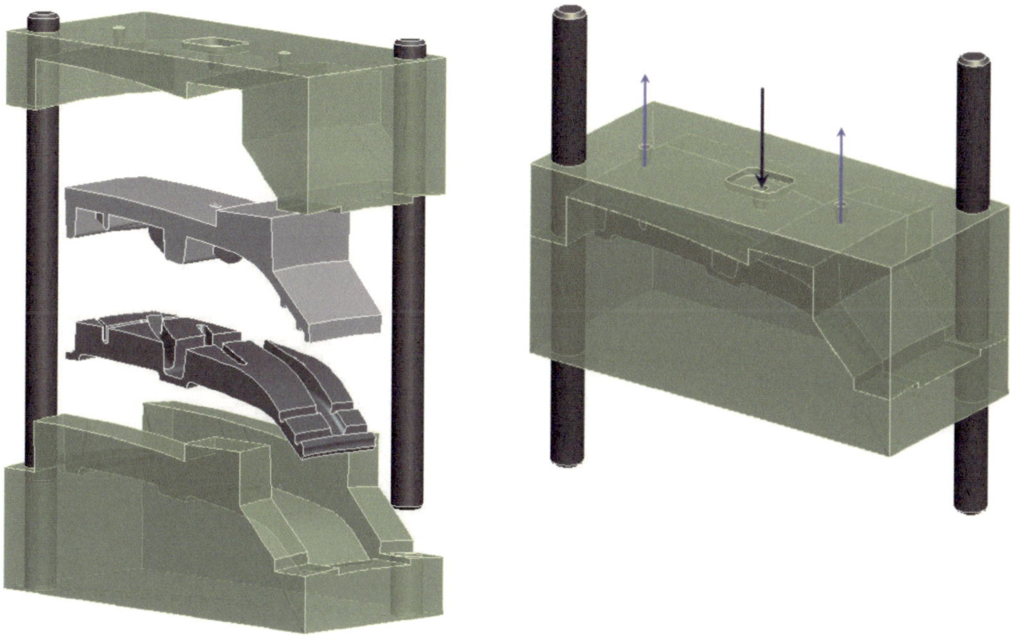

Fig. 11. Die-casting mould.

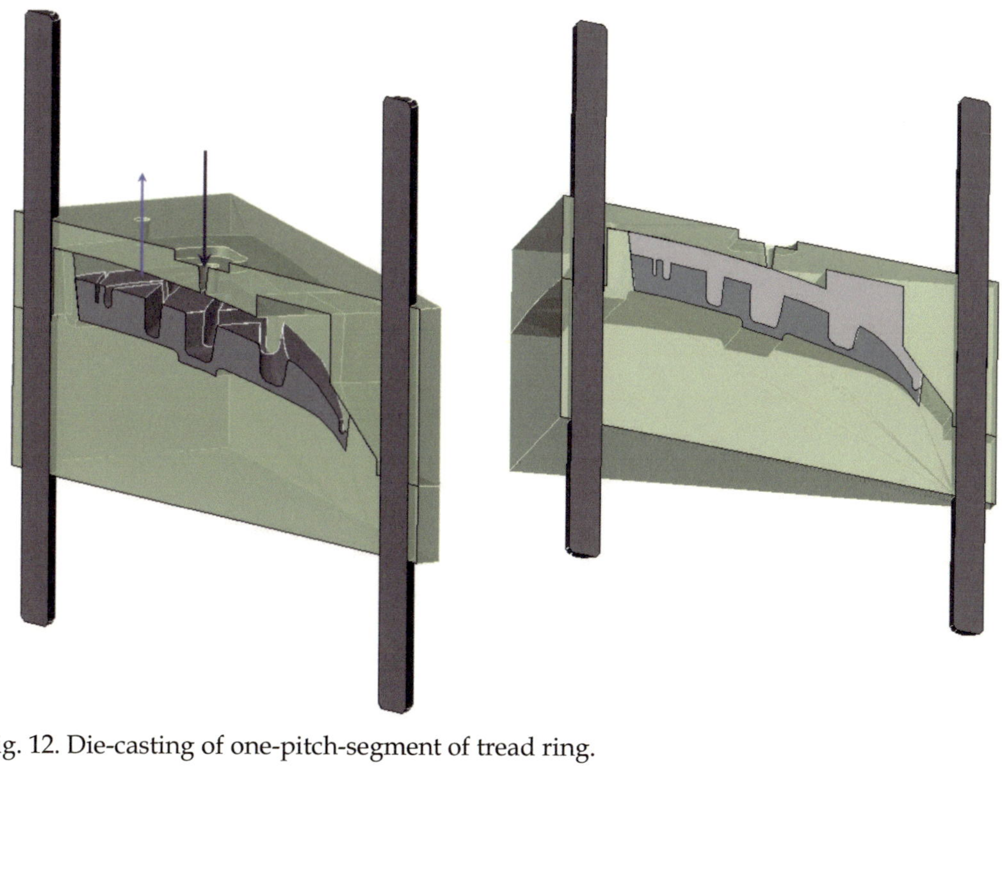

Fig. 12. Die-casting of one-pitch-segment of tread ring.

Finally, the one-pitch-segments are used to assemble the tread ring of the tyre vulcanization mould (Fig. 14). In this tooling strategy, the tread ring is assembled from AlSi one-pitch-segments as a large jig-saw puzzle tool, thus the utilization features are the same as they are for the tread rings that are produced by conventional three-steps casting procedure.

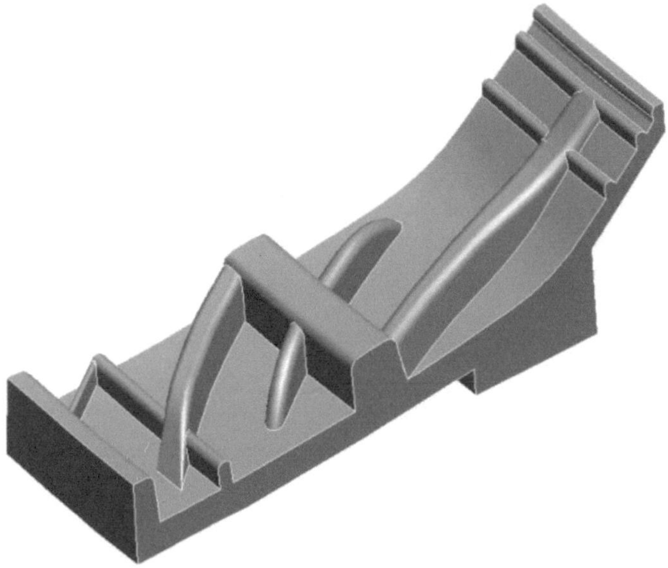

Fig. 13. Casted one-pitch-segment of tread ring.

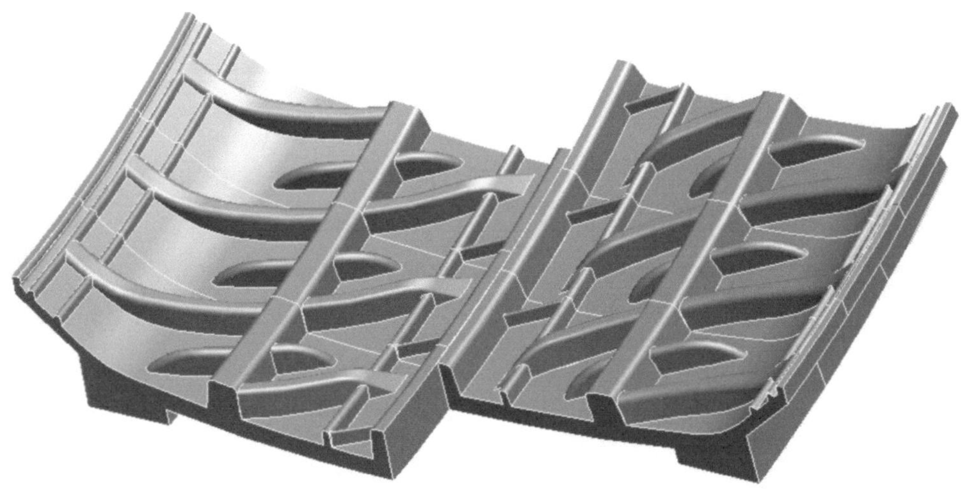

Fig. 14. Assembly of one-pitch-segments makes tread ring (segments).

5.2.1 Limitations on semi-direct RT strategy

Considering that the tread ring of the tyre vulcanization mould is of toroidal shape, it is very difficult to cast larger segments of the ring. If the segments are large (like 40° or 1/9 segments), the surfaces of the ribs that are presented at the tread ring are more slanted. This can cause opening of the mould impossible. Another constraint, which prevents from casting larger segments of the tread ring, is very specific disposition of the so-called pitches of the tyre tread. Usually, tyre tread has three or five different types of pitches that are characterized by slightly different geometry (Stojkovic & Trajanovic, 2001; Stojkovic et al., 2003; Stojkovic et al., 2005a, 2005b; Chu et al., 2006; Lee, 2008). These different pitches are repeated on the tyre tread by very specific disposition. Actually, this disposition has crucial influence on tyre vibration and noise (Sandberg & Ejsmont, 2002). Thus it is economically inappropriate to cast several different larger segments that include different combination of pitches. In addition, the larger segments take more metal powder and more production time, which finally affects on the higher production costs for material and maintenance of the RP system. In the case where it is necessary to produce 360 of 1/9 segments of tread rings (average series of 40 tyre moulds with 9 segments), the costs become unacceptable. The optimal solution is tread ring assembled from one-pitch-segments, where each mould segment corresponds to appropriate tyre tread pitch.

5.2.2 Time and costs consideration

The time and costs in semi-direct RT strategy are much more reduced as compared to the fully-direct strategy. In the case of five different pitches, DMLS system sinters them for about 50 working hours (Table 6).

	Sintering of 1 one-pitch-segment	Post-processing time for 1 one-pitch-segment	Set of 5 one-pitch-segments
Time (min)	600 (10h)	30	3150 (52,5h)
Processing cost [8]*($)*	120	30	750

Table 6. Time and costs for direct metal sintering of one-pitch-segments.

After the post processing of five master models, that takes 3h, 5120 of AlSi-alloy segments (128 jigsaw puzzles × 40 moulds) is moulded in fast die-casting process (Table 7).

	1 segment ($V \approx 21$ cm^3) ($m \approx 59$ g)	Segment post-processing (removing the gates, runners and burrs)	128 segments (one mould set)	5120 segments (set for series of 40 production moulds)
Time (min)	0.35 [9]	1.45	230	9200
Processing cost ($)				≈ 6000

Table 7. Time and costs for die-casting of series of mould segments.

[8] Processing cost does not include H20 material costs.

[9] The time for die-casting of five puzzles is 22 sec., in the tool that has 5 nests.

Duration of the die-casting process depends on number of the nests in the die as well as on the number of available die-cast machines. Whatever, we can claim that semi-direct RT strategy could make significant savings as compared to CNC engraving strategy, and conventional CNC-three-step-casting strategy.

6. Conclusion

The results of the first part of application study, in which utilization features were in focus, clearly showed that DMLS is the best choice among other MLS technologies for rapid tooling of the tread ring. The second part of the study, which took into consideration economic issues of tooling process in whole, leads to the conclusion that just small and thin i.e. skinny pieces (small volume and mass) with high complexity of geometry should be considered as candidates for using DMLS. Having that in mind as, well as the size and complexity of the mould, a conclusion was imposed that DMLS in this very particular case is worthwhile to be used just for fabrication of the set of inserts for die-cast mould that will be used for casting of one-pitch tyre tread ring segments. In the comparison to the conventional tooling processes, this, so-called semi-direct RT strategy is more direct, faster, simpler, more accurate and cheaper. At the same time, semi-direct RT strategy that utilizes DMLS technology is the most flexible procedure for manufacturing of prototype tread ring where the design changes are usually frequent.

7. References

Campanelli S. L.; Contuzzi N., Angelastro A. & Ludovico A. D. (2010). Capabilities and Performances of the Selective Laser Melting Process, In: *New Trends in Technologies: Devices, Computer, Communication and Industrial Systems*, Meng Joo Er (Ed.), 233-252, ISBN: 978-953-307-212-8, InTech

Chu C. H.; Song M. C. & Luo V. C. S. (2006). Computer aided parametric design for 3D tyre mould production, *Computers in Industry*, Vol.57, No.1, (September 2005), pp.11-25, ISSN 0166-3615

Ferreira J. C. (2004). Rapid tooling of die DMLS inserts for shoot-squeeze moulding (DISA) system, *Journal of Materials Processing Technology*, Vol. 155-156, pp. 1111-1117, ISSN 0924-0136

Fleming R. A. (1995). Tyre mould technology in die casting and venting: an overview, *Die Casting Engineer*, Vol.39, No. 5, pp.111-117, ISSN 0012-253X

Ilyas I.; Taylor C.; Dalgarno K. & Gosden J. (2010). Design and manufacture of injection mould tool inserts produced using indirect SLS and machining processes, *Rapid Prototyping Journal*, Vol.16, No.6, pp.429 - 440, ISSN 1355-2546

Khaing M. W. , Fuh J. Y. H. & Lu L. (2001). Direct metal laser sintering for rapid tooling: processing and characterisation of EOS parts, *Journal of Materials Processing Technology*, Vol.113, No.1-3, (June 2001), pp.269-272, ISSN 0924-0136

King, D. & Tansey T. (2003). Rapid tooling: selective laser sintering injection tooling, *Journal of Materials Processing Technology*, Vol. *132*, *No.* (1-3), pp. 42-48. ISSN 0924-0136

Knedla D. (2000). Tyre moulds: development, production, quality, *Tyre technology International*, pp. 150-153, ISSN 0969-7217

Kruth J.P.; Mercelis P.; Van Vaerenbergh J.; Froyen L. & Rombouts M. (2005) Binding Mechanisms in Selective Laser Sintering and Selective Laser Melting, *Rapid Prototyping Journal*, Vol. 11, No.1, 26-36, 2005.

Lee C. S. (2008). Geometric Modeling and Five-axis Machining of Tire Master Models, *International Journal of Precision Engineering and Manufacturing*, Vol. 9, No. 3, pp.75-78, ISSN 12298557

Milovanovic J. (2006). *Possibilities of using rapid prototyping technologies in tyre mould manufacturing*, MSc thesis, University of Nis, Serbia

Milovanovic J.; Stojkovic M. & Trajanovic M. (2009) Rapid tooling of tyre tread ring mould using direct metal laser sintering, *Journal of Scientific & Industrial Research*, Vol. 68 No. 12, pp. 1038-1042, ISSN 0975-1084

Milovanovic J.; Trajanovic M. & Stojkovic M. (2005) Possibilities of using selective laser melting for tyre mould manufacturing, *Proceedings 2nd International Conference on Manufacturing Engineering ICMEN*, pp. 187-193, Halkidiki, Greece, October 05-07, 2005

Panjan P.; Dolinšek S.; Dolinšek M.; Čekada M. & Škarabot M. (2005). Improvement of laser sintered tools with PVD coatings, *Surface and Coating Technology*, Vol. 200, No. 1-4, (October 2005), pp. 712-716, ISSN 0257-8972

Pessard E.; Mognolj P. Y.; Hascoët Y. & Gerometta C. (2008). Complex cast parts with rapid tooling: rapid manufacturing point of view, *The International Journal of Advanced Manufacturing Technology*, Vol. 39, No. 9, (November 2007), pp. 898-904, ISSN 1433-3015

Salaorni E. & Pizzini D. (2000). Constructing a Mould, *Tyre technology International*, pp. 154-158, ISSN 0969-7217

Sandberg U. & Ejsmont J. A., (2002). *Tyre/Road Noise Reference book*, INFORMEX, Harg, SE-59040 Kisa, ISBN 91-631-2610-9, Sweden

Simchi A.; Petzoldt F. & Pohl H. (2003). On the development of direct metal laser sintering for rapid tooling, *Journal of Materials Processing Technology*, Vol.141, No. 3, (November 2003), pp.319–328, ISSN 0924-0136

Stojkovic M. & Trajanovic M. (2001). Parametric design of automotive tyre, *Proceedings of First National Conference on Recent Advances in Mechanical Engineering*, ASME - Greek Section, ANG1/P046, September 17-20, 2001

Stojkovic M., Trajanovic M & Korunovic N. (2005). Computer Aided Tyre Design, *IIPP Journal*, No.8, year III pp. 19-32 (in Serbian), ISSN 1451-4117

Stojkovic M.; Manic M. & Trajanovic M. (2005). Knowledge-Embedded Template Concept, *CIRP - Journal of Manufacturing Systems*, Vol. 34, No 1, pp.71-79, ISSN 1581-504

Stojkovic M.; Manic M.; Trajanovic M. & Korunovic N. (2003). Customized Tyre Design Solution Based on Knowledge Embedded Template Concept, *22nd Annual Meeting and Conference of The Tyre Society*, Akron, Ohio, U.S., September 23-24, 2003.

Voet A.; Dehaes, J.; Mingneau J., Kruth J. & Van Vaerenbergh J. (2005). Laser sintered injection moulds, case studies made in Belgium, *Proceedings of the International Conference Polymers & Moulds Innovations PMI 2005*. International Conference Polymers & Moulds Innovations PMI 2005. Gent, Belgium, Apr 20-23, 2005

Xu, X.; Sachs, E. & Allen, S. (2001). The design of conformal cooling channels in injection molding tooling. *Polymer Engineering And Science*, Vol. 41, No. 7, (July 2001), pp. 1265–1279, ISSN 0032-3888

5

Sintering of Supported Metal Catalysts

José Luis Contreras[1] and Gustavo A. Fuentes[2]
[1]*Universidad Autónoma Metropolitana-Azcapotzalco, México City,*
[2]*Universidad Autónoma Metropolitana –Iztapalapa, México City, México*

1. Introduction

Supported metal catalysts are composed of small metal crystallites (nanoparticles) deposited on high surface area supports. As the size of the crystallites decreases, the dispersed metal is employed more efficiently because a larger fraction of the metal atoms can be made accessible at the surface of the crystallites. The supports are commonly metal oxides, but other materials can be used, such as carbides, carbon, silicates, clays, etc.

Use of a support allows the physical separation of the nanoparticles and thereby hinders their agglomeration into larger crystallites [Wanke & Flynn, 1975]. When it occurs, agglomeration causes a decrease in the number of surface metal atoms per unit mass of metal and therefore decreases the number of active sites of the catalyst. This phenomenon is called sintering or ripening, and it involves different processes that cause changes in the metal particle size distribution (PSD) over the support with an increase in the mean particle size. It is also possible to decrease the mean particle size, a phenomenon called redispersion, dealt with briefly in this work.

Although the support usually stabilizes the metal nanoparticles, in practice it is common to observe the shift of the PSD towards large particle sizes, especially when the catalyst is operated at elevated temperatures. Under those conditions the loss of active surface area by the agglomeration of small crystallites into larger ones is in some cases accompanied by the collapse of the pore structure and loss of internal surface area either for supported or unsupported catalysts [Butt & Petersen, 1988].

The loss of active surface area produced by sintering is an important cause of catalyst deactivation during industrial operation and this is very important in the case of noble metal catalysts. Textural or structural promoters, which modify either the support or the metallic phase to stabilize the metallic particles, are commonly employed in the development of industrial catalysts.

Sintering is a complicated phenomenon because it involves different mechanisms and the metal-support interaction plays an important role. There are several reviews about sintering of supported metal catalysts but in this case we will refer to three of them when dealing with the older literature: the work of Wanke and Flynn [Wanke and Flynn, 1975], that of Lee and Ruckenstein [Lee H.H. & Ruckenstein ,1983], and the last one from Butt and Petersen [Butt & Petersen,1988].

The goals of the chapter are:

1. To review the experimental results about sintering of supported metal catalysts, restricted to supported Pt, Pd, Rh, Ni and bimetallic systems.
2. To present the methods used to correlate sintering data.
3. To briefly analyze the mechanistic models proposed to explain sintering of supported metal catalysts.
4. To discuss stabilization strategies of supported metal catalysts against sintering.

2. Experimental methods to measure the metal dispersion

The metal surface area and the average metal particle size are obtained as a function of treatment conditions of supported metal catalysts.

The metal particle size will can be expressed in terms of metal dispersion, which is defined as the ratio of active surface atoms to total metal atoms. Several experimental methods have been developed to measure this property, they are: (i) gas chemisorption, (ii) Transmission Electron Microscopy (TEM) (iii) X-Ray Diffraction and (iv) Other methods.

2.1 Gas chemisorption

This method measures the amount of gas chemisorbed by the metal on the surface of the support which can be converted to a metal dispersion by assuming an adsorption stoichiometry. This stoichiometry is defined as the ratio of the number of adsorbed atoms or molecules adsorbed per surface metal atom.

Before the review of this technique was made by Wanke and Flynn [Wanke & Flynn,1975] others reviews were made [Whyte,1973,Dorling,1971, Muller,1969, Schlosser, 1967, Gruber, 1966,Dorling,1970,Flynn,1974,Prestridge,1971,Paryjczak,1974,Maat,1965,Renuoprez,1974 as cited in Wanke & Flynn,1975] and therefore only important points will be cited here.

The conversion of gas chemisorption uptakes to metal dispersion requires an assumption of the adsorption stoichiometry and in order for the adsorption technique to be an absolute measure of the metal surface area, the "correct" adsorption stoichiometry has to be used. In this respect Wanke [Wanke & Flynn,1975] said that this was not a problem because as long as the adsorption stoichiometry is constant, it does not depend on metal particle size, catalysts pretreatment conditions, etc., the calculated surface areas are correct on a relative basis.

There is experimental evidence that adsorption stoichiometries may vary considerably with metal particle size [Wilson,1970, Wanke,1972,Dalla Betta,1972,Kikuchi, 1974, as cited in Wanke & Flynn,1975] and this could produce serious difficulties since the changes in adsorption uptake are not directly proportional to the changes in dispersion if the adsorption stoichiometry is function of the dispersion. It is known that the adsorption stoichiometry for oxygen atoms on supported Pt catalysts has been reported to be less than 0.5 for small Pt crystallites (less than 15 Å) and approximately 1 for larger Pt crystallites (more than 20 Å) [Wilson,1970, Dalla Betta,1972, as cited in Wanke & Flynn,1975].

2.2 Transmission electron microscopy

According with Butt and Peterson [Butt &Petersen,1988] transmission microscopy is, the most direct method of examination and in recent years the resolution claimed has increased to observe individual particles (1 nm) and also the particle size distributions can be directly determined.

These distributions are based on three implicit assumptions: (i) The size of a particle is the same as its image recorded on the micrograph (corrected for magnification), (ii) Detection of a given size of particle implies that all particles of that size and larger are being detected and (iii) Contrast of the metal particles is distinguishable from the contrast arising from the support material.

The first assumption deals with the fact that the photographic image is two-dimensional, and of course crystallites which are not equiaxed will have apparent different dimensions depending on their orientation.

The second assumption is related with a question, is the sampling a valid representation of the true distribution in both size range and size frequency?. It is known that small metal particles are very difficult to see and sometimes these small particles are important for certain reactions.

The third assumption has to do with the contrast that is achieved between the support and the metal particles. As it is known [Anderson & Pratt,1985] the metal particles in a supported catalyst can be observed and measured directly by using an electron microscope in transmission. In most cases 100 keV electrons should be used and there is some advantage in using higher energies for thicker specimens.

It is important to keep in mind that particle size distributions determined by TEM are less reliable when the particle size extends below 1 nm.

2.3 X-Ray Diffraction (XRD)

Small crystalline particles on a catalyst, being analyzed by XRD, causes a broadening in the diffraction lines and this broadening can be related to the size of the particles [Wanke & Flynn, 1975]. This technique has been described by Klug and Alexander [Klug,1954, as cited in Wanke & Flynn,1975] and Dorling [Dorling, 1970, as cited in Wanke & Flynn,1975]

For the characterization of supported metal catalysts this technique has limited capacity to detect crystalline particles with a size below 2 to 4 nm [Spindler, 1972, Moss, 1967, as cited in Wanke & Flynn,1975] although Adams et al.,[Adams, 1962, as cited in Wanke & Flynn,1975] detected much smaller particles using a special spectrometer.

In general, for well dispersed catalysts the evaluation of dispersion with XRD can be erroneous because very small particles may go undetected.

Another technique, called Small-Angle X-ray Scattering (SAXS), sometimes is used. This technique has been used for several decades to determine particle size in the range of 10 to 50 Å [Guinier, 1969 as cited in Wanke & Flynn,1975].

Use of this technique with supported metal catalysts has been questioned because the low-angle scattering by the porous support obscures scattering by the metal particles. Somorjai [Somorjai,1968, as cited in Wanke & Flynn,1975] eliminated the support interference by compressing the catalyst samples at extremely high pressures to collapse the pores. Furthermore other technique use pore maskants of electron density similar to the support (for example CH_2I_2) in order to eliminate support interference [Renouprez, 1974, as cited in Wanke & Flynn,1975].

2.4 Others experimental methods

The benzene hydrogenation has been used to measure the metal dispersion [Narayanan & Sreekanth,1989] Oxides such as SiO_2, γ-Al_2O_3, TiO_2 (anatase and rutile), ZrO_2 and MgO with different properties have been used as supports for loading nickel by the pore volume impregnation method. Surface area and acidity measurements, X-ray diffraction and H_2 and O_2 adsorption measurements along with hydrogenation of benzene were used as tools for characterizing the catalysts.

The turnover number (TON) for benzene hydrogenation was dependent on the available metal area and on the crystallite size. A smooth correlation between TON and the Ni dispersion (S_{metal}/S_{BET}) was observed.

Within the limitation of X-ray Photoelectron Spectroscopy (XPS) sampling depth, this method is useful in studying sintering and redispersion of Pt on alumina. Pt/Al_2O_3 analysis by XPS is complicated by overlapping Al2p and Pt4f lines [Shyu, J.Z., Otto, K, 1988]. The interference problem is eliminated when the $Pt4d_{5,2}$ lines are used instead. Although these lines are relatively broad, they can be used to discriminate between particulate and dispersed phase Pt on alumina. The discrimination is based on the Pt binding energy measured after an oxidized sample has been exposed to hydrogen at 150°C.

3. Sintering of supported Pt/Al_2O_3 catalysts

The main factors affecting the rate of sintering of a specific supported metal are the temperature, time and type of atmosphere.

In some studies [Spindler, 1972, Mills, 1961,Armstrong,1972, Emelianova,1969, as cited in Wanke & Flynn,1975] thermal treatments were employed to change the metal crystallite size in order to determine the effect of metal crystallite size on adsorption uptakes and /or rates of reaction. In these studies, however, the rate of metal particle growth as a function of treatment conditions was not of interest, and hence the conditions used are often not described in detail.

In many studies the change of metal dispersion as a function of treatment conditions could be difficult to interpret because:

1. The experimental determination of metal dispersion may not be accurate.
2. The support may undergo changes such as collapse of the pore structure and the trapping of metal within the support.
3. The chemical state of the metal during the treatment may be ill-defined.

One of the first works to study sintering in Pt/Al_2O_3 catalysts was made in N_2 [Herrmann,1961, as cited in Wanke & Flynn,1975] at several temperatures and periods of time. H_2 chemisorption at 200°C and 9 torr was used to measure the Pt area. The H_2 uptakes were such that a dispersion greater than unity resulted for fresh catalysts if an adsorption stoichiometry of one H_2 atom per surface Pt atom was used.

3.1 Effect of air, O_2, H_2 and inert gases

After the previous study, Somorjai [Somorjai, 1968, as cited in Wanke & Flynn,1975] reported the effect of treatment in oxidizing and reducing atmospheres at 600 and 700°C for periods of up to 96 h on the average metal particle size for a $5\%Pt/Al_2O_3$ catalyst. The results of SAXS showed that the average Pt particle diameter increased as the treatment time increased and an inverse behavior for dispersion was found.

Other study changing the support was made by Bett et al., [Bett,1974, as cited in Wanke & Flynn,1975]. They study the sintering of Pt/carbon catalysts in N_2 and H_2 at temperatures of 600,700 and 800°C. The dispersion decreased as the temperature and time increased. The experiments showed that for the samples with high concentration of Pt (20wt%) the dispersion was lower than the dispersion for samples with low concentration of Pt (5wt%).

In other study Hughes et al.,[Huges,1962, as cited in Wanke & Flynn,1975]measured the changes in Pt dispersion of a $0.4\%Pt/Al_2O_3$ catalyst which were caused by treatment in H_2 at 900 and 1000°F for treatment times up to 1000 h. They found that the CO adsorption uptake U as a function of treatment time, t could be correlated by an exponential function, as follows:

$$U = a\,(t)^{b} \tag{1}$$

Where a and b were constants at a fixed temperature. They found a correlation between Dispersion (D) and time (t) assuming a stoichiometry of 1 for CO/Pt.

$$D = 0.73\,(t)^{-0.13} \tag{2}$$

With (t) in hours, and these equations are valid for $2 < (t) < 1000$ h.

Again, Wanke and Flynn [Wanke & Flynn,1975] made a review of several investigations where the sintering was evaluated changing the temperature sintering of supported Pt when the time was constant. The effect of treatment temperature on dispersion of supported Pt catalysts showed that for Pt catalysts having low Pt concentration (less than 3.7wt%Pt) the dispersion decreased as treatment temperature increased under several atmospheres (vacuum, air and H_2) and both Al_2O_3 and SiO_2 supports were studied using H_2PtCl_6, $Pt(NH)_4(OH)_2$, colloidal PtS deposited on $Al(OH)_3$ in aqueous suspension.

In the case of Pt/SiO_2 at 1% Pt the dispersion decreased near 54% after air calcination from 400 to 700°C. In the case of Pt/Al_2O_3 the dispersion decreased 97 % after air calcination from 550 to 700°C[Jaworska- Galas, 1966, as cited in Wanke & Flynn,1975]. Other experiments with these type of catalysts showed similar results [Renouprez,1974, Benesi, 1968, Wilson, 1972, as cited in Wanke & Flynn,1975].

Another procedure to vary the metal dispersion of supported catalysts is to treat catalysts at an elevated temperature for various periods of time. Results of such studies using a Pt/η-Al_2O_3 catalyst showed that after 70 h in air at 700°C the dispersion of Pt decreased 97%, whereas in the case of H_2 reduction after 70 h at 500°C the dispersion decreased only 36% [Gruber, 1962, as cited in Wanke & Flynn,1975]. This last experiment was extended to 1200 h at 500°C and the Pt dispersion decreased 61%. Then the Pt dispersion decreased more with air than with H_2.

In the case of Pt/Al_2O_3 Gruber [Gruber, 1962, as cited in Wanke & Flynn,1975] carried out studies with two catalysts prepared with $Pt(NH_3)_4(OH)_2$ and with H_2PtCl_6 having 0.7% and 0.6%Pt respectively the Al_2O_3 has 200 m^2/g. The thermal treatment was carried out in H_2 at 500°C for 1 to 82 days. The Pt dispersion (D) in function of time (t) in hours was correlated by the following equations:

$$D = 0.38\,(t)^{-0.1} \qquad \text{for } 0.6\%\ Pt/\eta\text{-}Al_2O_3 \tag{3}$$

and

$$D = 0.465\,(t)^{-0.073} \qquad \text{for } 0.7\%\ Pt/\eta\text{-}Al_2O_3 \tag{4}$$

These equations are valid between: $24 < (t) < 2000$ h.

In the case to study large Pt crystals supported on various crystal faces of Al_2O_3 by scanning electron microscopy (SEM) at 900°C in air at 0.5 and 1 atm, Huang and Li [Huang,1973, as cited in Wanke & Flynn,1975] measured Pt particle size changes as a function of treatment time. Treatment times of up to 4 days were studied and their results were well correlated by a power-law function as follows.

$$đ^4 - đo^4 = k\,(t) \tag{5}$$

Where đ is the average Pt particle diameter at time (t) and đo is the average Pt particle diameter at (t) = 0. The rate constant k varied by more than a factor of 10, depending on which crystal face of the Al_2O_3 was used as the support. The rate constant decreased by as much as a factor of 3 when the sintering atmosphere was charged from air at 1 atm to air at 0.5 atm pressure.

From experiments run at varying temperatures for the same amount of time is the second type of experimental data. Examples of this were given for Wanke and Flynn (Wanke &Flynn,1975) for Pt/SiO_2 as can be seen in Fig. 1.

From these results, it appears that ion exchange provides a means of obtaining higher initial dispersions, however it is not clear from these data whether and to what extent the ion-exchange materials are more resistant to sintering. Perhaps the most common type of

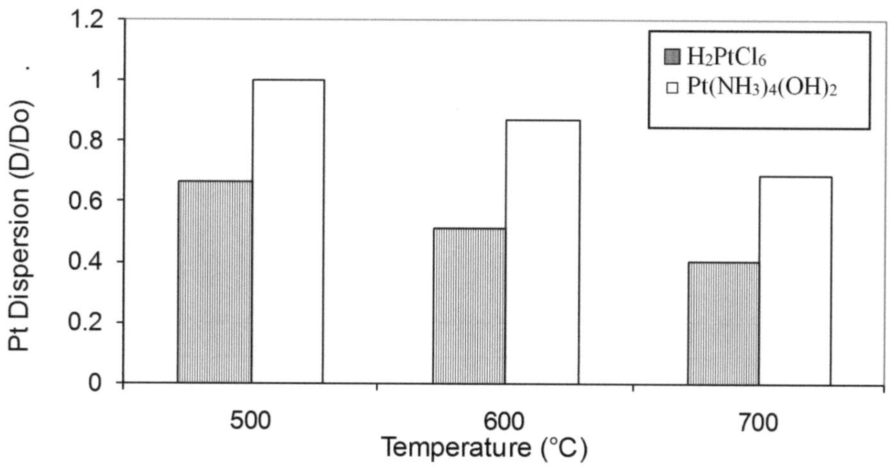

Fig. 1. Sintering of a catalyst of 1.6%Pt/SiO_2 in H_2 atmosphere (2 h) prepared by cationic and anionic precursor (Benesi et al.,1968).

experimental data on sintering comes from constant temperature-variable time studies (Fig. 2). In these experiments the Pt dispersion decreased very fast in air atmosphere (in less than 72 h), while in H_2 it did not decrease as much. Analysis of this type of sintering is simple, and gives immediate answers about the evolution of Pt particles.

In the literature there are many sintering studies of Pt on Al_2O_3 and other supports (MgO, SiO_2, zeolites, SnO_2, CeO_2, TiO_2, carbon, and others), but Al_2O_3 is the classical support, with the largest number of references.

The effect of different atmospheres, such as N_2, O_2, and H_2 on the sintering of diluted samples of Pt/Al_2O_3 reforming catalyst was studied in the temperature range 300-800 °C [Hassan,et al.,1976]. The results of sintering in these atmospheres at temperatures below 400 °C were very similar to those previously obtained in the case of unsupported platinum.

In the case of bimetallic catalysts we will group them taking into account the presence of Pt. Various samples of Pt-Al_2O_3-Cl reforming catalyst were subjected to a series of heat treatments and studied by several characterization techniques [Herrmann et al., 1961] and Pt was present in the original catalyst in a highly dispersed, perhaps ionic, form and that heat treatment causes the formation of Pt crystallites, and not merely their growth.

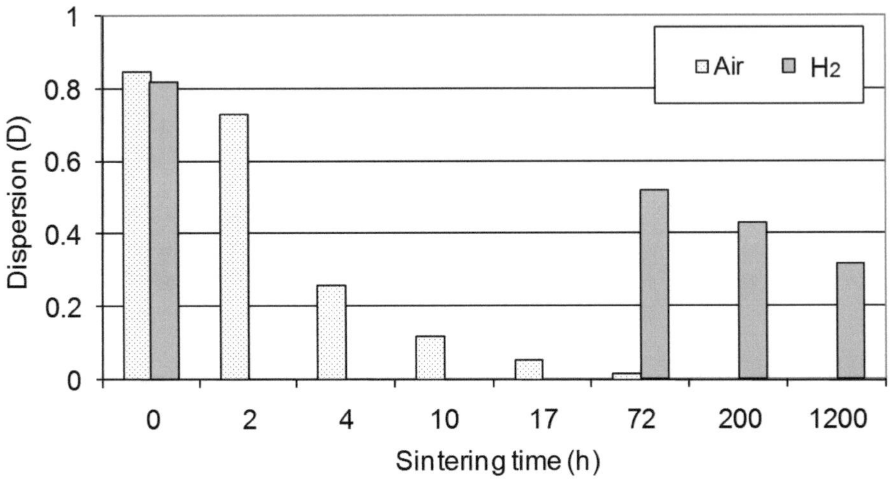

Fig. 2. Sintering of 0.6%Pt/ Al_2O_3 in air (at 780°C) and H_2 (at 500°C) as a function of sintering time. [from Maat & Moscou,1965 and Gruber, 1962].

The treatment of the catalyst samples at temperatures higher than 400 °C in both N_2 and H_2 could lead to an activation; the phenomenon which was observed previously by heating in vacuum. However, the treatment in O_2 showed a continuous deactivation over the whole range of studied temperatures.

A model catalyst suitable for examination in the transmission electron microscope was used to study the sintering of Pt crystallites supported on Al_2O_3 [Chu & Ruckenstein,1976]. The model catalyst consisted of a thin layer of γ-Al_2O_3 upon which Pt was deposited. The samples were subjected to heating in various chemical atmospheres for various lengths of time and temperatures. The change in size, shape, and position of each Pt crystallite was followed by examining the same region after each step in the treatment. Crystallite migration was identified as a cause of sintering. Evidence for the migration of crystallites of size larger than about 10 nm over large distances was presented. The degree of mobility of Pt crystallites depended on sintering conditions. Increase in temperature, periodic oxidation and reduction, reaction of Pt with the substrate, and burning off of coke accumulated on the surface all enhanced crystallite migration. The authors found that phase transformation occurring in the Al_2O_3 substrate was also associated with severe sintering of Pt crystallites. Heating of the model catalyst in pure O_2 or wet N_2 did not promote Pt sintering appreciably and significant sintering occurred during heating in wet H_2.

In other interesting work, Lietz et al. [Lietz et al, 1983] studied several Pt γ-Al_2O_3 catalysts treated in O_2 between 100 and 600 °C and in H_2 at 500 °C using uv-vis reflectance spectroscopy. The formation of different oxidized Pt surface species previously indicated by temperature programmed reduction (TPR) studies was confirmed by characteristic uv-vis spectra. The results are used as the basis for a model describing the types of surface reactions and details of the Pt surface species formed in O_2 and in H_2, and for a model of the sintering in O_2. The amount of soluble Pt was found to correspond with the amount of highly dispersed Pt. Hence, only surface Pt atoms were soluble.

In other work related with the effect of several treatments of O_2 and H_2 on the Pt/Al_2O_3 catalyst the authors studied the sintering of these catalysts following ageing-regeneration cycles typical of industrial use and after extreme treatments in "oxidizing" or "reducing" environments [White, et al., 1983]. Details of particle morphology and the statistics of particle size distribution (PSD) were obtained by high-resolution electron microscopy, with complementary data from H_2 chemisorption measurements. It was found that catalyst sintering during a simulated industrial cycle was minimal, whereas "reducing" or "oxidizing" treatments at high temperature (600 °C) caused loss of accessible Pt surface area: the PSDs and particle geometries, however, then differed greatly, despite similar chemisorptive capacities. Conversely, it was also observed that no appreciable particle growths resulted from severe reducing treatment despite a reduced chemisorptive capacity.

Sintering and redispersion of Pt oxide on alumina have been the subjects of many studies. Some authors [Yao et al.,1980] showed experimental results on sintering and redispersion phenomena which are interpreted on the basis of a thermodynamic formalism and direct formation of Pt sintering during heating of Pt catalysts which have been observed by TEM [Harris, 1986] to investigate Pt particle sintering in specimens of a Pt/Al_2O_3 catalyst heat-treated under oxidizing conditions. The specimens were prepared by sol-gel process. It was found that heating in air at 700 °C produced very rapid sintering, the mean particle diameter increasing from 50 to 300 Å after 8 h. The form of the particle size distributions, and the types of particle structure observed in sintered catalysts, suggested that the mechanism involved a combination of migration and coalescence of whole particles and Ostwald ripening.

In other interesting investigation related with the Pt sintering, the authors used a model of Pt/Al_2O_3 catalyst after heat treatment in H_2, and alternately in H_2 and O_2 at different temperatures (300 and 800 °C) by means of TEM [Sushumna & Ruckenstein,1988]. At both relatively low (500 °C) and high temperatures (750 °C), short-distance migration (usually 1-3 particle diameters) and coalescence of particles was observed to contribute considerably toward sintering.

At 500 °C and following two or three cycles of alternate heating in H_2 and O_2, significant migration of particles (up to 8 nm particles migrating over 25 nm) was observed on exposure of the oxidized particles to H_2 and/or of the reduced particles to O_2. Sintering was fast and pronounced in the initial 4 to 6 h of heating of a fresh sample; then it becomes slow. However, further continuous heating for extended periods of 12 h or more causes additional significant sintering by particle migration and coalescence among other mechanisms.

This suggests that investigations based on only 1 or 2 h of heating, as often reported in the literature, yield incomplete information. The present results indicate that at both 500 and 750 °C, sintering by ripening (apparently only of a localized kind) also occurs.

In other investigation [Sushumna & Ruckenstein,1988] the authors showed evidences of the occurrences of a variety of phenomena such as short-distance migration and coalescence of particles, migration toward or away from another particle, decrease in size and disappearance of larger particles near unaffected smaller particles, decrease in size or disappearance of both small and larger particles, decrease in size or disappearance of small particles near larger particles (ripening), decrease in size and subsequent migration of particles or vice versa, collision without coalescence, and collision-coalescence-separation into two particles again. The particles seem to feel the presence of other nearby particles via long-range inter-particle forces that induce the migration of a particle toward another or emission of atoms toward a nearby particle.

The two major mechanisms of sintering of supported metal crystallites appear to be (i) short-distance, direction-selective (in contrast to random) migration of particles followed by (a) collision and coalescence or (b) direct transfer of atoms between the two approaching particles, or (ii) localized ripening (direct ripening) between a few stationary, adjacent particles.

From literature, Pt sintering is produced mainly by temperature, type of atmosphere and time (Table 1). But there are other causes such as: type of support, impurities in the support and level of metal loading. Unfortunately, there is a disparity in the literature on such presumably factors in Pt/ Al_2O_3 catalysts [Butt, 1988].

In the literature some authors have gave attention to the factors of strong H_2 adsorption and strong metal-support interactions (SMSI) as being potentially important effects in the determination of metal exposure of crystallite size as related to studies of sintering [Butt, 1988].

In this point, it is important that when metal dispersion results are discussed, it will be understood that possible influences of strong chemisorption or SMSI have not been considered, then some results can be suspect because interpretation of sintering was based on dispersion measurements following such treatments. Although there is some evidence that high-temperature treatment in H_2 does not promote sintering, this issue is not clear in terms of the effect of strongly bound H_2.

From Table 1, air calcination of the 0.5%Pt/ Al_2O_3 catalyst at 500°C produced low dispersion (D=0.15). An increase of air temperature (600°C for 24 h) produced sintering, decreasing dispersion (D =0.06). The change of atmosphere from H_2 to O_2 in the 2.75%Pt/ Al_2O_3 catalyst decreased the dispersion from 0.67 to 0.5. This behavior is similar in the next other types of catalysts.

The presence of HCl in the 2%Pt/ Al_2O_3 catalyst after O_2 treatment at 520°C showed that the Pt dispersion increased from 0.75 to 1.08 and then HCl contributed to the Pt redispersion.

Some authors found that after heat treatments of Pt/ Al_2O_3 catalysts in H_2 and O_2 at high temperatures, drastic changes both in H_2 and O_2 chemisorption occurred in spite of no change of Pt particle size.

The great effect of the type of atmosphere has been investigated by Guo et al.[Guo et al.,1988], they studied the influence of thermal treatments in O_2, H_2 and inert atmospheres on the dispersion of Pt/ γ-Al_2O_3 and the Pt sintering was followed by H_2 chemisorption and wide-angle x-ray diffraction (XRD). Catalysts containing about 1.0 and 5.0 wt% Pt were treated in atmospheres containing H_2, He, N_2, O_2 and various chlorine compounds, XRD was used to obtain Pt crystallite sizes and size distributions.

The results showed that the treatment atmosphere had a significant influence on the Pt crystallite sizes and size distributions; treatment in He at 800°C resulted in relatively small

Pt crystallites, while treatment in O_2 usually resulted in bimodal size distributions. Treatment in chlorine-containing atmospheres resulted in sintering or redispersion of the Pt depending in the nature and concentration of the chlorine-containing species.

In other studies, the effect of an oxidative atmosphere is studied on fresh and sintered unchlorinated naphtha reforming catalysts containing 0.6-1% Pt.

Catalyst	Atmos-phere.	Tempera-ture (°C)	Time (h)	Pt Dispersion	Analytic Method	Reference
0.5%Pt/ Al_2O_3	Fresh	500	1	0.15	H_2 Chemi-sorption -TEM	(Harris,1983)
	Air	600	2	0.10		
	Air	600	8	0.09		
	Air	600	24	0.06		
2.75%Pt/ Al_2O_3	Fresh H_2	500	8	0.67	H_2 Chem.	(Apesteguia 1984)
	Air	590	0.5	0.35		
	Air	600	0.7	0.3		
0.93%Pt/ Al_2O_3	Fresh H_2	500	8	0.65	H_2 Chem.	(Apesteguia 1984)
	Air	600	5	0.48		
0.5%Pt/ Al_2O_3	Fresh H_2	500	1	1.1	H_2 Chem.	(Lieske,1983)
	O_2	800	Seq.	0.41		
	O_2	550	11	0.60		
2%Pt/ Al_2O_3	Fresh H_2	400	12	0.90	H_2 chem.	(Lee, 1984)
	H_2	700	3	0.39*		
	O_2	520	2	0.74		
	O_2+HCl	520	2	1.08		
	O_2+HCl	600	2	0.95		
	H_2+HCl	600	2	0.29*		
	O_2	300	2	0.69		
	O_2+HCl	520	2	0.92		
	O_2	800	2	0.07		

* The H/Pt ratios for these runs probably do not correspond to Pt dispersion

Table 1. Dispersion of Pt/Al_2O_3 catalyst in presence of several atmospheres, temperature and time using H_2 chemisorption to measure the Pt dispersion (stoichiometry H/Pt=1).

The TPR profiles show that only one of species is formed under given experimental conditions, regardless of the mean crystallite size of the metal particles [Borgna,et al.,1992]. The structural information supplied by extended X-ray absorption fine structure (EXAFS) compared with cuboctahedral particle modeling, implies that such species was a surface platinum oxide, the structure of which was close to that of PtO_2, (Figure 3) but largely distorted.

The same authors extended their study several years later and a main observation of their study is that chlorine was present in the first shell around platinum during the entire sintering process [Borgna et al., 1999]. The experimental data were well fitted only by including chlorine in the first coordination sphere of Pt. Since the oxidizing mixture used here did not contain chlorided compounds, this result indicates that chlorine originating from the support was always in the immediate surroundings of Pt during the metal sintering. The authors used in situ EXAFS spectroscopy for identifying the surface species involved in the sintering of alumina-supported Pt catalysts under dried oxidizing atmospheres.

A Pt/Al_2O_3 catalyst (0.62 wt% Pt, 0.88 wt% Cl) was heated in a 2% O_2/N_2 gaseous mixture from 300 to 525°C for about 120 min and then kept at this temperature for up to 720 min. The main observation, which is in good agreement with ex situ TPR experiments, was that chlorine was always present in the surrounding of Pt during the oxidizing treatment. The metal sintering process involved three successive steps during which the chlorine and oxygen coordinations passed through a maximum, whereas Pt coordination exhibited a minimum. Formation of $Pt(OH)_4Cl_2$ species was detected at the end of the first step, i.e., when the temperature reached 500°C. After about 4 h of treatment, they deduced that Pt species are made up of a metal platinum core surrounded by a double coating of oxychlorinated species (Fig. 3).

More precisely, EXAFS experiments suggested that surface platinum oxide is made of rigid PtO_6 octahedra, but their assembly led to a largely disordered structure. The absence of a long-range order allows the location of residual chlorine species either between the PtO_6 octahedra or at the Pt-oxidized surface shell.

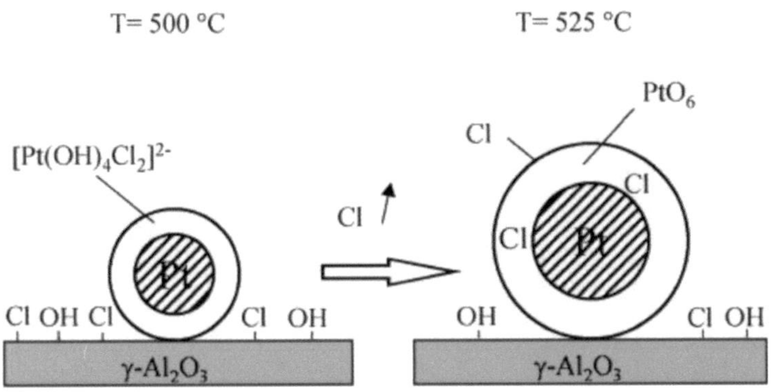

Fig. 3. Pt species are made up of a metal platinum core surrounded by a double coating of oxychlorinated species.

The effect of H_2O on the Pt sintering also has been studied by Barbier et al. [Barbier,et al.,1992]. They made a direct reduction by H_2, without precalcination, of Pt/Al_2O_3 activated by H_2PtCl_6. They studied several series of catalysts with different metal and chloride loadings. It appears that H_2O acts as a kinetic inhibitor of the reduction reaction but can also induce Pt sintering. HCl has no kinetic effect on the reduction but is able to increase the metallic accessibility. The higher the H_2 pressure, the lower the Pt dispersion. The extent to which the metal is accessible is considered to be resultant of two parallel reactions, namely polymerization and reduction of partly reduced Pt^{2+} complexes.

In other study, the sintering mechanism of Pt nanoparticles dispersed on a planar, amorphous Al_2O_3 support as a model system was studied by Simonsen,et al [Simonsen,et al.,2010] By means of in situ transmission electron microscopy (TEM), the model catalyst was monitored during the exposure to 10 mbar air at 650 °C. Time-resolved image series unequivocally reveals that the sintering of Pt nanoparticles was mediated by an Ostwald ripening process.

A statistical analysis of an ensemble of Pt nanoparticles shows that the particle size distributions change the shape from an initial Gaussian distribution via a log-normal distribution to a Lifshitz-Slyozov-Wagner (LSW) distribution. Furthermore, the time-dependency of the ensemble-averaged particle size and particle density is determined. A mean field kinetic description captures the main trends in the observed behavior. However, at the individual nanoparticle level, deviations from the model are observed suggesting in part that the local environment influences the atom exchange process.

3.2 Effect of sintering over the catalytic activity

The Pt particle size has an important effect over the catalytic activity and selectivity for different reactions. For example, Maat and Moscou [Maat and Moscou ,1964, 1965], correlated the sintering of a commercial reforming catalysts of 0.6%Pt/Al_2O_3 with 0.5% chloride, with its activity for n-heptane reforming. Variable time-constant temperature sintering experiments were carried out at 780°C in an inert atmosphere and the resulting Pt surface areas determined by H_2 chemisorption and electron microscopy. Corresponding crystallite sizes (average) ranged from about 1 nm for the fresh catalyst to 45 nm for the most severely sintered case (780°C, 72 h) and good correlation of sintering kinetics was obtained via equation:

$$dD / dt = - k D^n \qquad (6)$$

with n=2, where D is the Pt percentage exposed or Pt dispersion and k the sintering kinetic constant.

The integration of this equation yields:

$$k = 1/t \ln [Do/D] \qquad \text{where } n = 1 \quad (7)$$

$$k = D_o^{1-n} / t(n-1) [(Do/D)^{n-1} - 1] \qquad n=2 \quad (8)$$

where Do is the initial dispersion and D the dispersion at time t.

The authors found a significant change in activity, although it is not as large as one might have expected. The total change in surface area is from 233 to 5 m^2/g Pt a 98% decreases, yet total conversion decreases only from about 95 to 75%, roughly 25%. What is probably of more ultimate importance in these data is the change in product distribution. The selectivity for the various products changes markedly with sintering and the dehydrocyclization activity is severely affected, while the sum of isomerization plus cyclization remains approximately constant.

The isomerization selectivity increases, from 9 to 25 mol% product, with increasing sintering. Since the aromatics produced in dehydrocyclization reactions would be prominent in determining product octane, this change represents a severe decline in a desired selectivity. The reason for the alteration in selectivity is found in the bifunctional nature of the catalyst; aromatization reactions are strongly dependent on the metallic function, while the isomereization activity is predominantly a function of the acid Al_2O_3.

Maat and Moscou do not report any notable sintering of the support in these experiments, so presumably the acidic function is relatively unchanged and the apparent increase in isomerization activity on sintering is the result of increased reactant availability in competition with the metallic function.

In another type of catalysts, for catalytic reduction of NOx, the redox behavior of the fresh and sintered catalysts was investigated using a pulse reactor. A highly dispersed Pt/Al_2O_3 catalyst was used for the selective catalytic reduction of NOx using propene (HC-SCR) [Vaccaro et al.,2003]. Contact with the reaction gas mixture led to a significant activation of the catalyst at temperatures above 523 K. According to CO chemisorption data and TEM analysis, Pt particles on the activated catalyst had sintered. The authors observed that if Pt particles were highly dispersed (average size below ~2nm), only a small part (~10%) of the total number of Pt surface sites (as determined by CO chemisorption (Pt_{surf})) participated in the H_2/O_2 redox cycles ($Pt_{surf,redox}$). Then for a sintered catalyst, with an average particle size of 2.7 nm, the number of superficial Pt (Pt_{surf}) and $Pt_{surf,redox}$ sites were in good agreement.

In other work, volatile organic compounds (VOC) were burn by catalytic combustion using a Pt catalyst and the activity was promoted with the sintering of Pt crystallite, which was supported on the anodized aluminum plate with the electrolysis supporting method[Wang et al.,2005]. The effect of atmosphere and the sintering temperature were studied. Results of the pre-treatment under the respective combustion of toluene, benzene, or ethylbenzene showed that the linear relationship between the ultimate average particles size and VOC molecular weight could be concluded.

In order to make the link between sintering of a Pt/Al_2O_3 catalyst and its activity for the CO oxidation reaction Yang, et al., [Yang, et al., 2008] treated the catalyst (1.6% Pt) by thermal aging for different durations ranging from 15 min to 16 h, at 600 and 700 °C, under 7% O_2. It was observed a shift of the Pt particle size distributions due to sintering towards larger diameters. These distributions were studied by TEM. The number and the surface average diameters of Pt particles increase from 1.3 to 8.9 nm and 2.1 to 12.8 nm, respectively, after 16 h aging at 600°C.

The catalytic activity for CO oxidation under different CO and O_2 inlet concentrations decreases after aging the catalyst. The light-off temperature increased by 48 °C when the catalyst was aged for 16 h at 600 °C. The CO oxidation reaction is structure sensitive with a catalytic activity increasing with the platinum particle size.

Another investigation where the reforming reactions were used to know the state of metal particles was made with Pt/Al_2O_3. Consecutive oxidation-reduction cycles involving successive oxidation treatments of $0.3\%Pt/Al_2O_3$ catalyst were made at various temperatures, each followed by a reduction treatment in H_2 to produce an active catalyst, lead to changes in catalytic activity for the reactions of n-hexane with H_2. Decreased activity is paralleled by changes in selectivity brought about by the increased tendency of large Pt particles towards coking [Anderson,et al.,1989]. Results show that the reforming reactions of n-hexane can be used to gain information about the state of metal particles following sintering/redispersion of Pt on alumina.

In order to evaluate Pt catalysts using an accelerated test, a novel method was developed using a periodic pulse technique in which a catalyst is exposed to forced redox cycles. It was applied to the oxidation of C_2H_4 on Pt/Al_2O_3 catalysts [Murakami,et al.,1991]. It was found that the catalysts are sintered at relatively low temperature, i.e., at 588 K, in reaction atmosphere, and that the periodic pulse technique actually accelerates the sintering. The sintering in the continuous flow reaction for ca. 1000 h could be reproduced within ca. 100 h, indicating the possibility of rapid estimation of catalyst life. It is suggested that the sintering in reaction atmosphere is governed by the chemical factors rather than the thermal ones.

In another interesting study of effect of Pt sintering was made using a Pt/Al_2O_3 catalyst which was sintered in O_2 and H_2 atmospheres using two metal loadings of the catalyst: 0.3% and 0.6% Pt [Susu & Ogogo,2006]. After sintering, the aromatization selectivity was investigated with the reforming of n-heptane as the model reaction at a temperature of 500 °C and a pressure of 391.8 kPa. The primary products of n-heptane reforming on the fresh platinum catalysts were methane and toluene, with subsequent conversion of benzene from toluene demethylation. To induce sintering, the catalysts were treated with O_2 and temperatures (500 and 800°C). The 0.3%Pt/Al_2O_3 catalyst exhibited enhanced aromatization selectivity at various sintering temperatures while the 0.6%Pt/Al_2O_3 catalyst was inherently hydrogenolytic.

3.3 Redispersion of Pt on Al_2O_3

Changes in the dispersion of supported Pt/Al_2O_3 catalysts following reduction and a variety of thermal treatments have been monitored by gas uptake and electron microscopy [Flynn & Wanke,1975]. Evidence of redispersion was found after sintering of one catalyst in oxygen at 450 to 600 °C. Sintering was found to be sensitive to gas atmosphere and metal loading. Addition of a portion of presintered catalyst containing large Pt particles increased the rate of sintering of a catalyst. From electron micrographs of the same catalyst area before reduction and after reduction and various thermal treatments, it was concluded that Pt agglomeration occurs during all these steps. Some Pt crystallites remain in a fixed location during reduction and thermal treatments.

Other investigation of redispersion of Pt of thermally sintered Pt/Al_2O_3 catalysts used in the simultaneous oxidation of CO and propene has been achieved by an oxychlorination treatment [Cabello Galisteo et al.,2005]. The catalyst can be considered to model the active component of the catalytic converter fitted to diesel driven cars. Platinum crystallites redispersion was verified by XRD, H_2 chemisorption, TEM and Fourier transform infrared spectroscopy (FTIR). The extent of regeneration reflects the platinum particle redispersion achieved by such a treatment. Oxychlorination also introduced electronic effects in the Pt particle caused by the presence of chlorine at the Pt-Al_2O_3 interface but no detrimental result of this was observed in the oxidation reactions. The results indicate that the deactivation of the diesel oxidation catalysts can be reverted by this simple treatment resulting in a remarkable recovery of the catalytic activity.

4. Sintering of supported Pd catalysts

In the literature most of the sintering studies reported are for Pt catalysts. Less information is available on the sintering of other noble metal catalysts, such as Pd, Rh, Ni etc. Wanke and Flynn [Wanke & Flynn,1975] made a review of the sintering of these metals.

In the case of Pd, it was observed an unexpected growth rates of Pd particles supported on charcoal at temperatures below 50°C [Pope, 1971, as cited in Wanke & Flynn,1975]. In that study, it was observed a 30% loss of Pd surface due to reduction of a 10%Pd/charcoal catalysts at 25°C for 2 h. The reduction temperature required for the H_2 reduction of Pd precursors was very low.

In other case, Brownlie et al., [Brownlie, 1969, as cited in Wanke & Flynn,1975] prepared a Pd/charcoal catalysts by vapour deposition of Pd, for the hydroisomerization of 1-butene. The authors observed by TEM a change in average Pd crystallite size from 140 Å for the fresh catalyst to 1300 Å for the catalyst used for reaction and the reaction temperature was 43°C. Little particle growth was observed for another Pd/charcoal sample which was prepared by impregnation with palladous chloride and exposed to the same reaction condition.

In the case of Pd/Al_2O_3 at 5% Pd the dispersion decreased near 85% after H_2 reduction from 400 to 800°C[Aben, 1968, as cited in Wanke & Flynn,1975]. The same authors studied the Pd/SiO_2 catalyst and the dispersion decreased 80 % after H_2 reduction from 400 to 900°C.

In relation of the metal-support interaction in the Pd/SiO_2 system some authors [Lamber et al, 1990] have studied the effect of support pretreatment using TEM and electron diffraction. The authors found that heating of the Pd/SiO_2 system in a H_2 atmosphere may lead to chemical (strong) interaction between metal and support and growth of palladium salicide (Pd_2Si). Microdiffraction analysis showed an oriented growth of the Pd_2Si precipitate with respect the Pd matrix. The magnitude of metal-support interaction, the agglomeration of metal particles and the formation of an intermetallic compound are strongly influenced by the thermal pre-treatment of the SiO_2 substrate.

In other investigation, catalysts, containing Pd and Pt on a carbon support, were studied by the temperature-programmed reduction, in situ X-ray photoelectron spectroscopy, and X-ray absorption spectroscopy (XAFS)[Stakheev et al.,2004]. The reduction of Pd and Pt species in samples 2%Pd/C and 2%Pt/C calcined in an air flow at 370°C was studied. Reduction of the 2%Pd/C sample begins at 50 - 60°C and is completed at 250 - 300°C. Particles of various dispersions are formed during reduction. Long-distance peaks observed in the EXAFS spectra point to the presence of a fraction of relatively large crystallites. The average Pd - Pd coordination number (~5) at 200°C gives evidence that a number of very small Pd nanoparticles, oligomeric clusters, is present. Reduction at $T > 200$°C results in sintering of a small fraction of the Pd particles. Reduction of Pt in 2%Pt/C sample begins at 120 - 150°C and is completed at 300 - 350°C. The sintering-resistant monodispersed Pt particles are formed under these conditions.

5. Sintering of supported Rh catalysts

In the case of Rh, several samples supported on Al_2O_3 and SiO_2 were prepared and reduced in H_2 prior to the treatment in air [Wanke, 1972, Hughes,1962,Wanke, 1969, as cited in Wanke & Flynn,1975].

For a 5%Rh/Al_2O_3 catalyst, the dispersion of Rh decreased 25% at 600°C after 3 h in air (initial dispersion 0.79), when the temperature increased at 800°C for 8 h the dispersion decreased 84% [Wanke, 1969,1972, as cited in Wanke & Flynn,1975].

In the case of a catalyst of 5%Rh/SiO_2 the dispersion decreased 84% (initial dispersion 0.53) after calcinations in air at 800°C for 4 h[Yates, 1967, as cited in Wanke & Flynn,1975].

In the case of a catalyst of 0.3%Rh/Al_2O_3 the dispersion decreased 67% (initial dispersion 0.77) in presence of N_2 at 900°C [Hughes, 1962, as cited in Wanke & Flynn,1975]. The Rh in the 5%Rh/SiO_2 catalyst [Yates, as cited in Wanke & Flynn,1975] was present as $RhCl_3$ prior to the air treatment since the dried impregnated catalyst was not reduced prior to the sintering. The comparison between the catalysts supported on Al_2O_3 was difficult because the authors did not know the state of the metal prior to the treatment in N_2.

In a comparative study made by Fiedorow, et al[Fiedorow at al.,1978] changes in dispersion of alumina supported Pt, Ir, and Rh catalysts due to thermal treatment (250-800 °C) in O_2 and H_2 atmospheres were measured. In O_2 atmospheres the sequence of thermal stability was found to be Rh > Pt > Ir, while in H_2 atmospheres the sequence was Ir > Rh > Pt. Increases in dispersion due to treatment in O_2 were observed for Pt and Ir catalysts. The observed relative stabilities are compared to qualitative predictions based on sintering mechanisms. The extent of the changes was found to be quite different among the Al_2O_3 supports used.

6. Sintering of supported Ni catalysts

The Ni in the 10%Ni/ Al_2O_3-SiO_2[Carter, 1963, as cited in Wanke & Flynn,1975] and 6.7%Ni/SiO_2[Van Hardeveld, 1966, as cited in Wanke & Flynn,1975] catalysts was present as the nitrate at the beginning of the treatment. Complete reduction was obtained after H_2 reduction at 370°C.

Low Ni dispersion was found for the 3%Ni/Al_2O_3 catalyst (initial dispersion 0.021) and after 900°C in N_2 for 72 h the dispersion was 0.002 [Hughes, 1962, as cited in Wanke & Flynn,1975]. The sintering of Ni in these catalysts was difficult to measure.

The sintering of Ni in H_2 was evaluated in the 6.7%Ni/SiO_2 catalyst [Van Hardeveld,1962, as cited in Wanke & Flynn,1975]. The dispersion of Ni decreased from 0.11 to 0.028 after 1 h at 700°C, the dispersion decreased 74%. In these catalysts the methods used for measuring the dispersion were: H_2 chemisorption (static system at -78°C), CO adsorption (in He flow system) and X-ray line broadening.

7. Sintering of supported bimetallic catalysts

Some studies of supported bimetallic catalysts have been made using Pt,Pd Ir, Rh and Ru, for example: Pd-Ru, Pd-Pt, Pd-Ir, Pt-Ir, Pt-Ru and Pt-Rh. The characterization studies were difficult to interpret using the metal dispersion since significant alterations in the supports occurred at the elevated treatment temperatures [Armstrong, 1972, as cited in Wanke & Flynn,1975]. In this study it is interesting to note that in some of the catalysts (Pt-Rh and Pt-Ru) the use of steam between 1000 and 1200°C resulted in an increase in activity.

The sintering in bimetallic catalysts as Pt-Ir/Al_2O_3 has been studied [Graham & Wanke, 1981] with focus on the effects of thermal treatments in O_2 (300 to 600 °C) and H_2 (500 to 800 °C) on the metal dispersions of 1% Pt, 1% Ir, and 1% Pt-1% Ir catalysts. Treatment in O_2 at 400 to 600 °C resulted in an increased dispersion for the Pt catalyst. Small increases in dispersion for the Ir catalysts were observed after treatment in O_2 at 300 °C; at higher temperatures the Ir dispersion decreased significantly. Segregation of the metals occurred in the bimetallic Pt-Ir catalysts during treatment in O_2. The largest decreases in dispersion due to treatment in H_2 were observed for the Pt catalyst. Surprisingly, H_2 treatment of the Ir and Pt-Ir catalysts resulted in essentially the same relative decreases in metal dispersions

One of the beginning investigations of Pt-Ir catalyst was made by Graham &Wanke [Graham &Wanke,1981] They studied the effect of parameters such as temperature, gas flow rate, O_2 concentration and oxychlorination on the sintering of an industrial bimetallic reforming catalyst (Pt-Ir/ Al_2O_3). The time taken for carbon removal decreased with increasing gas flow rate (from 0.75 to 3 l/h/cc. cat.) and good metal dispersion and high BET surface area were achieved at a flow rate of 3 l/h/cc.cat. Increasing the initial combustion temperature from 370 to 450° C led to considerable sintering of the metal. However, it did not significantly change the BET surface area. Better metal dispersion and higher BET surface area were restored using O_2 concentrations in the range 0.3 to 0.6% above which value considerable sintering of both metal and support occurred. In sintered catalysts, metal redispersion could be improved to some extent by oxychlorination treatment. Dispersion of the metallic phase was found to be higher at higher chlorine levels in the catalyst.

Bimetallic catalysts have shown good catalytic stability. The research on bimetallic catalysts has had a major impact in the reforming of petroleum naphtha fractions to produce high octane number components for gasoline [Sinfelt, 1983]. Two industrial catalysts for naphtha reforming were developed in 1969 for different companies (Exxon and Chevron). Pt-Ir and Pt-Re supported on Al_2O_3 [Jacobson et al., as cited in Sinfelt , 1983]. During the 1970´s bimetallic catalysts largely replaced traditional Pt catalysts in reforming.

The research in bimetallic catalysts has received particular attention: (a) the investigation of selectivity effects in catalysis by such bimetallic materials, (b) the preparation and characterization of highly dispersed bimetallic catalysts and (c) the metallic stability.

Since the ability to form bulk alloys was not a necessary condition for a system to be of interest as a catalysts, it was decided not to use the term alloy in referring to bimetallic catalysts in general. Instead, terms such as bimetallic aggregates of bimetallic clusters have been adopted in preference to alloys. Bimetallic clusters refer to bimetallic entities which are highly dispersed on the surface of a carrier.

In an investigation of Pt-Ir catalysts, the effects of thermal treatments in O_2 (300 to 600 °C) and H_2 (500 to 800 °C) on the metal dispersions of 1% Pt, 1% Ir, and 1% Pt-1% Ir catalysts were studied by Graham & Wanke [Graham & Wanke,1981] . Treatment in O_2 at 400 to 600 °C resulted in increases in dispersion for the Pt catalyst. Small increases in dispersion for the Ir catalysts were observed after treatment in O_2 at 300 °C; at higher temperatures the Ir dispersion decreased significantly. Segregation of the metals occurred in the bimetallic Pt- Ir catalysts during treatment in O_2. The largest decreases in dispersion due to treatment in H_2 were observed for the Pt catalyst. Surprisingly, H_2 treatment of the Ir and Pt-Ir catalysts resulted in essentially the same relative decreases in metal dispersions.

Studying the same bimetallic catalyst, Bishara et al., [Bishara et al.,1983] investigated the effect of temperature, gas flow rate, O_2 concentration and oxychlorination on the sintering of an industrial bimetallic reforming catalyst (Pt-Ir/Al_2O_3). The time taken for carbon removal decreased with increasing gas flow rate from 0.75 to 3 l/h/cc. cat. and good metal dispersion and high BET surface area were achieved at a flow rate of 3 l/h/cc.cat. Increasing the initial combustion temperature from 370 to 450° C led to considerable sintering of the metal. Better metal dispersion and higher BET surface area were restored using O_2 concentrations in the range 0.3 to 0.6% above which value considerable sintering of both metal and support occurred. In sintered catalysts, metal redispersion could be improved to some extent by oxychlorination treatment. Dispersion of the metallic phase was found to be higher at higher chlorine levels in the catalyst.

In other study by Deng et al., [Deng et al.,1988] the effect of addition of Ba on sintering of Pt clusters on Pt/γ-Al_2O_3 catalysts was examined. During the preparation and thermal treatment of the catalysts, two new species - $BaCO_3$ and $BaAl_2O_4$ - were formed from Ba acetate ($Ba(AC)_2$). Ba acts as an inhibitor for sintering at either a low Ba/Pt atomic ratio range or a high ratio range, whereas in-between those two ranges it acts as a promoter. The inhibiting effect of Ba $(AC)_2$ on sintering of Pt clusters cannot be related to the neutralization of support acidity. The inhibiting and promoting effects of Ba can be ascribed to the presence of $BaCO_3$ at a low Ba/Pt atomic ratio and $BaAl_2O_4$ at a high Ba/Pt atomic ratio, respectively.

One of the most common bimetallic catalyst is Pt-Re. Pieck et al. [Pieck et al., 1990] studied the influence of temperature and gas flow rate on total metallic dispersion and specific surface area of Pt-Re/Al_2O_3 reforming catalyst during the burning of coke deposited on its surface. A laboratory catalyst coked to different coke concentrations and a commercial catalyst coked in a commercial plant were used. The chloride level during oxychlorination and its influence on total metallic dispersion were also studied. The actual temperature inside the catalyst during coke burning is the main parameter affecting the total metallic dispersion. During catalyst oxychlorination the metallic phase is redispersed and the total dispersion at values of 0.9-1% chloride on the catalyst is the same, independent of the initial dispersion. The total metal dispersion of Pt-Re/Al_2O_3 is fixed by the platinum complexes in either sintering or redispersion. This is because, on reduction of Pt complexes, metallic Pt forms crystallites where Re is reduced producing the Pt-Re clusters.

Shinjoh et al.,[Shinjoh et al.,1991] studied both sintering and activity behaviors over noble metal catalysts aged in oxidative atmospheres with various O_2 contents at 1000°C for 5 h. With increasing O_2 contents, catalytic activities over aged Pt, Rh, and Pt/Rh catalysts decreased, and, in contrast, those over aged Pd and Pd/Rh catalysts increased. While, an order of sintering for the noble metal particles on aged catalysts agreed closely to that of the each percentage conversions as well as to that of vapor pressures of respective catalyst species, such as noble metals or their oxides. It is confirmed through the above results that the performance of aged catalysts are tightly governed by the properties of noble metals and the selectivity data are also important for exploring the catalytic activities, in particular, over multi-functional catalysts such as automotive exhaust ones.

In other study, a series of Pt-Rh/Al_2O_3 catalysts designated PtRhx, where x is the atomic percentage %Rh/Rh+Pt were prepared by successive impregnation of Al_2O_3 pretreated in H_2 at 850°C, with aqueous solutions of H_2PtCl_6 and $RhCl_3$[Kacimi & Duprez,1991]. They were calcined at 450°C (fresh catalysts) and then sintered (2% vol. O_2/Ar for 2h) at 700, 800 and 900°C. The catalysts were dechlorinated at 400°C. This treatment induces a decrease of the metal area of Pt while there is no change for Rh. It was observed a significant change in the surface composition of Pt and Rh for the sintered catalysts which are strongly enriched in Rh at high Rh content and in Pt at low Rh content. Two models are proposed to explain their results, they take into account the structural and morphological changes of Rh and Pt in oxidizing atmosphere as well as the degree of interaction between the two components of the bimetallics.

The same authors [Kacimi, et al., 1993] studying the sintering of Pt-Rh/Al_2O_3 catalysts in O_2 atmosphere at 700 and 900°C, found a strong enrichment in Rh at high Rh content while, at low Rh content, the bimetallics would be enriched in Pt. All the states of Rh (including Rh^{3+} in strong interaction with Al_2O_3) were analyzed by XPS $^{18}O/^{16}O$ and oxygen storage capacity (OSC) techniques and also they analyzed Rh° and Rh_2O_3.

In other investigation on Pt-Sn/Al_2O_3 reforming catalysts Yaofang et al, [Yaofang et al.,1994] studied the kinetics of coke combustion, platinum sintering in a nonhydrogen atmosphere, and redispersion on Pt-Sn catalyst. Using kinetic results the authors carried out mathematical simulation of coke burning in a radial continuous regenerators. Good agreement was obtained between the temperatures calculated from the simulation and measured values. Gaseous O_2 may enhance resistance against sintering at high temperatures. The platinum sintered in the catalyst without a decrease in support surface area and it could be completely redispersed through chlorination and oxidation.

The same authors [Yaofang, et al.,1995] studied the kinetics of coke combustion, Pt sintering in a non-H_2 atmosphere, and redispersion on Pt-Sn reforming catalyst. By using the kinetic results, mathematical simulation of coke burning in a radial continuous regenerator was established. Good agreement was obtained between simulation-calculated and measured temperatures. The O_2 in the gas may enhance resistance to sintering at high temperatures in the presence of water. The sintered Pt on the catalyst, if there had been no decrease of the support surface area, can be completely redispersed through chlorination and oxidation..

Platinum sintering and redispersion were systematically investigated with Pt-Re and Pt-Sn reforming catalysts by Yaofang et al.,[Yaofang, et al.,1995]. It was found that platinum might sinter both in catalyst dehydration during the start-of-run and in coke burning during regeneration and had the same sintering mechanism as the following: $PtO_2 \rightarrow Pt + O_2$, except melt of alumina supporter. The sintered platinum could be completely redispersed through regeneration if the alumina had no decrease in its surface area. Otherwise the loss in the platinum dispersion or dehydrogenation activity of the catalyst after regeneration would be proportional to the loss in surface area of alumina. The sintered Pt redispersed quite fast, taking about 10 minutes of chlorination and 30 minutes of oxidation.

In other investigation of Pt-Rh/Al_2O_3 and Pt-Rh/CeO_2-Al_2O_3 bimetallic catalysts (Pt+Rh≈60 μmol g^{-1}) the samples were prepared via chlorine-free precursors[Martin et al.,1995]. Oxygen storage capacities (OSC) were measured on the fresh (calc. 723K) and on the sintered catalysts (1%O_2+10%H_2O, 2h, 973K and 1173K). On Al_2O_3 catalysts, only Rh can promote OSC which is extremely sensitive to sintering. OSC values are higher on Al_2O_3-CeO_2 catalysts, but do not depend on the composition of the bimetallics. Moreover CeO_2

renders the catalysts resistant to sintering. Pt-Rh/Al_2O_3 and Pt-Rh/CeO_2-Al_2O_3 were modified by Cl, SO_4^{2-} and K. On Al_2O_3, OSC variations due to the additives follow the same trend as the variations of O_2 mobility (deduced from $^{18}O/^{16}O$ isotopic exchange). Chlorine and sulfur are inhibitors of OSC while K, at low content, is a promoter.

The addition of W to 0.3 wt % Pt/γ- Al_2O_3 catalysts has been found to stabilize and even increase Pt dispersion after high reduction temperature (up to 1073 K)[Contreras &Fuentes,1996]. This effect appears to be caused by formation of Pt species bound to WOx during drying and calcination and by strong interaction between Pt crystallites and the mixed oxide surface after H_2 reduction. UV-vis spectra show the formation of WOx and PtOy species on the surface of alumina in the calcined state. Interaction between Pt^{4+}- and W^{6+}-bearing structures seems to enhance UV absorption with minor variations in the location of the band maxima. Pt/WO_3- Al_2O_3 were in general more stable towards deactivation than monometallic Pt catalysts (reduced at 773 K) during hydrogenation of benzene. Reduction at 1073 K produced catalysts with lower initial activity but with increased resistance towards deactivation. Impregnation of Pt on WOx-(γ- Al_2O_3) gave the best overall catalysts, apparently because of an improved interaction of crystallites with the modified surface of Al_2O_3. This result is important in order to control deactivation and sintering of the metallic phase if high operation temperatures are involved.

In other investigation bimetallic Pd-Pt/Al_2O_3 catalysts were prepared by controlled surface reactions [Micheaud et al.,1996] . The reduction of $PtCl_6^{2-}$ can be made by using as reductant hydrogen preadsorbed on a parent monometallic Pd/Al_2O_3 catalyst nothing or metallic palladium itself (direct redox reaction). Two precursors ($Pd(NH_3)_4Cl_2$ and $Pd(Acetate)_2$) were used to prepare the (1 wt % Pd) parent catalyst. The platinum loading was within the range of 0.13-0.14 wt %. The two preparation methods lead to different kind of Pt deposit, whatever the Pd parent catalyst. By using preadsorbed H_2, Pt can be deposited in decoration on low coordination sites (corners, edges) of Pd particles. By direct redox reaction, the deposit of Pd atoms occurs preferentially on the (111) faces of Pd particles. However some sintering can occur, especially during hydrogen treatment in aqueous medium and that sintering is enhanced by chloride ions.

In other study made by Sakamoto et al.,[Sakamoto, et al.,1999], the addition of Fe to the Pt/Al_2O_3 catalyst was made, with focus on the catalytic activities of stoichiometric and lean mixture simulated exhaust from automotive engines on bimetallic Pt-Fe/Al_2O_3 catalysts after the O_2, O_2-H_2, and O_2-H_2-O_2 treatments and compared to that of Pt/Al_2O_3 catalysts. The state of the Pt particles of the Pt-Fe/ Al_2O_3 catalysts was also investigated for several techniques. The activity of the Pt-Fe/Al_2O_3 catalyst was greater than that of the Pt/Al_2O_3 catalyst after the O_2-H_2-O_2 treatment for the stoichiometric mixture. Also, the activity of the Pt-Fe/Al_2O_3 catalyst after the O_2-H_2 treatment was greater than that after the O_2 and O_2-H_2-O_2 treatments. Pt particles were found to react with Fe additives to form homogeneous Pt-Fe alloy particles on Al_2O_3 under reducing conditions. Also, the Pt-Fe alloy particles on Al_2O_3 were found to segregate into Pt and Fe_2O_3 and to form a Fe_2O_3 coverage layer on Pt particles so that Pt particles were prevented from sintering when heated at 800 °C for a lean mixture. On the other hand, the activity of the Pt/Al_2O_3 catalyst was greater than that of the Pt-Fe/Al_2O_3 catalyst after the O_2-H_2-O_2 treatment for lean mixture. The layer of Fe_2O_3 on Pt is responsible for the low activity.

Hayashi et al., [Hayashi et al.,2002] investigated the sintering behavior of Pt metal particles by supporting them on silica-coated alumina. Silica coating was found to be effective for the retention of a large surface area of alumina even after calcination at elevated temperatures. Before sintering, the size of Pt metal particles on all the silica-coated aluminas, including the uncoated alumina, was identical, while the particle size was larger on silica than on alumina. After sintering the Pt catalyst at 1073 K, the particle size increased on uncoated alumina as well as on alumina coated with thicker silica layers, especially on the supports previously calcined at > 1473 K. On the other hand, the size of Pt metal particles did not increase much on alumina coated with monolayer silica. The observed suppression of sintering of Pt metal particles resulted from the retention of a large surface area of alumina with a thinner silica layer. In the case of a thicker silica layer, although a large surface area was maintained after calcination at elevated temperatures, the existence of a bulk silica-like property of the support did not favor the suppression of sintering of Pt metal particles.

In other investigation made by Mazzieri et al. [Mazzieri, et al.,2008] the deactivation, by coke deposition and sintering, and the regeneration of the metal function of Pt-Re-Sn/ Al_2O_3-Cl and Pt-Re-Ge/ Al_2O_3-Cl catalysts were studied. The analysis of the carbon deposits and of the final state of the metal and acid functions were performed by means of temperature programmed oxidation, temperature programmed reduction and temperature programmed pyridine desorption (TPO, TPR and TPD-Py) respectively. The degrees of deactivation and activity recovery of the metal function were measured by means of the cyclohexane dehydrogenation and cyclopentane hydrogenolysis reactions. It was found that the Pt-Re-Sn catalysts were more stable than the Pt-Re-Ge ones. Regeneration produced a segregation of the metal phase on both kinds of catalysts. Metal particle sintering at 650 °C modified the metal function severely and Pt was segregated from the other components. All the rejuvenation treatments (Cl, air and high temperature) were unable to restore the original state of the metal function.

In other work of bimetallic catalysts, Kaneeda et al.[Kaneeda, et al.,2009] studied the Pt-Pd/ Al_2O_3 catalysts for the NO oxidation after severe heat treatments in air. For this purpose, the addition of Pd has been attempted, which is less active for this reaction but can effectively suppress thermal sintering of the active metal Pt. Various Pd-modified Pt/ Al_2O_3 catalysts were prepared, subjected to heat treatments in air at 800 and 830 °C, and then applied for NO oxidation at 300 °C. The total NO oxidation activity was shown to be significantly enhanced by the addition of Pd, depending on the amount of Pd added. The Pd-modified catalysts are active even after the severe heat treatment at 830 °C for a long time of 60 h. The optimized Pd-modified Pt/ Al_2O_3 catalyst can show a maximum activity limited by chemical equilibrium under the conditions used. From their characterization results as well as the reaction ones, the size of individual metal particles, the chemical composition of their surfaces, and the overall TOF value were determined for discussing possible reasons for the improvement of the thermal stability and the enhanced catalytic activity of Pt/ Al_2O_3 catalysts by the Pd addition. The Pd-modified Pt/ Al_2O_3 catalysts should be a promising one for NO oxidation of practical interest.

In other investigation, Contreras et al., [Contreras et al.,2009] found that W^{6+} ions inhibited sintering of the Pt crystallites once they were formed on the Al_2O_3 after reduction at 1073 K. Sequential samples impregnating Pt on WOx-Al_2O_3 were more active and stable during benzene hydrogenation. W^{6+} ions promoted high thermal stability of Pt crystallites when sequential catalysts were reduced at 1073 K and decreased their Lewis acidity. The same authors studied the thermal stabilization of γ-Al_2O_3 using W^{+6} ions, which has been found useful to the synthesis of Pt/Al_2O_3 catalysts [Contreras et al.,2010].

The sequential impregnation method was also used by the same authors to study the n-heptane hydroconversion. They found that the W^{6+} ions delayed reduction of a fraction of Pt^{+4} atoms beyond 773 K and that W^{6+} ions inhibited the sintering of the metallic crystallites once they were formed on the surface. After reduction at 1073 K, sequential samples impregnating Pt on WOx/Al_2O_3 were more active and stable during n-heptane hydroconversion than monometallic Pt/Al_2O_3 catalyst. Selectivities for dehydrocyclization, isomerization and hydrocracking changed significantly when the W/Pt atomic ratio and reduction temperature increased. Initial and final reaction rates were more sensitive to reduction temperature. W^{6+} ions promoted high thermal stability of Pt crystallites when sequential catalysts were reduced at 1073 K and deactivation of bimetallic catalysts reduced at 773 K and 1073 K was less than the deactivation of Pt/Al_2O_3 catalyst.

In other paper of Pt-Rh catalyst Fernandes et al., [Fernandes et al.,2010] studied the influence of thermal ageing effects on catalytic activity for a Pt-Rh three-way catalyst. Deep textural, structural and physicochemical changes in the catalyst washcoat were caused by exposure to high temperatures (900 °C and 1200 °C). The catalytic activity was evaluated in terms of CO and propane oxidation and the reduction of NO by CO. In general, the conversions were lower after the ageing.

In other paper of Pt-Rh catalyst Hirata et al.[Hirata et al.,2011] studied the behaviors of sintering and reactivation of Pt and Rh on various metal oxide supports by TEM, CO pulse chemisorption and XAFS analysis. The results suggested that the phenomenon of reversible sintering and re-dispersion to reduced active metallic sites is related to the electron density of O atoms in support and to the crystal structure of support. As a result of in situ XAFS and in situ TEM analysis, Pt reversible sintering and re-dispersion phenomenon was observed on CeO_2 based metal oxide.

In a new investigation of Pt-Ni, and Ru-Ni bimetallic catalysts, Guo et al.,[Guo et al., 2011] studied that a trace amount of noble metal (Ru or Pt <0.1 wt%) was doped onto an anodic Al_2O_3-supported Ni catalyst, to know its performance in the steam reforming of methane, especially during the daily startup and shutdown operation. Although the steam purge treatment at high temperatures seriously deactivated the Ni catalyst because of the oxidation of metallic nickel with steam into Ni^{2+}, trace Ru assisted the regeneration of active metallic nickel by hydrogen-spillover. And, the Ni sintering was largely alleviated by the addition of Ru, and it was probably due to the formation of Ru-Ni alloy. In comparison with the Ru-doped Ni catalyst, the Pt-doped Ni catalyst showed a more favorable tolerance to steam oxidation, even at 900 °C.

8. Proposed models of the sintering process

In accordance with Wanke and Flynn [Wanke & Flynn,1975] the factors that influence the sintering of supported metal catalysts are: temperature, type of atmosphere, nature of the support and the degree of metal loading appear to be of secondary importance.The experiments have shown that the rate of loss of dispersion increases with increasing temperature even in absence of O_2. In the case of presence of O_2 some investigators have observed an increase in metal dispersion at certain temperatures [Wanke & Flynn,1975].

Redispersion in O_2-containing atmospheres appears to be restricted to temperatures between 400 and 620°C. The Pt dispersion decreases at temperatures above this interval or for long periods of time.

The rate of loss of dispersion is larger in O_2 atmosphere than in H_2 atmosphere [Somorjai,1968, as cited in Wanke & Flynn,1975] and the rate of loss in dispersion decreased with decreasing O_2 partial pressure [Huang,1973, Wynblat,1973, as cited in Wanke & Flynn,1975].

The rate of sintering in N_2 atmosphere is approximately similar to the rate in H_2 atmospheres and the rate of sintering in vacuum is lower than if a gas phase is present [Boudart, 1968, as cited in Wanke & Flynn,1975].The effect of the support on the rate of sintering has a strong dependence however is difficult to establish. Some authors [Huang, 1973, as cited in Wanke & Flynn,1975] showed strong dependence on the support but the applicability of these results to supported metal catalysts is questionable because the size of metal crystallites in this work was much larger than those encountered in supported metal catalysts.

Wanke and Flynn [Wanke & Flynn,1975] compared sintering data for Al_2O_3, SiO_2 and carbon-supported catalysts and they did not find that it was not a possibility to determine the effect of the support on the sintering rate since sintering conditions, initial metal dispersions and metal loadings are different for all these samples.

The mathematical models based on postulated mechanism for the sintering of supported metal catalyst have appeared in the literature. The model developed by Ruckenstein and Pulvermacher [Ruckenstein & Pulvermacher,1973a, 1973b, as cited in Wanke & Flynn,1975] envisages the sintering of supported metal catalysts to occur by migration of metal crystallites over the surface of the support and the resulting collision and fusion of metal particles causes the loss in dispersion (crystallite migration model).

The second model proposed by Flynn and Wanke [Flynn,1974a, 1974b, as cited in Wanke & Flynn,1975] considers the sintering to occur by dissociation of atomic or molecular species from the metal crystallites. These atomic or molecular species migrate over the support surface and become incorporated into metal crystallites upon collision with the stationary metal crystallites (atomic migration model).

The crystallite migration model is similar to the mechanism proposed by Smoluchowski [Smoluchowski, 1917, as cited in Wanke & Flynn,1975] for the coagulation of colloidal suspension by Brownian motion. The atomic migration model is similar to the model proposed by Ostwald (Ostwald ripening) [Ostwald,1900, as cited in Wanke & Flynn,1975] for the change in mercury oxide particle sizes in solution. The transport of metal or metal compounds through the vapor phase is a third possible mechanism for the sintering of supported metal catalysts. The transport of metal by this mechanism in non-oxidizing atmospheres is negligible for Group VIII metals at temperatures below 1000°C.

Some authors have mentioned that the growth of supported metal crystallites are related with the interaction metal-support [Wynblatt, 1975, as cited in Wanke & Flynn,1975], in fact it could be so important to determine which mechanism will predominate.

In other studies, Fuentes & Salinas [Fuentes & Salinas,2001] have showed that apparent inconsistencies between leading models and experimental data were caused by the use of limited solutions to the models. In particular, the authors have studied solutions to the Ostwald ripening model and they found that the classical analytical solution, known as the Lifshitz, Slyazov and Wagner solution (LSW), is only a particular solution of a family of solutions. The important implication of that analysis is the observation that the particle size distribution predicted by their solution can fit adequately in the experimental distributions observed during sintering of Pt, Pd and Ni supported in a variety of supports.

8.1 Atomic migration model

The model presented by Flynn and Wanke [Flynn &Wanke, 1974a, 1974b, as cited in Wanke & Flynn,1975] envisages sintering to occur as a three-step process: (i) Escape of metal atoms or molecules such as metal oxides in an oxygen atmosphere from the metal crystallite to the support surface, (ii) Migration of these atoms along the support surface and (iii) Capture of these migrating atoms by metal crystallites upon collision of these migrating atoms with stationary metal crystallites.

These processes result in the growth of the large metal crystallites and in a decrease in size of the small metal crystallites since the rate of loss of atoms is smaller than the rate of capture for large crystallites, while for small crystallites the rate of loss is greater than the rate of capture. This occurs because large crystallites are in equilibrium with a lower concentration of migrating atoms than small crystallites (this is the two-dimensional analogy of Ostwald ripening).

8.2 Crystallite migration model

The crystallite migration model postulates that metal crystallites migrate as entities along the surface of the support. Rapid diffusion of metal atoms on the surface of a metal crystallite will cause metal atoms to accumulate on one side of the crystallite by random fluctuations. This rapid, random surface diffusion will cause Brownian type motion of the particles on the support [Wynblatt, 1975, as cited in Wanke & Flynn,1975].

Direct evidence of this crystallite migration is difficult to obtain. The usually cited references in support of crystallite migration [Phillips, 1968,Thomas, 1964, Sears,1963, as cited in Wanke & Flynn,1975] use microscopy to investigate this phenomenon, and the results show either very slight crystallite motion, such as small rotations, or they are anomalous in that large particles appeared to move faster than small particles.

If appreciable particle migration occurs, it is restricted to metal crystallites smaller than 50 Å in diameter [Whnblatt,1975, as cited in Wanke & Flynn,1975]. The model presented by Ruckenstein and Pulvermacher [Ruckenstein and Pulvermacher,1973, as cited in Wanke & Flynn,1975] is hence restricted to the early stages of sintering when crystallites are < 50 Å in size. Two limiting cases of the crystallite migration model were developed: (i) Surface diffusion controlled (i.e. the rate of migration of crystallites is the rate-determining process) and (ii) Sintering controlled (i.e. the merging of two metal crystallites coming into physical contact by collision is the rate-determining process).

For both of these cases the authors were able to reduce the kinetic equations of the sintering process to power-law equations where the value of the power-law order, n depends on the rate-determining step. For the sintering controlled case (i.e. merging of particles), n is equal to 2 or 3 and for the surface diffusion controlled case $n > 4$. The magnitude of n for the surface diffusion controlled case is related to the dependence of the surface diffusion coefficient on the average metal crystallite size.

The authors Pulvermacher [Pulvermacher, 1974, as cited in Wanke & Flynn,1975] described mathematical techniques for differentiating between sintering and diffusion controlled cases. The application of these techniques to sintering data obtained by TEM, SAXS, X-ray broadening and chemisorption was presented.

Other authors [Wynblatt 1975, as cited in Wanke & Flynn,1975] showed on the basis of theoretical predictions that the merging of metal crystallites cannot be the rate-determining process at temperatures above 500°C, i.e., the sintering controlled case is generally not realizable under normal sintering conditions. This is supported by observations that metal blacks sinter is appreciable (fusion of particles in physical contact) at temperatures below 200°C [McKee, 1963, Khassan, 1968, as cited in Wanke & Flynn,1975].

It is important to notice that during the latter stages of sintering a mechanism other than crystallite migration causes the observed growth of metal crystallites.

Experimental attempts to determine which of the two mechanisms (atomic migration model and crystallite migration model) is predominantly responsible for the sintering of supported metal catalysts have not been conclusive. In large part this inconclusiveness arises from the complexity of the sintering process.

In another study, Handa & Matthews, [Handa & Matthews,1983] proposed a model, which utilizes a Monte Carlo simulation, of the sintering and the redispersion of supported metal catalysts is presented and compared with experimental data from a Pt/Al_2O_3 system. The model is based on an atomic migration mechanism, but includes instantaneous diffusion and coalescence of crystallites of size less than 1 nm in reducing and inert atmospheres. It was found that behavior in reducing, oxidizing and inert atmospheres could be predicted.

9. Metal stabilization

The stabilization of Pt/Al_2O_3 catalysts by addition of other oxides, preparation methods or metals has great importance in the preparation of catalyst because the sintering of Pt particles is reduced. One example of stabilization of Al_2O_3 by addition of oxides was done by Mizukami et al.,[Mizukami et al, 1991].

In this study thermostable high-surface-area Al_2O_3 was obtained by investigating the influence of preparation procedures, raw materials and additives on sintering of Al_2O_3. The complexing agent-assisted sol-gel method to obtain Al_2O_3 with large surface area and high durability at around 1000°C as compared with those prepared by conventional methods.

Upon addition of BaO, La_2O_3 and SrO to Al_2O_3, the sintering of Al_2O_3 was furthermore retarded and the effect was maximum when 10wt% BaO was incorporated to alumina. The activity order of the three catalysts for the CO oxidation was as follows: (commercial Al_2O_3) < (sol-gel Al_2O_3) < (sol-gel BaO-Al_2O_3). This sequence could be explained by the differences in the Pt crystal growth and α- Al_2O_3 appearance rates in the three catalysts.

In other study of stabilization, Pt/SiO_2 catalysts with well defined pore size distributions and metal particle sizes were prepared by the sol-gel method [Zou & Gonzalez, 1993]. Tetraethoxysilane (TEOS) and Pt(Acetyl acetonate)$_2$ were used as precursors. Pore size distributions were controlled by varying the H_2O/TEOS ratio during synthesis. It was found that when the particle size was matched to coincide with the average pore diameter the resulting Pt/SiO_2 catalysts were resistant to sintering in O_2 at temperatures up to 675°C.

In other work, catalysts of Pt supported on MgO, with Pt loadings of 0.5 to 5.9 wt%,[Ademiac, et al., 1993] were characterized by H_2 chemisorption, wide-angle x-ray diffraction and TEM after various high temperature treatments in O_2 and H_2. It is concluded that the H/Pt ratios for Pt/MgO do not, correspond to the Pt dispersion, even for catalysts with very low levels of impurities. Reduction at temperatures ≥400°C (HTR), of sulfur-containing Pt/MgO causes large decreases in H_2 adsorption, but the H_2 adsorption capacity can be restored by O_2 treatment at 500 - 550°C followed by reduction in H_2 at 250 to 300°C (LTR). Treatment of Pt/MgO in O_2 at 550 to 800°C results in decreases in average Pt crystallite size (redispersion); no significant Pt particle growth (sintering) occurs during O_2 treatment at temperatures as high as 800°C. Chlorine increases the rate of Pt redispersion, but chlorine is not required for redispersion of Pt on MgO

In other investigation, Pt clusters smaller than 1.5 nm on γ-Al_2O_3 support were stable against thermal sintering in H_2 at temperatures as high as 500°C[Zhang & Beard,1999]. The stability of small Pt clusters against thermal sintering can be ascribed to the anchoring effect of Pt by the acid sites on the surface.

The linearly adsorbed CO on Pt is characteristic of small Pt clusters. A band at 2112cm^{-1} suggests the presence of charged Pt atoms. Bridged CO infrared bands, 1825 and 1661cm^{-1}, which are absent from bulk Pt, are also evidence for the formation of small Pt carbonyl clusters. It was discovered in this work that the small Pt clusters coalesce readily upon H_2 reduction at 320°C when the catalyst is pretreated with a solution of a chloride salt such as NH_4Cl. The agglomeration of small Pt clusters in the presence of a chloride salt under a reducing environment is ascribed to the deanchoring effect of the chloride, which weakens the interaction of Pt with the anchoring Lewis acid sites. As temperature increases, the formation of mobile Pt chloride complexes that leads to Pt particle agglomeration is proposed .

The Pt particle sizes after the reductive agglomeration are distributed within a narrow range of 5-8 nm. The electronic character of the large Pt particles is identical to that of bulk Pt metal. It was further demonstrated that Pt cluster agglomeration is independent of the heating rate and the length of heating period, indicating that the predominant mechanism of Pt particle agglomeration is due to rapid coalescence of chlorinated primary small Pt particles under an H_2 atmosphere.

In other study of stabilization by supported bimetallic catalysts, highly dispersed Pt-W structures were formed on γ- Al_2O_3 from bimetallic cluster precursors with Pt-W bonds[Alexeev, et al.,2000]. The interactions of W cations with O_2 atoms of the support, on one hand, and with Pt clusters, on the other, stabilized Pt dispersion in clusters. Highly dispersed bimetallic structures highly resistant to sintering under high-temperature oxidation-reduction conditions were evident. Increases in the amount of W initially bonded to Pt, increased resistance of the supported Pt to migration and aggregation on the support. The resistance to aggregation was ascribed to the oxophilic metal-noble metal interactions. The interaction of W cations with Pt decreased the chemisorption capacity of Pt clusters for H_2 and CO, and the catalytic activity for toluene hydrogenation.

In other paper, one interesting investigation related with different preparation method of Pt/Al_2O_3 was made by Ikeda, et al.[Ikeda et al., 2001]. They found that the catalyst prepared by microemulsion had a higher resistance to sintering than the catalysts prepared by the sol-gel and impregnation methods. The resistance to sintering in all the catalysts was improved by pressing. The pressed microemulsion catalyst was little deactivated in the NO-CO reaction by thermal treatment at 700 °C for 12 h, and had a high activity relative to that of the sol-gel and impregnation catalysts.

In other stabilization work, sintering behavior of Pt/γ- Al_2O_3, Pt/ZrO_2 and Pt/CeO_2 catalyst was studied using an originally developed 3D sintering simulator [Shinjoh,, et al.,2009]. While Pt on the γ-Al_2O_3 sintered significantly, Pt on CeO_2 presented the highest stability against sintering. On the other hand, grain growth of supports was significant in the order; $ZrO_2 > CeO_2 >$ γ- Al_2O_3.

In other interesting work made by Romero et al., [Romero, et al., 2002] a series of Pt/Al_2O_3 catalysts were prepared using a sol-gel method. It was found that the BET surface area decreased with an increase in the platinum content. A surface area of 500 m^2/g was obtained following calcination at 773K. The structure of fresh samples corresponded to a pseudoboehmite structure. Samples prepared using a water/alkoxide ratio (H_2O/ammonium ter-butoxide (ATB)) of 9 showed a well-defined, uniform pore size distribution following calcination at 773K. Aging studies (calcination at 873K for 24h) performed on these catalysts, exhibited sintering behavior which were similar to Pt/Al_2O_3 catalysts prepared by other methods. The sample prepared using a H_2O/ATB ratio of 9 had the highest surface area and was more thermally resistant towards metal sintering. A bimodal metal particle size distribution was observed: some particles exhibited sintering while others of similar size showed a greater thermal stability to sintering. The sample having the largest surface area and the highest thermal stability following thermal treatment was a consequence of a more condensed structure and a higher pore roughness obtained after drying the gel.

The use of zeolite as catalysts with noble metals was proved by Kanazawa [Kanazawa, 2006] in automotive exhaust systems. These catalysts were subjected to sintering at extreme temperatures, leading to deterioration of catalytic activity. Zeolite with the zeolite MFI structure having mesoporosity (ZSM5) structure is examined as a support for Pt particulate catalysts. The MFI structure is composed of agglomerates of single-crystal zeolite with interstitial mesoporosity. Pt fixed within these mesopores is shown through high-temperature aging tests in air to be highly resistant to sintering due to the mechanical constraints on particle size imparted by the mesoporous structure. The deterioration of catalytic activity after aging is significantly lower than that for comparable γ-alumina supported catalyst.

In other work, the stabilization of Pt particles was studied by Liotta et al.,[Liotta et al.2009]. The authors studied a structured 1wt%Pt/ceria-zirconia/alumina catalyst and the metal-free ceria-zirconia/alumina was prepared, by dip-coating, over a cordierite monolithic support. In the Pt supported catalysts ceria-zirconia were present as a $Ce_{0.6}Zr_{0.4}O_2$ homogeneous solid solution and that the deposition over the cordierite does not produce any structural modification. Moreover no Pt sintering occurs and TEM investigations of the redox cycled Pt/ceria-zirconia/alumina catalyst detected ceria-zirconia grains with diameter between 10 and 35 nm along with highly dispersed Pt particles (2-3 nm) strongly interacting with ceria. The effect of redox aging on the NO reduction by C_3H_6, in lean conditions, was investigated over the Pt/ceria-zirconia/alumina monolith. The catalyst shows at low temperature (290 °C) good NO removal activity and appreciable selectivity to N_2.

In another stabilization work, the addition of W^{6+} ions to Al_2O_3 has been found useful to the synthesis of Pt/Al_2O_3 catalysts. The simultaneous and sequential methods were used to study the effect of W^{6+} upon Pt/γ-Al_2O_3 reducibility, Pt dispersion, and benzene hydrogenation [Contreras et al.,1996]. The same authors studied the thermal stabilization of γ-Al_2O_3 using W^{+6} ions in the synthesis of Pt/Al_2O_3 catalysts applied to the n-heptane hydroconversion. The sequential impregnation method was used to study the effect of W^{6+} upon Pt/γ-Al_2O_3 reducibility and n-heptane hydroconversion. The authors found that the W^{6+} ions delayed reduction of a fraction of Pt^{+4} atoms beyond 773 K. At the same time, W^{6+} inhibited sintering of the metallic crystallites of Pt once they were formed on the surface. After reduction at 1073 K, Pt on WOx/γ-Al_2O_3 were more active and stable during n-heptane hydroconversion than monometallic Pt/γ- Al_2O_3 catalyst.

In other study, the stabilization of Pt was obtained when the authors supported Pt on carbon black but their Pt nanoparticles were covered with microporous silica layers using successive hydrolysis of 3-aminopropyl-triethoxysilane and methyltriethoxysilane [Nakagawa, et al., 2010]. The Pt/carbon nanoparticles covered with microporous silica layers showed high sintering resistance of Pt metal particles to thermal treatment at 973 K in a H_2 atmosphere as compared with Pt/carbon. Furthermore, this catalyst showed higher catalytic activity for cyclohexane decomposition than Pt/carbon after thermal treatment.

An interesting paper made by Suzuki et al. [Suzuki, et al. 2010] studied multi-scale theoretical methods for predicting sintering behavior of Pt on various catalyst supports. In this regard, the capability of theoretical durability studies to offer an efficient alternative methodology for predicting the potential performance of catalysts has improved in recent years and various types of Pt diffusions depending on support were confirmed by the micro-scale ultra-accelerated quantum chemical molecular dynamics method. With this purpose, macro-scale sintering behavior of Pt/Al_2O_3, Pt/ZrO_2 and Pt/CeO_2 catalysts were studied using a developed 3D sintering simulator. Experimental results were well reproduced. While Pt on Al_2O_3 sintered significantly, Pt on ZrO_2 sintered slightly and Pt on CeO_2 demonstrated the highest stability against sintering.

A new procedure to stabilize Pt particles was developed by Lee et al., [Lee,et al., 2011]. The authors encapsulated supported Pt nanoparticles in heterogeneous catalysts to prevent their sintering. Model catalysts were first prepared by dispersing ~3-nm Pt nanoparticles on ~120-nm silica beads. These were then covered with a fresh layer of mesoporous silica, a few tens of nanometers thick, and etched to re-expose the metal surface to the reaction mixtures.. The resulting encapsulated platinum nanoparticles were shown to resist sintering during calcination at temperatures as high as 1075 K, whereas the unprotected catalysts were seen to sinter by 875 K.

In a recent investigation Li et al., [Li et al., 2011] prepared a support, composed by CeO_2-ZrO_2-La_2O_3-Al_2O_3(CZLA) with different contents of CeO_2, ZrO_2 and La_2O_3 by the peptizing method and characterized by XRD and N_2 adsorption-desorption. The low Pt-Rh content catalysts were prepared using $Ce_{0.4}Zr_{0.4}Y_{0.1}$ $La_{0.1}O_{1.9}$ and CZLA samples as the support, and the catalytic performance showed that both La, Ce and Zr can inhibit the Al_2O_3 sintering and then suppress Al_2O_3 transition to α-phase. The CZLA sample with 2%La_2O_3 maintained a highest specific surface area (97 m^2/g) after aged at 1100°C for 5 h. The catalyst with CZLA and 2%La_2O_3 had an excellent catalytic performance. Its light-off temperature was 249°C, and full conversion temperature was 286°C for C_3H_8. Both results show that Ce-Zr-La co-modified alumina has great potential as support.

The stabilization of Pt nanoparticles supported on Al_2O_3 with conventional and perovskite based materials was reported by Dacquin et al.[Dacquin, et al.,2011]. Successive thermal treatments under reductive and oxidative atmospheres induced bulk and surface reconstructions. These modifications considerably alter the catalytic behavior of Pt in interaction with $LaFeO_3$ or γ- Al_2O_3 in terms of activity and selectivity toward the selective transformation of NOx. It has been found that oxidic Pt^{4+} species initially stabilized on $LaFeO_3$ lead after subsequent H_2 reduction to the formation of metallic nano-Pt particles in stronger interaction than on γ-Al_2O_3 support and then become more resistant to sintering during thermal aging in 1000 ppm NO, 1000 ppm N_2O, 3 vol % O_2, 0.5 vol % H_2O, and 0.5 vol % H_2 at 500 °C.

10. Conclusion

Based on the data reviewed from the literature, it is possible to draw some general conclusions about the factors that influence sintering of supported metal catalysts. Temperature and the characteristics of the atmosphere are the most important factors, although the nature of the support and type of structural promoters play an important role, too. Bimetallic catalysts have shown good catalytic stability for several industrial applications. Sintering can be slowed down by the use of additives to the support together with alternate preparation methods. These new procedures, particularly in the case of Pt nanoparticles, have shown to be effective in attaining more stability, and hence activity and selectivity, in several commercial reactions.

11. Aknowledgements

The authors acknowledge Universidad Autónoma Metropolitana and Instituto Mexicano del Petróleo for their support.

12. References

Adamiec, J. , Szymura, J.A., Wanke, S.E. (1993) Sintering, Poisoning and Regeneration of Pt/MgO, *Stud. In Surf. Sci. and Catalysis,* Volume 75, Issue C, , Pages 1405-1418 ISSN: 01672991.

Alexeev, O.S. , Graham, G.W., Shelef, M. , Gates, B.C. (2000), γ-Al_2O_3-supported Pt catalysts with extremely high dispersions resulting from Pt-W interactions, *J. of Catal.* Volume 190, Issue 1, 15 February, Pages 157-172 ISSN: 00219517.

Anderson J.R. and Pratt, K.C. (1985), Introduction to characterization and testing of catalysts" *Academic Press* Australia, p.355.

Anderson, J.A., Mordente, M.G.V., Rochester, C.H.(1989) Effects of oxidation-reduction treatments of Pt/Al_2O_3 on catalytic activity and selectivity for hexane reforming, Volume 85, Issue 9, Pages 2991-2998 ISSN: 03009599.

Barbier, J., Bahloul, D., Marecot, P. (1992), Reduction of Pt/Al_2O_3 catalysts: Effect of hydrogen and of water and hydrochloric acid vapor on the accessibility of platinum. *J. of Catal.,* Volume 137, Issue 2, October, Pages 377-384 ISSN: 00219517.

Bishara, A., Murad, K.M., Stanislaus, A., Ismial, M., Hussian, S.S (1983) Factors controlling the sintering of an industrial bimetallic reforming catalyst during regeneration, *Appl. Catal.*Volume 7, Issue 3, 15 September, Pages 351-359, ISSN: 01669834.

Borgna, A. , Le Normand, F. , Garetto, T.F. , Apesteguia, C.R. , Moraweck, B. (1999),X-ray Absorption Spectroscopy: A powerful tool to investigate intermediate species during sintering-redispersion of metallic catalysts, *Stud. In Surf. Sci. and Catal.*Volume 126, Pages 179-186 ISSN: 01672991.

Borgna, A., Garetto, T.F., Apesteguia, C.R., Le Normand, F., Moraweck,B. (1999), Sintering of chlorinated Pt/γ-Al_2O_3 catalysts: An in situ study by X-ray absorption spectroscopy *J. of Catal.* Volume 186, Issue 2, Pages 433-441, ISSN: 00219517.

Borgna, A., Le Normand,F.,Garetto, T., Apesteguia, C.R., Moraweck,B. (1992) Sintering of Pt/Al_2O_3 reforming catalysts: EXAFS study of the behavior of metal particles under oxidizing atmosphere *J. of Catal.*Volume 13, Issue 3, September, Pages 175-188, ISSN: 1011372X.

Butt J.B. and Petersen, E.E. (1988) Activation Deactivation and Poisoning of Catalysts, *Academic Press, Inc.* U.K. Edition (London) Ltd.171-234

Cabello G. F., Mariscal, R. , López Granados, M. , Fierro, J.L.G. , Daley, R.A. , Anderson, J.A. (2005), Reactivation of sintered Pt/Al_2O_3 oxidation catalysts *Appl. Catal. B: Environmental,* Volume 59, Issue 3-4, 8 August, Pages 227-233 ISSN: 09263373.

Contreras, J.L. , Fuentes, G.A. (1996) Effect of tungsten on supported platinum catalysts, *Stud. In Surf. Sci. and Catalysis,* Volume 101 B, Pages 1195-1204 ISSN: 01672991.

Contreras, J.L. , Fuentes, G.A. , Zeifert, B. , Salmones, J. (2009),Stabilization of supported platinum nanoparticles on γ-alumina catalysts by addition of tungsten, *Journal of Alloys and Compounds,* Volume 483, Issue 1-2, 26 August Pages 371-373 ISSN: 09258388.

Contreras, J.L.,Fuentes, G.A., Salmones, J. , Zeifert, B.(2010) Thermal stability of Pt nanoparticles supported on WOx/Al_2O_3 for n-heptane hydroconversion,19th International Materials Research Congress 2010; Cancun; 15 August 2010 through 19 August 2010; Code 84177, Volume 1279, 2010, Pages 123-141 ISSN: 02729172.

Chu, Y.F., Ruckenstein, E. (1978),On the sintering of platinum on alumina model catalyst *J. of Catal.* Volume 55, Issue 3, December Pages 281-298 ISSN: 00219517.

Hassan, S.A. , Khalil, F.H. , El-Gamal, F.G.(1976) The effect of different atmospheres on the sintering of Pt Al_2O_3 reforming catalysts, *J. of Catal.* Volume 44, Issue 1, July, Pages 5-14 ISSN: 00219517.

Dacquin, J.P., Lancelot, C , Dujardin, C., Cordier-Robert, C., Granger, P. (2011) Support-induced effects of $LaFeO_3$ perovskite on the catalytic performances of supported Pt catalysts in DeNOx applications *J. of Phys. Chem.* Volume 115, Issue 5, 10 February 2011, Pages 1911-1921 ISSN: 19327447.

Deng, Yi, Wang, Junyu, Min, Enze (1988) Effect of $Ba(Ac)_2$ on sintering of Pt clusters on Pt/γ-Al_2O_3 catalyst, Shiyou Xuebao/*Acta Petrolei Sinica*,Volume 4, Issue 2, June Pages 57-65 ISSN: 02532697.

Fernandes, D.M. , Scofield, C.F. , Neto, A.A., Cardoso, M.J.B. , Zotin, F.M.Z. (2010)Thermal deactivation of Pt-Rh commercial automotive catalysts, *Chemical Engineering Journal,* Volume 160, Issue 1, 15 May, Pages 85-92 ISSN: 13858947.

Fiedorow, R.M.J., Chahar, B.S., Wanke, S.E.(1978), The sintering of supported metal catalysts. II. Comparison of sintering rates of supported Pt, Ir, and Rh catalysts in hydrogen and oxygen, J. of Catal. Vol.51,2, p. 193-202,ISSN: 00219517.

Flynn, P.C., Wanke, S.E (1975),. Experimental studies of sintering of supported platinum catalysts, *J. of Catal.* Volume 37, Issue 3, 6 June,p. 432-448, ISSN: 00219517.

Fuentes G.A. and Salinas R.E.(2001), Realistic particle size distributions during sintering by Ostwald ripening *"Catalysts Deactivation 2001"Stud.Surf.Sci.Catal.,(J.J.Spivey, G.W. Roberts, and B.H.Davis, eds.)*V.139,p.503-510 Elsevier, Amsterdam 2001, ISSN: 01672991.

Garetto, T.F., Borgna, A., Monzón, A. (1996), Modelling of sintering kinetics of naphtha-reforming Pt/Al_2O_3-Cl catalysts, *J. of the Chem. Soc. Faraday Trans.*Volume 92, Issue 14, 21 July Pages 2637-2640 ISSN: 09565000.

Graham, A.G., Wanke, S.E. (1981) The sintering of supported metal catalysts. III. The thermal stability of bimetallic Pt-Ir catalysts supported on alumina, *J. of Catal.* Volume 68, Issue 1, March, Pages 1-8 ISSN: 00219517.

Guo, Y. , Zhou, L. , Kameyama,H.(2011) Steam reforming reactions over a metal-monolithic anodic alumina-supported Ni catalyst with trace amounts of noble metal . International *Journal of Hydrogen Energy,* Volume 36, Issue 9, May, Pages 5321-5333 ISSN: 03603199.

Guo, I. , Yu, T.-T. , Wanke, S.E. (1988) Changes in Pt crystallite sizes as a result of treating Pt/Al_2O_3 catalysts in different atmospheres. *Stud. Surf. Sci. and Catal.* Volume 38, Issue C,Pages 21-32 ISSN: 01672991.

Handa, J.C. Matthews, P.K., (1983), Modeling of sintering and redispersion of supported metal catalysts. *J. of AIChE,* Volume 29, Issue 5, September, Pages 717-725, ISSN: 00011541.

Harris, P.J.F. (1986) The sintering of platinum particles in an alumina-supported catalyst: Further transmission electron microscopy studies, J. of Catal. Volume 97, Issue 2, February, Pages 527-542, ISSN: 00219517.

Hayashi, K. , Horiuchi, T. , Suzuki, K. , Mori,T (2002), Sintering behavior of Pt metal particles supported on silica-coated alumina surface , Catalysis Letters,Volume 78, Issue 1-4, March, Pages 43-47 ISSN: 1011372X.

He, W., Liu, D., Zhang, L.(1998), Huaxue Fanying Gongcheng Yu Gongi Chem. Investigation on GPLE sintering kinetics of platinum alumina *React. Eng. and Tech.* Volume 14, Issue 2, Page 25, ISSN: 10017631.

Herrmann, R.A., Adler, S.F., Goldstein, M.S., Debaun, R.M. (1961), The kinetics of sintering of platinum supported on alumina, *J. of Catal.,* Volume 65, Issue 12, , Pages 2189-2194 , ISSN: 00223654.

Hirata, H., Kishita, K., Nagai, Y., Dohmae, K., Shinjoh, H., Matsumoto, S. (2011) Characterization and dynamic behavior of precious metals in automotive exhaust gas purification catalysts , *Catal. Today* Volume 164, Issue 1, 30 April, Pages 467-473 ISSN: 09205861.

Ikeda, M., Tago, T., Kishida, M., Wakabayashi, K.(2001) Thermal stability of Pt particles of Pt/Al_2O_3 catalysts prepared using microemulsion and catalytic activity in NO-CO reaction *Catal. Communications,* Volume 2, Issue 8, , Pages 261-267 ISSN: 15667367.

Kacimi, S. , Kappenstein, C. , Duprez , D. (1993) Effect of Sintering in O_2 Atmosphere on the Surface Properties of PtRh/ Al_2O_3 Catalysts, *Stud. In Surf. Sci. and Catalysis,* Volume 75, Issue C, Pages 2519-2522 ISSN: 01672991.

Kanazawa, T. (2006) MFI zeolite as a support for automotive catalysts with reduced Pt sintering *Appl. Catal. B: Environmental,*Volume 65, Issue 3-4, 6 June, Pages 185-190 ISSN: 09263373.

Kaneeda, M. , Iizuka, H. , Hiratsuka, T. , Shinotsuka, N. , Arai, M.(2009) Improvement of thermal stability of NO oxidation Pt/Al_2O_3 catalyst by addition of Pd, *Appl. Catal.,B: Environmental,*Volume 90, Issue 3-4, 17 August, Pages 564-569 ISSN: 09263373.

Kacimi, S. , Duprez, D. (1991) Characterization of Bimetallic Surfaces by $^{18}O/^{16}O$ Isotopic Exchange. Application to the Study of the Sintering of PtRh/Al_2O_3 Catalysts. *Stud. In Surf. Sci. and Catalysis,* Volume 71, Issue C, , Pages 581-592 ISSN: 01672991.

Lamber,R., Jaeger N., Schulz-Erloff,G./(1990), Metal-support interaction in the Pd/SiO_2 system: Influence of the support pretreatmet, *J. of Catal.* 123, 285-297.

Lee H.H. and Ruckenstein E.(1983) *Catal. Revi. Sci. Eng.,* 25(4),475-550.

Lee, I., Zhang, Q., Ge, J., Yin, Y., Zaera, F. (2011) Encapsulation of supported Pt nanoparticles with mesoporous silica for increased catalyst stability, *Nano Research,* Volume 4, Issue 1, January Pages 115-123 ISSN: 19980124.

Lietz, G. , Lieske, H., Spindler, H. , Hanke, W.A , Völter, J.A. (1983), Reactions of platinum in oxygen- and hydrogen-treated Pt γ-Al_2O_3 catalysts. II. Ultraviolet-visible studies, sintering of platinum, and soluble platinum *J. of Catal.* Volume 81, Issue 1, May, Pages 17-25, ISSN: 00219517.

Liotta, L.F. , Longo, A. , Pantaleo, G. , Di Carlo, G. , Martorana, A. , Cimino, S. , Russo, G. , Deganello, G.(2009),Alumina supported Pt(1%)/$Ce_{0.6}Zr_{0.4}O_2$ monolith: Remarkable stabilization of ceria-zirconia solution towards $CeAlO_3$ formation operated by Pt under redox conditions, *Appl. Catal. B. Environmental,*Volume 90, Issue 3-4, 17 August Pages 470-477 ISSN: 09263373.

Li, X.-M., Peng, N., Chen, S.-H., Zhao, M., Chen, Y.-Q., Gong, M.-C.(2011) Preparation of Ce-Zr-La co-modified alumina by peptizing method and its catalytic performance, Gaodeng Xuexiao Huaxue Xuebao/*Chemical Journal of Chinese Universities*, Volume 32, Issue 1, January 2011, Pages 1-3 ISSN: 02510790.

Martin, D., Taha, R., Duprez, D. (1995) Effects of sintering and of additives on the oxygen storage capacity of Pt-Rh catalysts, *Stud. In Surf. Sci. and Catalysis*, Volume 96, Issue C, , Pages 801-811 ISSN: 01672991.

Micheaud, C., Guerin, M., Marecot, P., Geron, C., Barbier, J. (1996)Preparation and characterization of Pd-Pt/Al_2O_3 catalysts, J. *De Chimie Physique et de Physico-Chimie Biologique*, Volume 93, Issue 7-8, , Pages 1394-1411 ISSN: 00217689.

Mazzieri, V.A., Pieck, C.L., Vera, C.R., Yori, J.C., Grau, J.M. (2008), Analysis of coke deposition and study of the variables of regeneration and rejuvenation of naphtha reforming trimetallic catalysts, *Catalysis Today*, Volume 133-135, Issue 1-4, April Pages 870-878 ISSN: 09205861.

Mizukami, F. , Maeda, K. , Watanabe, M. , Masuda, K. , Sano, T. , Kuno, K..(1991), Preparation of Thermostable High-Surface-Area Aluminas and Properties of the Alumina-Supported Pt Catalysts, *Stud. Surf. Sci. and Catalysis*, Volume 71, Issue C, Pages 557-568 ISSN: 01672991.

Murakami, Y., Komai, S., Hattori, T. (1991) Accelerated Sintering of Pt Catalysts in Reaction Atmosphere for Rapid Estimation of Catalyst Life , *Stud. Surf. Sci. and Catal.* Volume 68, Issue C, , Pages 645-652 ISSN: 01672991.

Nakagawa, K., Tanimoto, Y. , Okayama, T. , Sotowa, K.-I. , Sugiyama, S. , Takenaka, S. , Kishida, M.(2010) Sintering resistance and catalytic activity of platinum nanoparticles covered with a microporous silica layer using methyltriethoxysilane, Catal. Letters,Volume 136, Issue 1-2, May, Pages 71-76 ISSN: 1011372X.

Narayanan, S., Sreekanth, G.(1989) *J. of the Chemical Society, Faraday Transactions1: Physical Chemistry in Condensed Phases* Volume 85, Issue 11, Pages 3785-3796.

Pieck, C.L., Jablonski, E.L., Parera, J.M. (1990) Sintering-redispersion of Pt-Re/Al_2O_3 during regeneration, Appl. Catal. , Volume 62, Issue 1, Pages 47-60 ISSN: 01669834.

Romero-Pascual, E. , Larrea, A. , Monzón, A. , González, R.D. (2002), Thermal stability of Pt/Al_2O_3 catalysts prepared by sol-gel *J. of Solid State Chemistry*, Volume 168, Issue 1, October, Pages 343-353 ISSN: 00224596.

Sakamoto, Y., Higuchi, K., Takahashi, N., Yokota, K., Doi, H., Sugiura, M.(2010) Effect of the addition of Fe on catalytic activities of Pt/Fe/γ-Al_2O_3 catalyst, *Appl. Catal.,B: Environmental,* Volume 23, Issue 2-3, 1 November, Pages 159-167 ISSN: 09263373.

Schaal, M.T., Rebelli, J., McKerrow, H.M., Williams, C.T., Monnier, J.R.(2010) Effect of liquid phase reducing agents on the dispersion of supported Pt catalysts, *Appl. Catal. A: General* Volume 382, Issue 1, 30 June, Pages 49-57 ISSN: 0926860X.

Shinjoh, H., Muraki, H., Fujitani, Y.(1991) Effect of Severe Thermal Aging on Noble Metal Catalysts, *Stud. In Surf. Sci. and Catalysis*, Volume 71, Issue C, Pages 617-628 ISSN: 01672991.

Shyu, J.Z., Otto, K.(1988) Identification of platinum phases on γ-alumina by XPS *Appl. Surf. Sci.*, Volume 32, Issue 1-2, June, Pages 246-252 ISSN: 01694332.

Simonsen, S.B. , Chorkendorff, I. , Dahl, S. , Skoglundh, M , Sehested, J. , Helveg, S. (2010) Direct observations of oxygen-induced platinum nanoparticle ripening studied by in situ TEM, *Journal of the American Chemical Society,*Volume 132, Issue 23, 16 June, Pages 7968-7975 ISSN: 00027863.

Stakheev, A.Yu. , Tkachenko, O.P. , Kapustin, G.I. , Telegina, N.S. , Baeva, G.N. , Brueva, T.R. , Klementiev, K.V. , Grunert, W. , Kustov, L.M. (2004)Study of the formation and stability of the Pd and Pt metallic nanoparticles on carbon support, *Russian Chemical Bulletin,*Volume 53, Issue 3, March, Pages 528-537 ISSN: 10665285.

Sushumna, I., Ruckenstein, E. (1988), Events observed and evidence for crystallite migration in Pt /Al_2O_3 catalysts, *J. of Catal.* Volume 109, Issue 2, February, Pages 433-462,ISSN: 00219517.

Susu, A.A., Ogogo, E.O. (2006),Sintering-induced aromatization during n-heptane reforming on Pt/Al_2O_3 catalysts, *Journal of Chem. Thechnology and Biotechnology.* Volume 81, Issue 4, April, Pages 694-705 ISSN: 02682575.

Suzuki, A., Sato, R., Nakamura, K., Okushi, K., Tsuboi, H., Hatakeyama, N., Endou, A., Takaba, H., Kubo, M., Williams, M.C., Miyamoto,(2010)A. Multi-scale theoretical study of sintering dynamics of Pt for automotive catalyst, SAE *International Journal of Fuels and Lubricants,* Volume 2, Issue 2, , Pages 337-345 ISSN: 19463952.

Suzuki, A. , Nakamura, K. , Sato, R. , Okushi, K. , Koyama, M. , Tsuboi, H. , Hatakeyama, N., Endou, A. , Takaba, H. , Del Carpio, C.A. , Kubo, M. , Miyamoto, A. , (2009),Development and application of sintering dynamics simulation for automotive catalyst, *Topics in Catal.*Volume 52, Issue 13-20, December Pages 1852-1855 ISSN: 10225528.

Vaccaro, A.R. , Mul, G. , Pérez-Ramírez, J. , Moulijn, J.A. (2003) On the activation of Pt/Al_2O_3 catalysts in HC-SCR by sintering: Determination of redox-active sites using Multitrack, *Appl. Catal. B: Environmental,* Volume 46, Issue 4, 15 December, Pages 687-702 ISSN: 09263373.

White, D.A , Baird, T.A , Fryer, J.R.A , Freeman, L.A.B , Smith, D.J.b , Day, M.C (1983) Electron microscope studies of platinum/alumina reforming catalysts, *J. of Catal.,* Volume 81, Issue 1, May, Pages 119-130 , ISSN: 00219517

Wang, L., Sakurai, M., Kameyama, H. (2005) Sintering effect of platinum catalysts supported on anodized aluminium plate on VOCs' catalytic combustion *7th World Congress of Chemical Engineering, GLASGOW2005, incorporating the 5th European Congress of Chemical Engineering,* Glasgow, Scotland; 10 July 2005 through 14 July 2005, p. 115-116 ISBN: 0852954948.

Yao, H.C., Wynblatt, P., Sieg, M., Plummer Jr., H.K.(1980) Sintering, redispersion and equilibrium of platinum oxide on gamma -alumina. Sintering Processes, *Proc of the Int Conf on Sintering and Relat. Phenom, 5th; Materials Science Research* 18 June 1979 through 21 June 1979,Volume 13, pp 561-571, ISSN: 00765201.

Wanke S. E. and Flynn, P.C. (1975) "The sintering of Supported Metal Catalysts" *Catal.Rev. Sci. Eng.* Vol. 12(1), 93-135.

Yang, J. , Tschamber, V. , Habermacher, D. , Garin, F., Gilot, P. (2008) Effect of sintering on the catalytic activity of a Pt based catalyst for CO oxidation: Experiments and modeling, *Appl. Catal. B :Environmental.,* Volume 83, Issue 3-4, 23 September, Pages 229-239 ISSN: 09263373.

Yaofang, L., Jiujing, Y., Guaghua, Y. (1994) Chemical characteristics and kinetics in regeneration of Pt-Sn/ Al_2O_3 reforming catalyst, *Khimiya I Tekhnologiya Topliv I Masel* Issue 9-10, September, Pages 12-15 ISSN: 00231169.

Yaofang, L., Jiujing, Y., Guanghua, Y. (1995) Chemical characteristics and kinetics in regeneration of Pt-Sn/Al_2O_3 reforming catalyst, *Chemistry and Technology of Fuels and Oils*, Volume 30, Issue 9-10, September, Pages 353-359 ISSN: 00093092.

Yaofang, Liu, Guoqing, Pan, Guanghua, Yang, (July 1995) Sintering and redispersion of platinum in reforming catalyst, *Proceedings of the 210th National Meeting of the American Chemical Society;* Chicago, IL, USA; 20 August through 25 August; Code 44301 Volume 40, Issue 3, , Page 437 ISSN: 05693799.

Zhang, Z.C., Beard, B.C. (1999) Agglomeration of Pt particles in the presence of chlorides *Appl. Catal., A, General,* Volume 188, Issue 1-2, 5 November, Pages 229-240 ISSN: 0926860X.

Zou, W. and Gonzalez, R.D.(1993) Stabilization and sintering of porous Pt/SiO_2: a new approach *Appl. Catal., A.General, Volume* 102, Issue 2, 31 August 1993, Pages 181-200 ISSN: 0926860X.

6

Sintering Characteristics of Injection Moulded 316L Component Using Palm-Based Biopolymer Binder

Mohd Afian Omar and Istikamah Subuki
AMREC, SIRIM Berhad, Lot 34, Jalan Hi Tech 2/3,
Kulim Hi Tech Park, Kulim, Kedah
Malaysia

1. Introduction

Metal injection moulding (MIM) has been widely recognised as an advanced technology for the fabrication of complex-shaped, low cost and high performance components. Fine powders, less than 20 micron in diameter, are mixed with suitable thermoplastic binder and formed into the desired shapes. The binder aids the flowability and formability of fine metal powders during moulding, and they have to be removed in the next stage to enable high density components to be produced. The removal of the binder is done either thermally in the furnace or by solvent extraction. Ideally, the removal of the binder would open up pore channels which allow accelerated removal of the higher boiling point components. The components are sintered following the debinding stage. This stage is crucial to the MIM process as appropriate sintering conditions would ensure pore-free structures that have good mechanical properties (German, 1990; German and Bose, 1997)

Theoretical studies of sintering treat the powder as a spherical particle and divide sintering into three stages. The early stage of sintering occurs at low temperatures and is characterised by neck growth at the contact points between the particles. The intermediate stage of sintering is characterised by an interconnected pore system having complex geometry. (German, 1996)

The final stage begins when the pore phase becomes closed and the shrinkage rate of the components slow down. This final stage characterised by pores on four-grain corners that shrink rapidly, and sphereodised powders that separate from grain boundaries and shrink slowly. When all pores on four-grain corners have been eliminated, sintering densification essentially ceases (German, 1990 ; German 1992, German and Hens, 1993; German, 1996)

There has been a considerable motivation to design and develop binders that are locally produced in order to minimize the manufacturing cost and to fulfill the requirements of the MIM process. Prior study by Johnson (1988) revealed that moulded specimens made with peanut and vegetable oil mixed with polymer binder of polyethylene (PE) and polypropylene (PP), substantially eliminate cracking phenomenon that is commonly observed in the moulding process. These peanut and vegetable oil were used to replace the function of wax and surfactant in a binder system to ensure good wetting of the powder. As the characteristics of palm oil are similar to those oils mentioned, which consists of glycerides used as a surface active agent in many binder systems, palm oil can be considered as a potential binder. Since Malaysia is one of the largest producers and exporter of palm oil, it can be an opportunity to design a new locally sourced binder system to achieve these goals.

Besides the suitable constituent, palm oil also has other attributes as a binder such as low viscosity, high decomposition temperature, lower molecular weight to avoid residual stress and distortion, environmentally acceptable, inexpensive and easily dissolved in organic solvent. Thus, Iriany (2002) investigated the possibility of palm oil as a binder via characterisation of feedstock. The results indicate that palm oil can be used as a binder component and was found to be compatible with polyethylene and stainless steel powder. However, the investigation was limited to the flow behaviour of the proposed system. To accomplish the investigations through all process of MIM, the new developed binder, palm stearin and feedstock prepared with different particle size, powder loading and binder formulation were studied by considering a variety of aspects. These aspects include torque evaluation curves, rheological behaviour, injection moulding, kinetic solvent extraction, thermal pyrolysis and effect of sintering when specimen is sintered at different temperature, atmosphere, soaking time and heating rate. The results on moulding behavior, rheological behavior and debinding study has been discussed elsewhere. (Omar, 2007; Omar, 2006a and 2006b; Omar 2001; Omar, 2002) However, in this chapter, the results reported was on sintering characteristics of 316L stainless steel powder using palm stearin (PS) as a major portion and a minor fraction of poly propylene (PP).

2. Method and materials

The 90 %-22 μm 316L stainless steel powder used in the present study was obtained from Sandvik. The mean particle size distribution was determined using HELOS Particle Size Analysis WINDOX 5 and around 9 micron A scanning electron micrograph showing the powder morphology is spherical. The powder was mixed with a natural polymer based binder (palm stearin) at a solid loading of 65-volume % for injection molding. The binder system consists of different percentage of palm stearin (PS) and polypropylene (PP), as tabulated in table 1. The mixing was carried out in a sigma blade mixer for 1 hour at 160 °C to produce feedstock. (Omar, 2007; Omar 2006) The mixing was left for 1 ½ hour. After mixing has completed, the heater was shut off and the feedstock was allowed to cool with the mixing blade still in motion. This procedure gives a granulated feedstock.

Label	Vol.% fraction	Composition, wt.%	
		PS	PP
PSPP65-2	65	50	50
PSPP65-3	65	60	40
PSPP65-4	65	70	30

Table 1. Feedstock made with different composition binder of PS/PP.

The granulated feedstock was then injected into tensile bars using a simple, vertically aligned and pneumatically operated plunger machine, MCP HEK-GMBH. Feedstock was fed into the barrel and then injected through the nozzle in the mold cavity. Test bars were successfully molded at temperature of 220°C at pressure 300 bar. The dimensions and weight including density were measured in order to determine the solvent removal and shrinkage after sintering. The densities of the specimens were measured using water immersion method.

The test bars were debound by a two-step process where at the first stage the samples were solvent debound in order to remove all the wax portion of the binder, in this case palm stearin which is consist the major fraction of the binder. Molded samples termed the green body were arranged in a glass container, which then immersed in heptane and held at temperature 60°C for 5 hours. The glass container was covered to prevent evaporation of the heptane during extraction Subsequent thermal pyrolysis was performed in Lynn Furnace. The thermal debinding cycle consisted of 3°C/min to 450°C and soaking for 1 hour before furnace cool. Sample that completely undergoes thermal debinding termed the brown body.

The components were sintered in vacuum furnace with the heating rate at 10°C/min to the sintering temperature 1360°C, and held for 1 hour at this temperature before cooled down by furnace cool. The dimensions, density and weight of the sintered specimens were measured to calculate sintered shrinkage and final density.Tensile properties of the sintered samples were determined using an Instron Series IX Automated Materials Testing System. The yield strength, ultimate strength and elongation were measured at strain rate of 0.1/s . Finally, the microstructure analysis was carried out using optical microscopy and scanning electron microscopy.

3. Results and discussion

3.1 Properties of sintered specimen made with various composition of PS/PP

Three different specimens made with different compositions of binder system and labelled as PSPP65-2, PSPP65-3 and PSPP65-4, were sintered at various temperatures. The specimens were sintered under vacuum conditions. The samples were heated to 450°C, soaked for 1 hour with the heating rate 3°C/min and then heated up at a rate of 10°C/min to sintering temperature of 1300°C, 1320°C, 1340°C and 1360°C with a soaking time of 1 hour.

3.1.1 Physical properties

Figure 1 shows the density of the sintered specimens made with PS/PP binder with various compositions as a function of sintering temperature. As the sintering temperature was increased from 1300 °C to 1360 °C, the density was improved for all specimens. An increase of 20 °C of the sintering temperature can improve the sintered density by up to 3%.

It clearly shows that an increase of 10 wt.% of PS resulted in 0.41% to 1.83% improvement in sintered density. The sample PSPP65-4 gave the highest density at all sintering temperatures compared to the other two compositions. It explains that higher PS content can provide better sintered specimen integrity. Table 2 gives the relative density for the samples. It clearly shows that the density data where higher than those of PS/PE (Istikamah, 2010) This circumstance might be due to a higher residual carbon of PS/PP compared to PS/PE system leaving the sample prior to sintering process. This residual carbon which left after debinding, diffused into the structure of SS 316L powder during sintering and eventually increased the density of the sintered specimens.

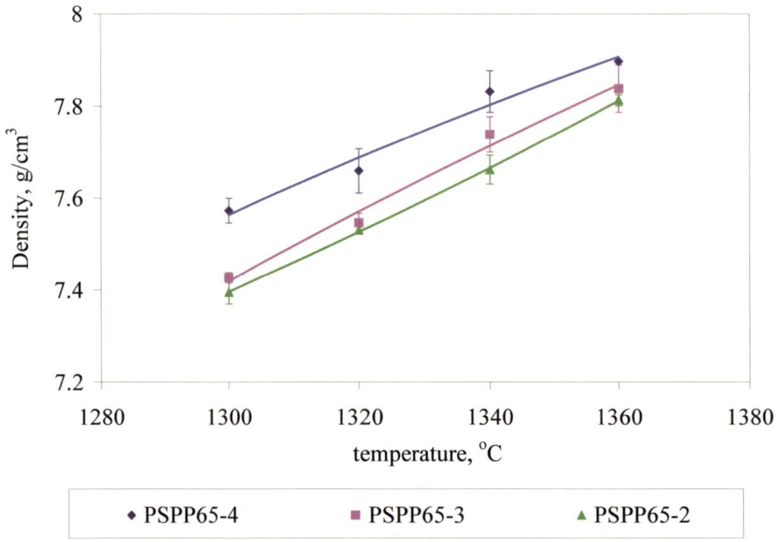

Fig. 1. Effect of PS content on the density of specimen sintered at different temperatures.

sintered specimen	Sintering temperature (°C)			
	1300	1320	1340	1360
PSPP65-2	± 93	± 95	± 96	± 98
PSPP65-3	± 93	± 95	± 97	± 99
PSPP65-4	± 95	± 96	± 98	± 99

Table 2. Relative density (%) of 316L sintered specimens at different wt% of PS/PP and sintering temperature.

Specimen shrinkage as a function of PS/PP contents and sintering temperature is shown in Figure 2. It clearly shows from the study that as the sintering temperature was increased from 1300 °C to 1360 °C there was a progressive increase in shrinkage percentage for all sintered specimens. The same relation was found in the density as discussed before. It shows that greater density results in a greater shrinkage. As the sintering temperature increased, it reduces the pore volume, leading to compact densification and thus increases the sintered density. Simultaneously, more active bulk transport that create a change in the interparticle spacing as neck growth, takes place and as a result increases the linear shrinkage.

This plot gives the highest sintering shrinkage of 15% resulted in 98% final density (Figure 2). The shrinkage is much higher for the specimen made with high content of PS. This may be due to the high shrinkage of brown body during thermal debinding of PP after high composition of PS was totally removed through solvent extraction. Sintering inherently involves substantial shrinkage, the process being characterized by a linear shrinkage of between 10 to 30%. From the result obtained, all specimen show linear contraction and produce good specimen without distortion.

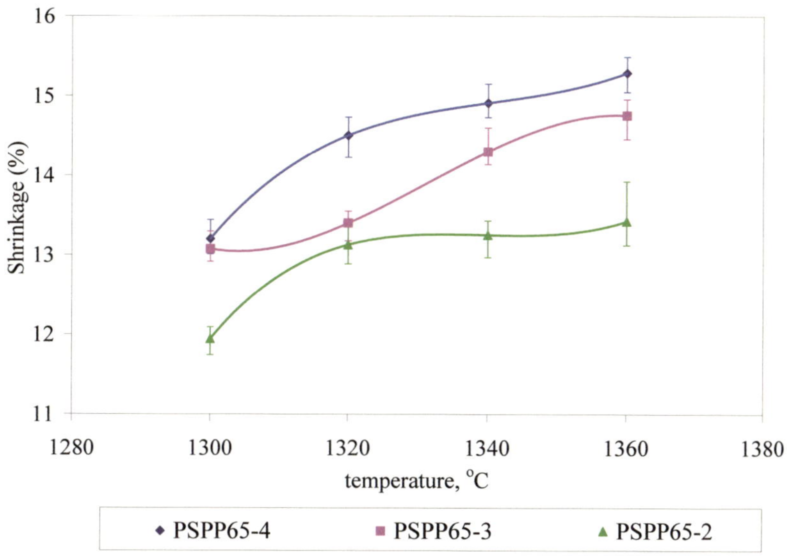

Fig. 2. Effect of sintering temperature on linear shrinkage of specimen made with different composition binder system.

Figure 3 shows the corresponding porosity for all compositions at various temperatures. It clearly shows that the porosity of the sintered specimens decreased as the sintering temperature increased for all specimen tested. Moreover, sintering temperature greater than 1320 °C gives much less deviation as compared to 1300 °C. These might be due to the changes and transformation it undergoes from intermediate to final stage of sintering to become dense and high strength, with attendant changes in the pore size and shape. At lower sintering temperature, there is insufficient fusion of metal powder in order to weld together (German, 1996). As a result, pore volume cannot be reduced as there is an insufficient kinetic energy. However, at higher temperatures especially beyond 1320 °C, this kinetic energy seems to be sufficient to remove free surface with the secondary elimination of grain boundary area via grain growth. (German, 1996)

At this final stage of sintering (1320 °C to 1360 °C), the elimination of isolated spherical pores was difficult, since vacancies must diffuse to distant grain boundaries, which is a very slow process. Also, with prolonged sintering, pore coarsening causes the mean pore size to increase while the number of pores decreases. As a consequence, there are slightly different porosity percentages at this stage of sintering. (Zhang and German, 1991)

Figures 4 to 6 show the optical micrograph of samples sintered at temperatures from 1300 °C to 1360 °C after etching. There is not much difference in grain size observed as temperature is increased. At that temperature, sintering is likely to slow down, as during this final stage of sintering, spherical pores shrink by a diffusion mechanism. The events lead to the isolation of a pore and spheroidisation due to rapid grain growth (German, 1996). At 1360 °C, it shows a change in the grain size as well as a decline in the total porosity during sintering as discussed before. The grain growth increased from 1300 °C to 1360 °C due to the annihilation of the pores on the grain boundaries.

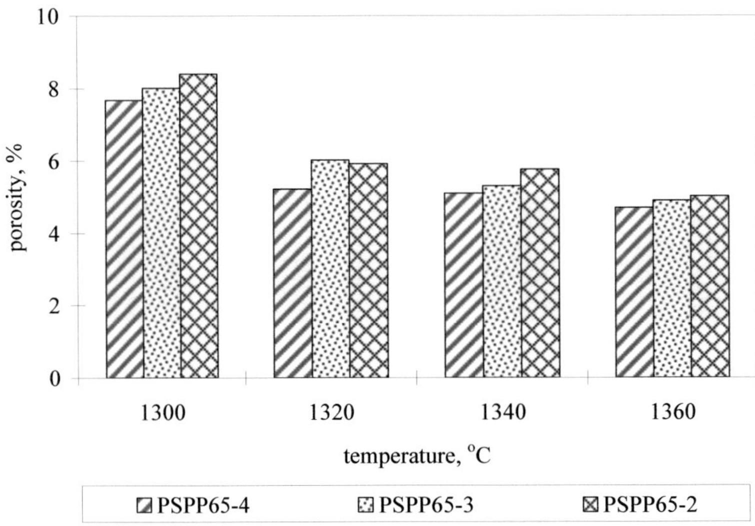

Fig. 3. The porosity of specimen sintered at different temperatures.

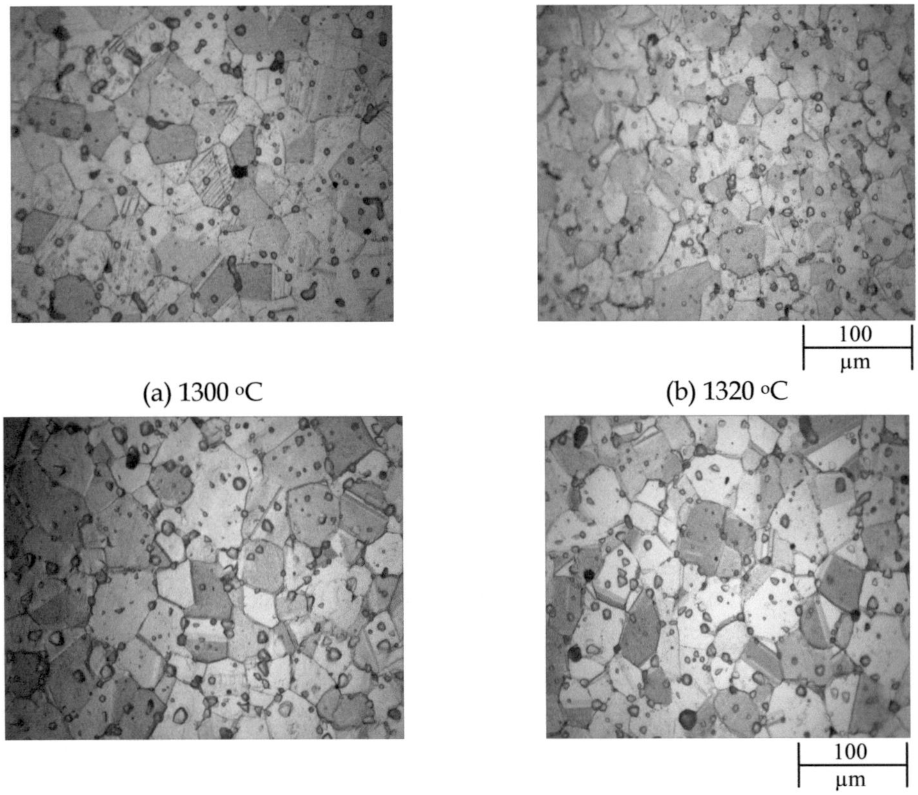

(a) 1300 °C (b) 1320 °C

(c) 1340 °C (d) 1360 °C

Fig. 4. Optical micrographs of PSPP65-4 specimen sintered at various temperatures. Magnification 500x.

As shown in the figure, the pore geometry is highly convoluted and the pores are located at grain boundary intersection. The microstructures show a progression from irregular pores along the particle boundaries a shown in Figure 4. (a), 5 (a) and 6 (a) to spherical pores as shown in Figure 4 (d), 5 (d) and 6 (d) within or on grain boundaries where densification occurs by decreasing the pore radius.

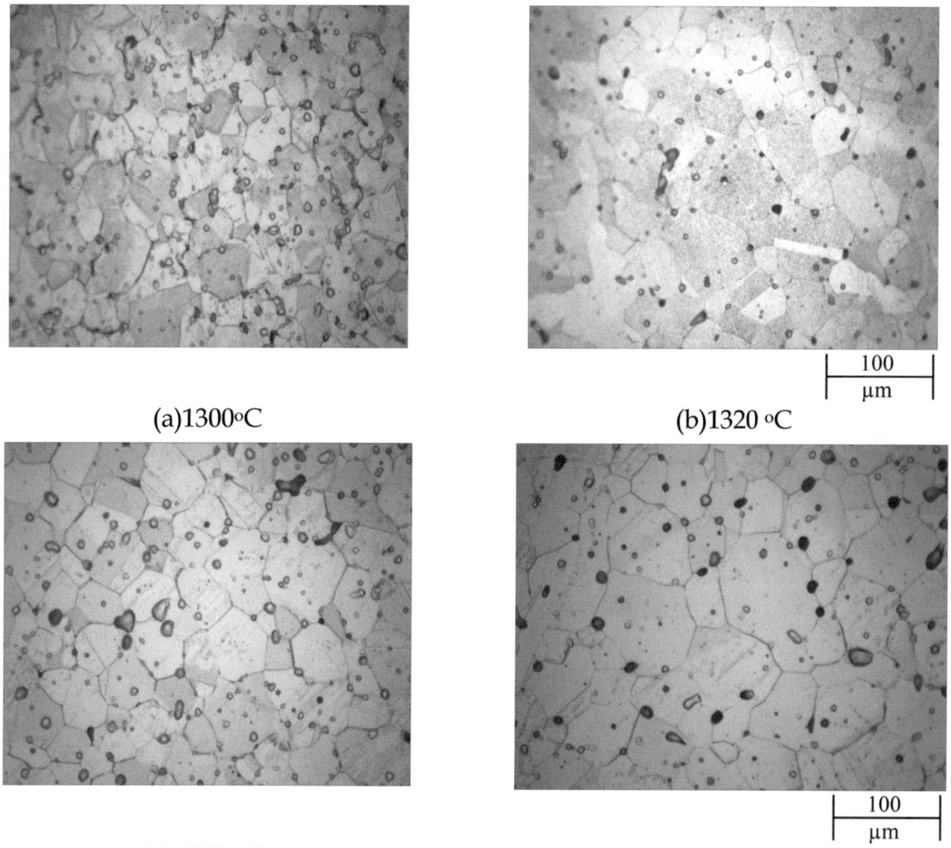

(a)1300 °C (b)1320 °C (c) 1340 °C (d) 1360 °C

Fig. 5. Optical micrographs of PSPP65-3 specimen sintered at various temperatures. Magnification 500x.

At high densities the pores are mostly associated with the largest grains. Consequently, there is a relation between grain sizeG, pore diameter d_p, and fractional porosity V_p,

$$\frac{G}{d_p} = \frac{K}{RV_p} \tag{1}$$

where R expresses the ratio of attached pores to randomly placed pores and K is a geometric constant. Value of R range from 1.7 to 5.7 for various sintering materials. It should be noted that the degree of boundary-pore contact remains essentially constant during sintering (German, 1996). Indeed, with the prolonged sintering, pore coarsening causes a number of pore size to decrease and that lead to increased grain size. Early in the sintering, the large pores are immobile and pin the grain boundaries, maintaining a small grain size. Late in sintering there are fewer pores, which are small due to shrinkage and the grains become large. (German, 1996).

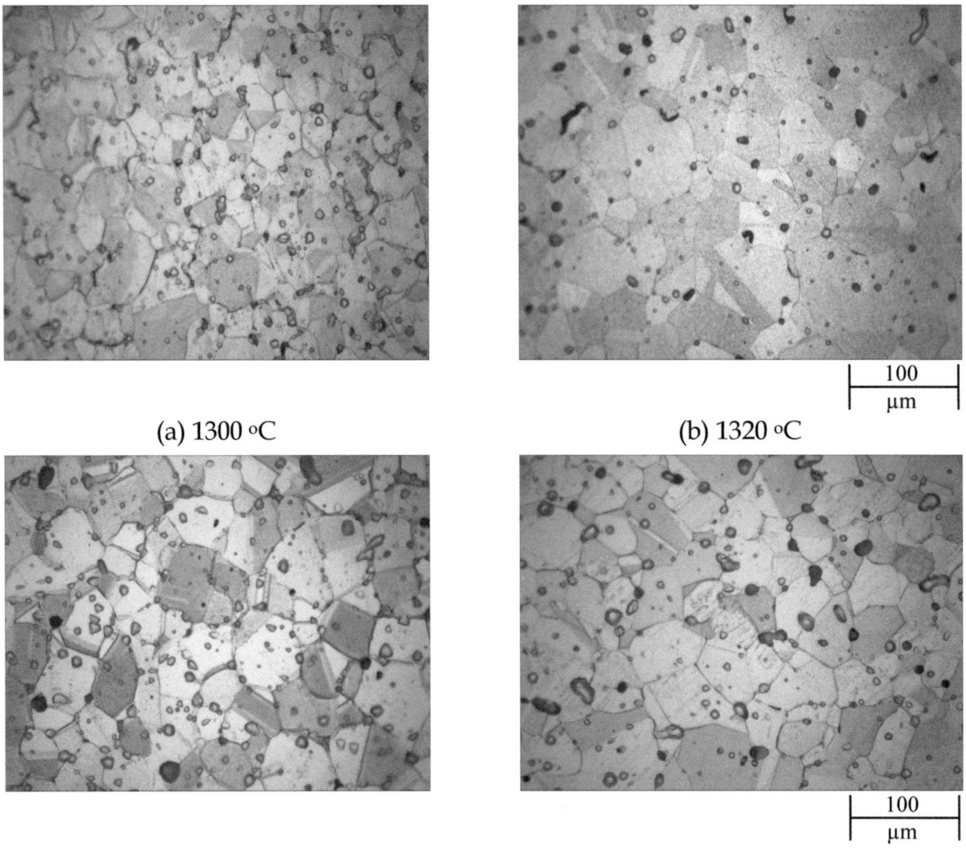

(a) 1300 °C (b) 1320 °C

(c) 1340 °C (d) 1360 °C

Fig. 6. Optical micrograph of PSPP65-2 specimen sintered at various temperatures. Magnification 500x.

3.1.2 Mechanical properties

The effect of PS content on tensile strength is shown in Figure 7. It clearly shows that, the specimen PSPP65-4 exhibited tensile strength of greater than 500 MPa for three different sintering temperatures (1320 °C, 1340 °C, 1360 °C). From the standpoint of sintering temperature, an increase in this parameter would increase the sintered strength for all composition tested.

To meet the MPIF Standard 35, the specimens made with 316L SS must have tensile strength of at least 500 MPa. As shown in the Figure 7, all compositions shows a significant linear increment of strength as a result on increasing sintering temperature. High strength greater than 500 MPa was observed in all specimens sintered at 1360 °C, thus complying with the international standard MPIF 35 MIM specimens.

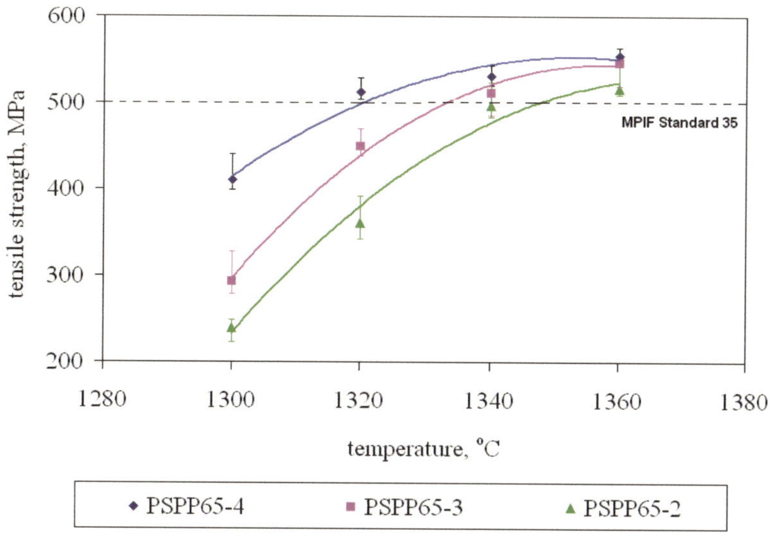

Fig. 7. Effect of the PS content on the sintered strength of specimen sintered at different temperatures.

The sudden increase in strength was obtained when specimens were sintered at 1320 °C especially for specimen of PSPP65-4. These might be due to the changes from intermediate to final stage of sintering that undergoes a transformation to become dense and to be high in strength, with attendant changes in the pore size and shape. Thus, sintering inherently involves substantial shrinkage with elimination of pores that leads to a dramatic increase in strength of specimen. As temperature is increased from 1320 °C to 1360 °C, small deviation changes the strength of all specimens. At this final stage of sintering, coarsening processes consume the surface energy that is responsible for densification, but do not reduce porosity.

The result of hardness of sintered specimens as a function of PS/PP content and sintering temperature is shown in Figure 8. From the results obtained, it can be seen that the hardness of the sintered specimens increases when the sintering temperature increases. The increase in hardness is due to the better densification at higher sintering temperature. It is evident that sintering temperature of 1360 °C gave rise to the highest hardness for all compositions (greater than 200 Hv) that complies with the MPIF Standard 35.

By comparing three properties of sintered density, tensile and hardness, it clearly shows that the higher content of PS has worked successfully in MIM process that complies with the MPIF Standard 35. The sintering temperature of 1360 °C was believed to be sufficient enough to obtain reasonable and acceptable sintered properties.

From the investigation, it clearly showed that the new locally developed binder system using PS as the base material mixed with PP was successful injection moulded with the 316L stainless steel powder. A maximum 70 wt.% of PS was believed to be a promising amount to give sufficient properties that meets the requirement of MPIF Standard 35. This residual carbon which left after debinding, diffuse into the structure of SS 316L powder during sintering and eventually promotes the significant changes of mechanical properties of the sintered specimens.

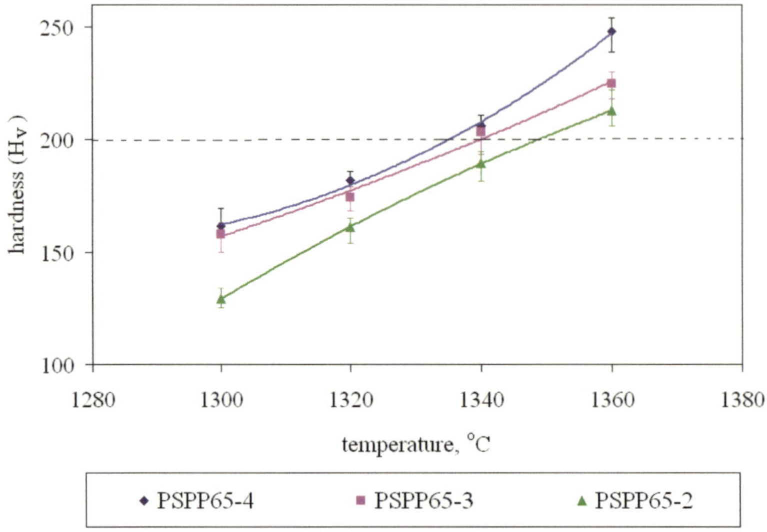

Fig. 8. Effect of the PS content on the hardness (Hv) of specimen sintered at different sintering temperatures.

3.1.3 Sintering study at different heating rate of sintering

The specimen made with 65 vol.% gas atomised 316L SS powder with the composition of binder system PSPP65-4 respectively was sintered at different time, temperature and atmosphere.

Heating rate is important in sintering practice. Too low heating rate at low temperature dissipates driving force and results in nearly no densification, while too high heating rate will result in distortion and warpage. In this study, 3 different heating rates ranging from 5 °C/min to 15 °C/min with soaking time of 1 hour were employed. The sintering temperature was kept at 1360 °C under vacuum atmosphere.

Figure 9 shows the effect of heating rate on the density of the sintered specimens. At the heating rate of 5 °C/min, the mean density was 7.69 g/cm^3, which is approximately 97% of theoretical density. At low heating rate, surface diffusion typically dominates in sintering. This provides bonding without densification, thus resulting in low sintered density of specimen.

From Figure 9, it can be seen that the density increases as the heating rate was increased from 5 °C/min to 10 °C/min. However, at heating rate of 15 °C/min, the density dropped to 7.67 g/cm^3, which is approximately 97% of theoretical value. High heating rates induce large thermal stress that accentuates sintering beyond that found in isothermal sintering (German, 1996). At higher temperatures, the densification dominates diffusion resulted in enhance densification without coarsening. Simultaneously, there has been an ability to reduce grain growth while achieving high sintered density.

High density can be achieved as the specimen is sintered at the heating rate of 10 °C/min. Densification observation of the sintered specimens indicates that the heating rate of 10 °C/min results in 98 % of theoretical density. Too low heating rate is not preferable for consideration of economy. From the results obtained, heating rate of 10 °C/min shows to be resulting in the optimum sintered density.

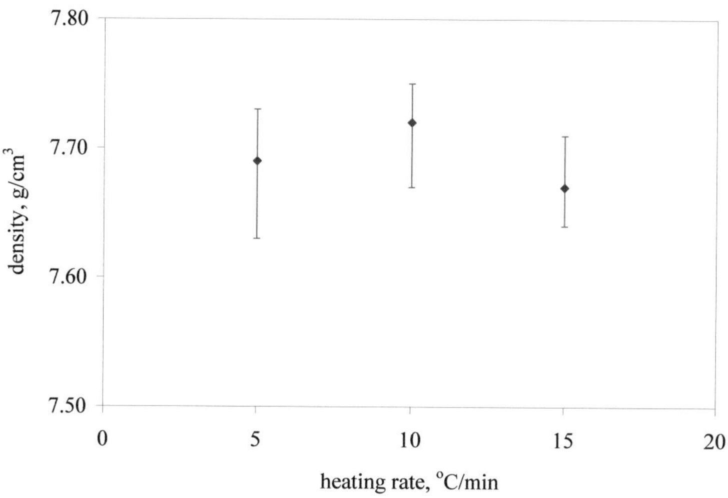

Fig. 9. Density as a function of heating rate.

Figure 10 shows the percentage of porosity at different heating rates. It can be seen that with an increase of heating rate from 5 °C/min to 10 °C/min, the porosity is lower. One effect of slow heating is that pore reducing occurs by surface diffusion. This tends to reduce the driving force for the densification mechanism at the higher temperature. It clearly shows that heating rate of 10 °C/min enabled for the complete diffusion to take place and reducing the number of pores as can be evidenced in Figure 11 (b). However, the numbers of pores increases as the heating rate increase to 15 °C/min. High porosity is due to too fast heating rate and this is probably due to some pores are still remaining prior to isothermal sintering. These pores are isolated and cannot be eliminated due to it being isothermal process. (German, 1996)

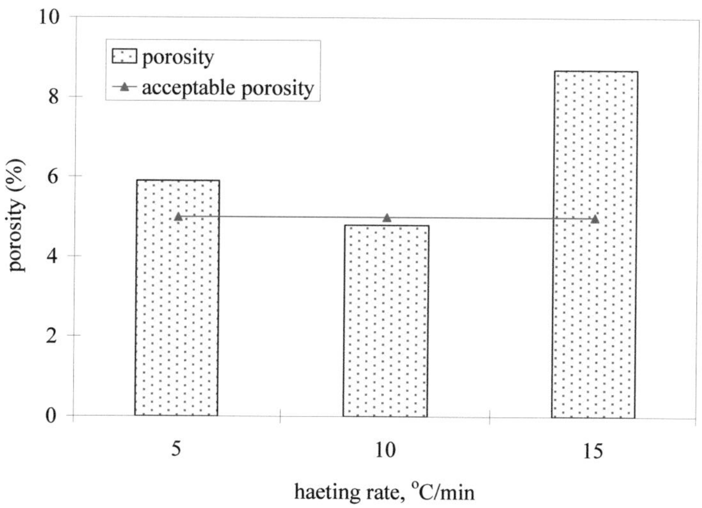

Fig. 10. Porosity as a function of heating rate.

Figure 11 shows the microstructure of the sintered specimens after etching. It can be seen that with an increase in the heating rate, the grain becomes finer. Normally a higher heating rate induces large thermal stresses that accentuate sintering beyond that are found in isothermal sintering (German, 1996). On the other hand, higher heating rate demonstrated an ability to reduce grain growth while achieving high final density. Although the grain boundaries tend to become finer with the heating rate of 15°C/min as shown in Figure 11 (c), the pores within or on the grain boundaries seems to be increasing. High porosity occurs due to too fast heating rate, the metal powder is unable to completely diffuse away. Some pores are still remaining prior to isothermal sintering. These pores are isolated and cannot be eliminated due to isothermal process. These circumstances result in lower sintered density.

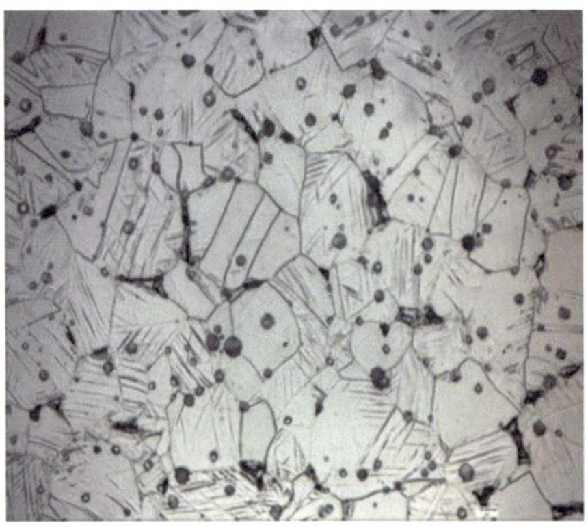

(a) 5°C/min

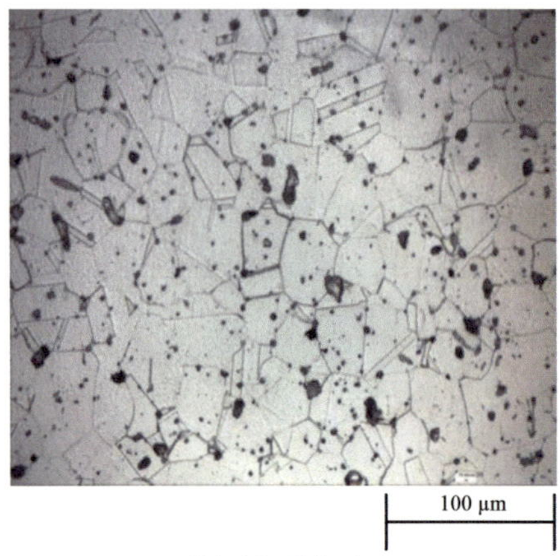

(b) 10°C/min

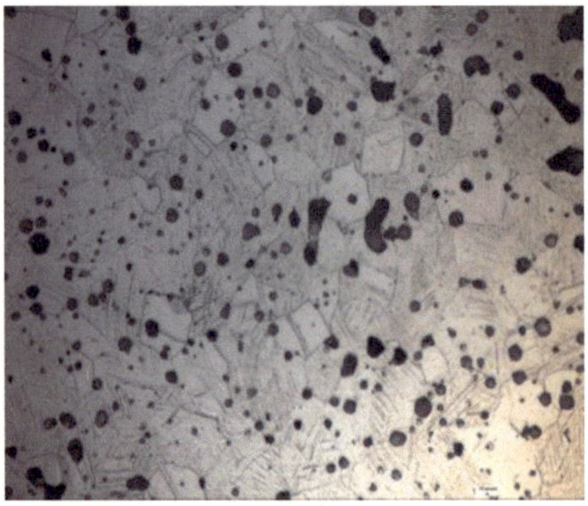

(c)15°C/min

Fig. 11. Microstructure of specimens sintered at different heating rate in vacuum conditions at 1360 °C. Magnification 500x.

The mechanical properties of the 316L SS samples prepared at various heating rates from 5 °C/min to 15 °C/min are compared as shown in Figure 12 and 13. The trends in tensile strength, elongation and hardness for the sintered specimen are similar with density. The tensile strength and elongation increased with the decrease of porosity. At a heating rate of

15 °C/min, the elongation and tensile strength of sintered specimen decreased due to the increasing number of porosity which can be evidenced from the microstructure shown in Figure (c). The greater pore induced the stress concentration that reduced the mechanical properties.

As shown in Figure 12 and 13, heating rates of 10 °C/min shows the highest tensile strength and hardness that can be considered as the optimum schedule for a sintering process. At this heating rate of sintering, near full density was obtained for the specimen tested.

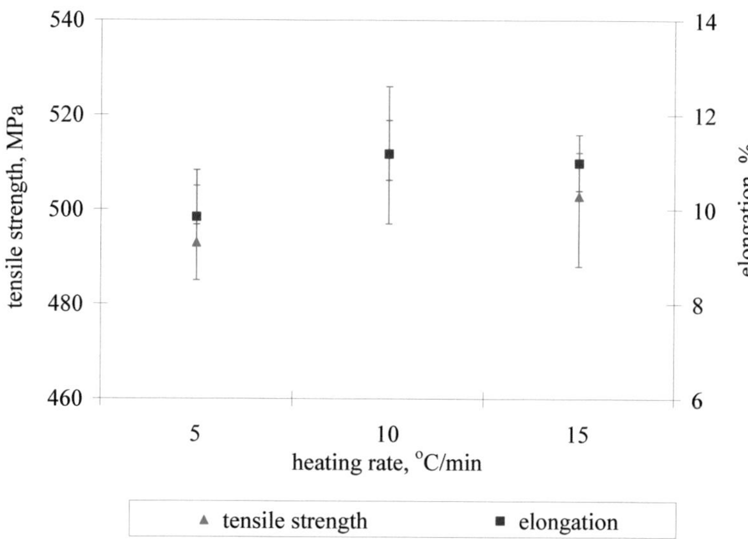

Fig. 12. Tensile strength and elongation as a function of heating rate.

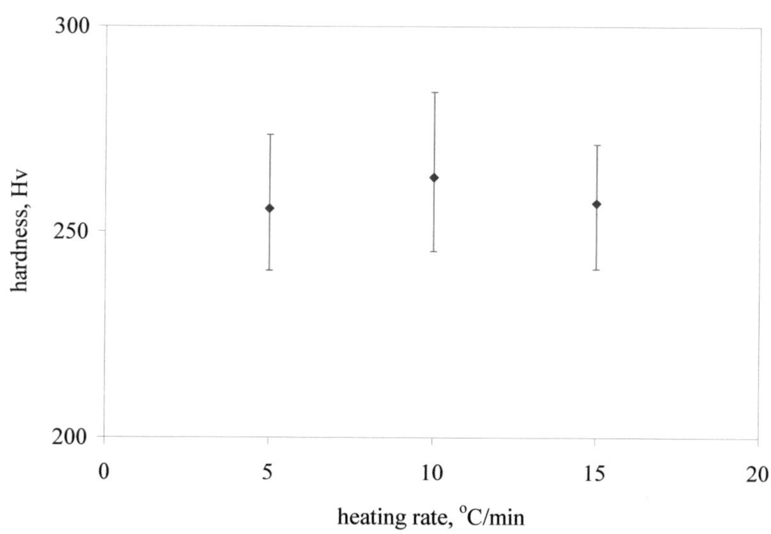

Fig. 13. Hardness as a function of heating rate.

The presence of pores prones to lowering tensile strength, elongation and hardness. This is because the large amount of pores corresponds to a lesser cross sectional area during tensile testing based on the equation of σ=P/A where *P* is pressure or force applied (German, 1996). The strength should decline since pore reduces the load-bearing cross sectional area (*P*). So that, the tensile strength, elongation and hardness are lowered as the porosity of sintered specimen becomes higher. As the consequence, the porosity is considered as an important factor affecting the tensile strength and elongation.

The 316L SS is considered partly as temperature sensitive, thus the optimum heating rate is useful in controlling microstructure evolution since it affects the mechanical properties of sintered specimen. From the results obtained, high physical and mechanical properties of 316L SS were achieved when sintered at 1360 °C with the heating rate of 10 °C/min.

3.1.4 Effect of temperature and atmospheres

Furnace atmosphere does play a critical role during sintering which affects the microstructure, pore size and properties of the sintered specimen. In order to investigate the effect of sintering atmosphere, 3 different atmospheres of vacuum, 95% N_2/5% H_2 and argon were used. Among these atmospheres, sintering in vacuum yields the highest density followed by sintering in 95% N_2/5% H_2 and argon respectively as shown in Figure 14. During heating, the residual gas atmosphere can be trapped in pores that inhibits full densification. As a consequent, high density can be obtained when sintered in vacuum since no gas trapped in the pores. From the result obtained, near full density (97% of theoretical density) was reached at sintering temperature of 1360 °C in the vacuum.

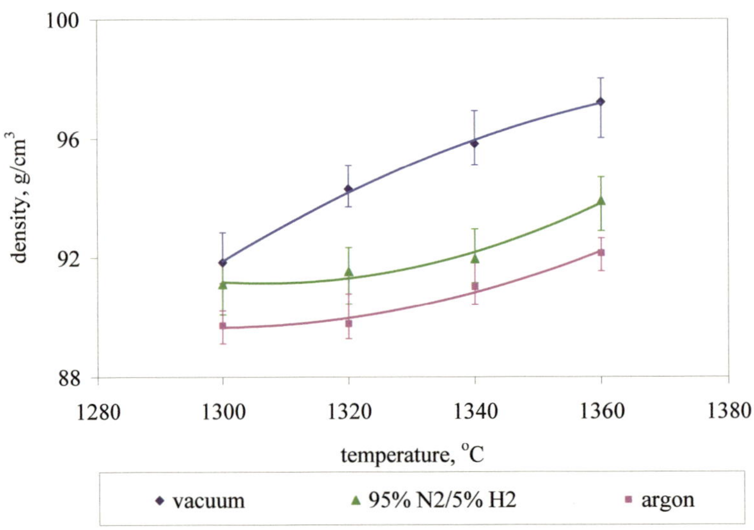

Fig. 14. Densification of 316L SS specimen after sintering in different atmosphere.

There were slight variations in density between sintering in 95% N_2/5% H_2 and in argon as shown in Figure 14. A little variation has been shown for the specimen sintered under 95% N_2/5% H_2 atmosphere in the duration of sintering temperature of 1300 °C to 1340 °C.

However, increasing the sintering temperature to 1360 °C shows a high density achieved which is 90 % of theoretical density.

The result of tensile strength as a function of sintering atmosphere is shown in Figure 15. The trends exhibit by tensile strength and density with temperature are nearly similar, which is increase with increasing sintering temperature. When sintered in vacuum, the maximum ultimate tensile strength of 510.37 MPa was achieved at the temperature of 1360 °C.

The results however show that the tensile strength of specimen sintered under 95% N_2/5% H_2 and argon atmosphere at the temperature in the range of 1300 °C to 1360 °C did not meet the specification of MPIF Standard.

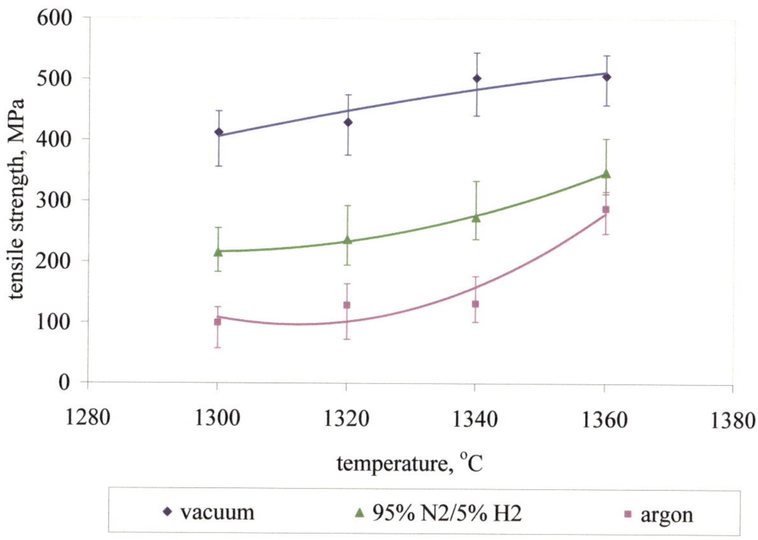

Fig. 15. Tensile strength of 316L SS specimen after sintering in different atmosphere.

The atmosphere (inert) protects surface against contamination during sintering. Additionally, a cleaning function is desired from the atmosphere to remove contamination. In many full-density systems best densification and properties are achieved using vacuum sintering (German, 1996; German and Bose, 1997).

The result obtained shows that the important factors in a sintering cycle are heating rate, temperature and furnace atmosphere. These factors can affect the microstructure, pore size and shape and final density of sintered stainless steel 316L. These in turn influence the mechanical properties of the sintered specimens.

3.1.5 Scanning electron microscopy observation: Microstructure evolution during sintering

The fracture surfaces of the specimen PSPE65-4 shown in Figure 16 reveal clearly the different morphology between the stages of pre-sintering process. The specimens were sintered in a tube atmosphere furnace at different temperatures ranging from 600 °C, to 1200 °C. Figure 16 (a) to (d) demonstrated the progressive microstructural development with increasing temperature during the sintering process.

Sinter bonding is evident as bonds grow at the particle contacts. Figure 16 (a) shows a formation of neck between sintering particles and almost all powder are discretely evident as sintered to 600°C. At the early stage of sintering at 800°C, the powder particles moved to fill the pores by the growth of several necks on each particle as shown in Figure 16 (b). The establishment of interparticles bonds by partial fusion is clearly seen. Increasing the temperature up to 1000°C extends the fusion bonding further with a substantial reduction in porosity. The particles take many paths to form the bonds. The original shape of the powders can still be discerned, although the particles have been fused together.

As the bonds become larger, they impinge on each other and form a network of pores as shown in Figure 16 (d). These pores are termed open pores that are accessible from the compact surface, so the sintering atmosphere can move in or out of the pores during sintering. The open pores are attributed to a lower elongation of sintered specimen corresponding to an intermediate stage of sintering.

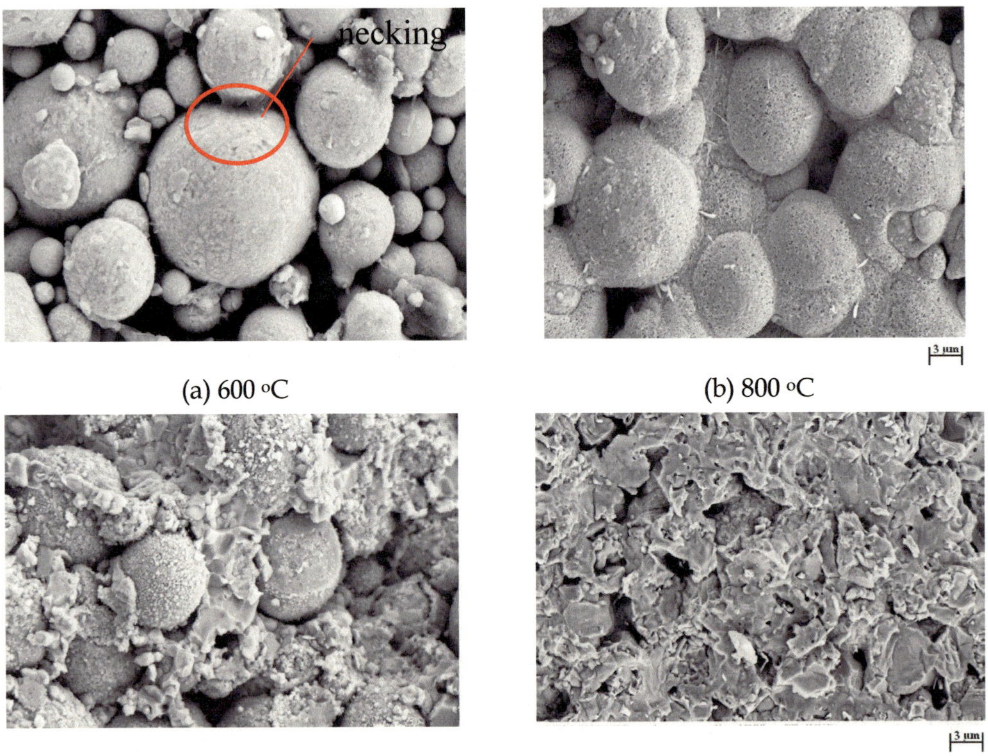

(a) 600 °C (b) 800 °C

(c) 1000 °C (d) 1200 °C

Fig. 16. Scanning electron micrograph of fracture surface of specimen after presintering at different temperatures.

During sintering, the initially loose powder compact undergoes a transformation to become a dense, polycrystalline structure with physical and mechanical properties similar to engineering materials. The final stage of sintering has a few pores sitting on grain boundaries.

Figure 17 depicts the SEM micrographs of vacuum sintered SS 316L specimens (PSPP65-4) at different temperatures ranging from 1300 °C to 1380 °C. At 1300 °C, it clearly shows that the powder boundaries were replaced by grain boundaries as shown in Figure 17 (a). As the temperature was increased, the microstructure began to coarsen that considerable reduces surface area, increases grain size and compact strengthening with attendant changes in the pore size and shape.

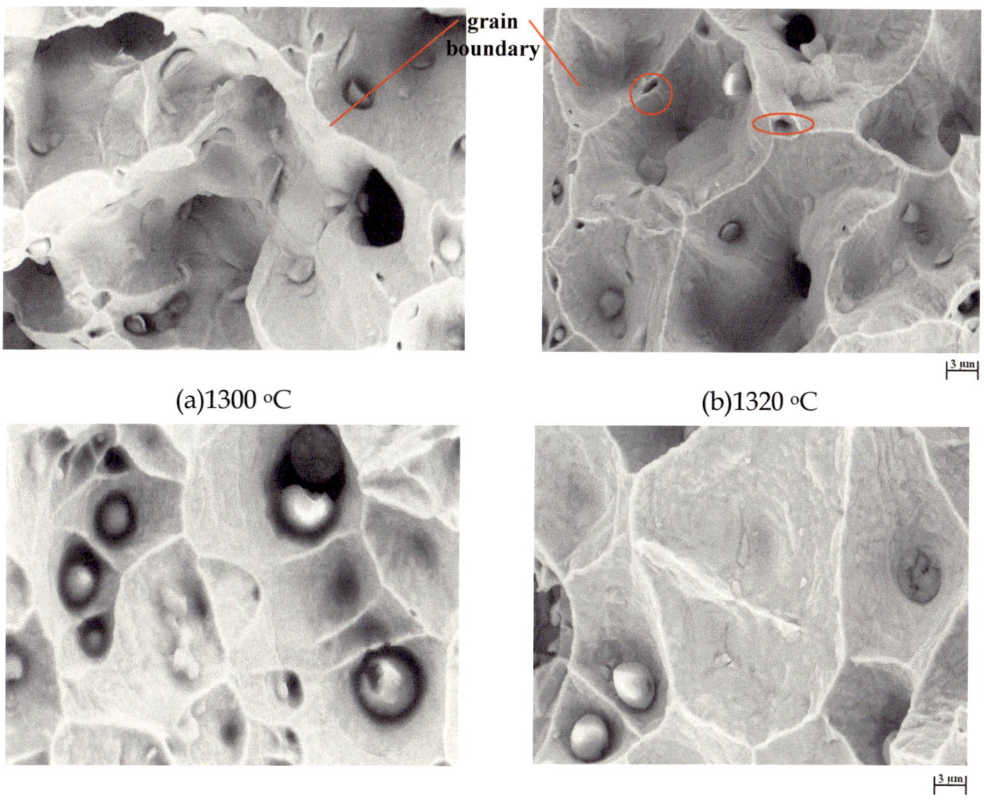

Fig. 17. Scanning electron micrograph showing the fracture surface of 316L stainless steel specimen sintered at various temperatures.

Sample sintered at 1320 °C shows spherodising and shrinking of pores , which are not connected to the compact surface (Figure 17 (b)). These pores are termed closed pores. The pore located on grain boundaries as shown in Figure 17 (a) disappeared as temperature increased, but pores disconnected from grain boundaries remain stable as shown in Figure 17 (c) and (d). The typical ductile fracture mechanisms is evidenced by dimples in the final stage of sintering as can be seen in Figures 17(c) and (d). A few isolated pores (closed pores) can also be observed in the Figures 17 (c) and (d) suggesting that closed pores play little part in the fracture. These closed pores are sealed and inaccessible via the sintering atmosphere.

In many literatures, sintering of stainless steel was carried out in the range of 1300 °C and 1380 °C (White and German, 1995; Cai and Geman, 1995; Loh et al., 1996; Afian, 1999). However, in this study, as the sintering temperature rose up to 1380 °C, separation has been noticed.

Besides that, sample experienced swelling defect and deteriorates. The specimen was to swell and changed the shape required. Figure 18 shows the fracture surface of specimen sintered at 1380 °C. At this stage, the sintering is faster as temperature approaches the melting point because of the increased number of moving atoms resulted in the atoms landing at the bond between particles where it helps annihilate surface area and surface energy (German, 1996).Figure 18 clearly shows the greater size of pores indicating that more energy had been activated. These large pores are unable to remain attached to moving grain boundaries and become stranded at the grain interior. The grain boundary tends to break away from the pores. The growth of the pores could cause a decrease in density during prolonged final stage sintering. As 316L SS has a melting temperature of 1441°C , it is preferable to sinter the specimen ranging between 1300 °C and 1360 °C.

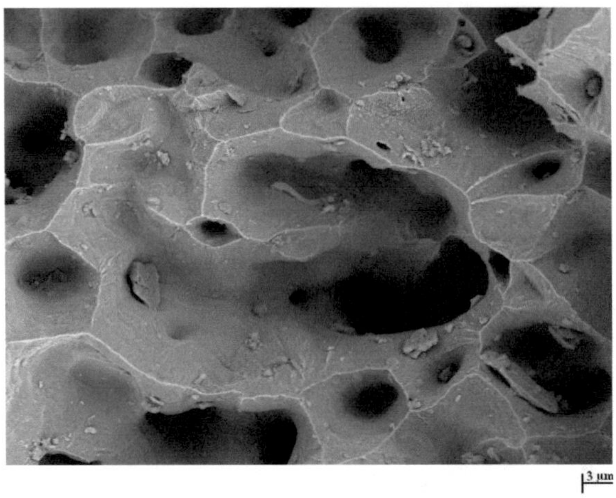

Fig. 18. Scanning electron micrograph of fracture surface of specimen sintered at 1380 °C.

There are many fine particles that appeared in the centre of a grain boundary. These particles have a dimension of approximately 3 µm. As shown in Figure 19, the energy dispersive spectroscopy (EDS) result shows that this is the particles which is not diffused and appeared exclusively. Table 3 shows the elements found in the particles.

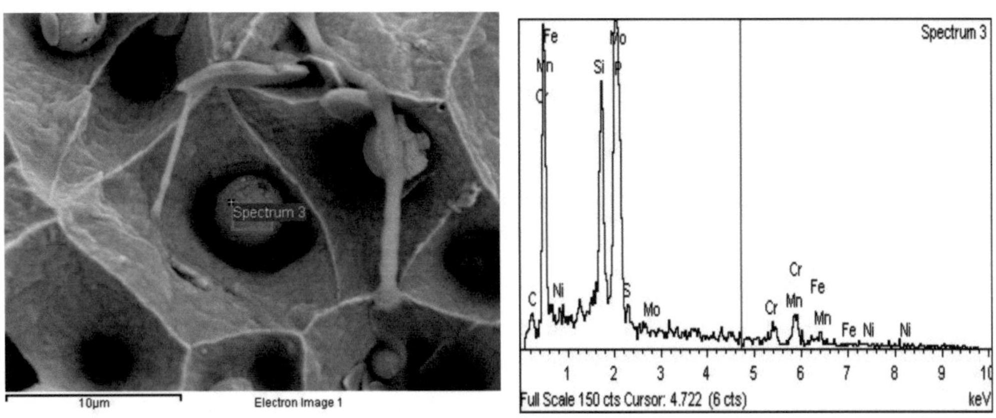

Fig. 19. Energy dispersive spectroscopy of sintered specimen.

Element	Weight %
C K	4.27
Si K	13.68
P K	5.74
S K	3.62
Cr L	61.39
Mn L	0.00
Fe L	11.30
Ni L	0.00
Mo L	0.00
Total	100%

Table 3. EDS results showing the elements present in the fine particles.

4. Conclusion

The new developed binder systems, palm stearin can be used as a binder system combined with polypropylene in the metal injection moulding process. The maximum metal powder loading in the powder/binder mixture for successful injection moulding was 65 vol.% for gas-atomised 316L stainless steel.

The sintering process of injection moulded 316L stainless steel specimen is clearly influenced by sintering temperature. With a high sintering temperature, the density shrinkage, tensile strength and the hardness of the sintered specimens increase due to the pore shrinkage. The closure of the pores enhanced the mechanical properties of the sintered samples. From this study, it can be concluded that the best sintering temperature for the 316L stainless steel powder using palm based binder is 1360°C which result in good properties for the sintered parts, and comply with the requirement for MPIF Standard 35 for Metal Injection Moulded Parts.

5. Acknowledgment

The authors would like to thank MOSTI for financial support under ScienceFund grant No. 03-03-02-SF0040, 03-03-03-SF0124 and TechnoFund 1208D168.

6. References

Berginc,B., Kampus, Z and Sustarsic, B. (2007). Influence of feedstock charateristics and process parameters on peroperties on MIM parts made of 316L stainless steel, *Powder Metallurgy*, 50 , 172-183

German, R. M. & Donal, K. (1993). Evaluation of injection moulded 17-4 PH stainless steel using water atomized powder. *The International Journal of Powder Metallurgy*, 29(1), 47-61

German, R.M, (1996), *Sintering Theory and Practice*, John Wiley and Sons, Inc, New York

German, R.M. & Hens, K.F. (1992). Identification of the effect of key powder characteristics in PIM. *Proc. of Powder Injection Moulding Symposium*, MPIF, Princeton, NJ, 1-16.

German, R.M. (1993). Technological barriers & opportunities in powder injection moulding. *Powder Metallurgy International*, 25(4), 165-168.

German, R.M. and Bose, (1997) *A. Injection Moulding of metal and Ceramis,* MPIF, Princeton , NJ.

German, R.M., (1990), *Powder Injection Moulding*, Princeton , New York.

Iriany, (2002), PhD Thesis, Universiti Kebangsaan Malaysia, Malaysia

Istikamah, S., (2010), PhD Thesis, Universiti Teknologi Mara, Malaysia

Li, Y., Li, L. and Khalil, K.A. (2007). Effect of Powder Loading on Metal Injection Moulding Stainless Steels, *Journal of Materials Processing Technology*, 183, 432-439

Omar, M.A., (2006) Metal injection Moulding- An Advanced Processing Technology", *Journals of Industrial Technology*, 15 (1) , 11-22

Omar, M.A., and Subuki, I., (2007) Rapid Debinding of 316L Stainless Steel Injection Moulded Using Palm Based Biopolymer Binder, Proceedings of 3rd Cooloquium on Postgraduate Research on Materials, Minerals and Polymers 2007, Vistana Hotel, Penang

Omar, M.A.,, .Subuki, I and Ismail, M.H., (2006) . Observation of Microstructural Changes During Solvent Extraction and Polymer Burnout Process of MIM Compact", Proceedings of 5th International Materials Technology Conference and Exhibition 2006 (IMTCE 2006), July, 2006, Kuala Lumpur

Omar,M.A**.**, Davies, H.A., Messer, P.F. and Ellis,B. (2001). The Influence of PMMA Content on the Properties of 316L Stainless Steel MIM Compact, *Journal of Materials Processing Technology*, 113 :477-481.

Omar,M.A**.**, Mohamad, M., Sidik, M.I. and Mustapha, M. (2002). A PEG-based Binder System for Metal Injection Moulding, *Solid State Science and Technology*, **10**, No. 1& 2 : 359-364

Zhang,H and German, R.M. (1991). Structural development During Sintering on Injection Moulded Fe-2Ni Steel, *Advances in Powder Metallurgy*, 2,:181-194

7

Liquid Phase Sintering of Fe-Cu-Sn-Pb System for Tribological Applications

Cristina Teisanu
University of Craiova
Romania

1. Introduction

Powder metallurgy technique has become increasingly interesting for engineering parts manufacturers due to its advantages which include high productivity, minimum consumption of raw materials and energy, high efficient use of the initial metals (95–98%), near net shape character and unique capability of porous material production.

This technique results in sintered parts that reach sufficiently high strength properties, for example, similar to cast iron, already at a porosity of 20-15%, and present remarkable physical, chemical and mechanical characteristics, which are determined by their chemical composition and the phase structure as well as the shape and mass distribution of the powder particles. Moreover, by powder metallurgy methods can be obtained materials of virtually any structure, composition and porosity and, thus, any mechanical and service properties.

The performance of equipments, machines, devices and mechanisms is strongly related to their subassemblies and component parts performance. These machines and mechanisms withstand various loads and, thus, complex processes during operating times, but friction, lubrication and wear are among the most frequently met processes.

Sintered antifriction materials are widely used in various tribological applications due to their more homogenous structure and controlled open porosity and grain size, which actually would be impossible to be fabricated by other manufacturing approaches. Besides the high wear resistance and low friction coefficient, sintered antifriction materials have the best possible volume and surface strength, which combines the high strength of the surface layer and the high conformability of the friction pair. Furthermore, they ensure self-lubrication of surfaces with oil from pores, which implies no additional external oil supply. Therefore it is essential to know the actual loading conditions of the part and modify alloying and the treatment conditions of the material on the basis of these conditions.

Sintered iron base antifriction materials were not developed until the last decade because of poor corrosion resistance and antifriction properties. Considering the low cost and availability of iron powders, the more homogenous structures obtained by sintering and the increased specific load-carrying capacities and sliding rates, the development of the iron base sintered alloys for tribological applications was continuously improved. Additions such as copper, graphite, manganese, lead, antimony and tin to iron have been attempted but improvement in one property was offset by a decrease in other properties (Kostornov & Fushchich, 2007; Kostornov et al., 2007; Shahparast & Davies, 1978; Verghese & Gopinath, 1989). To overcome weaknesses in existing alloy systems and to meet the challenging nature of newer machines, it is important to develop a modified alloy system which can succeed in dealing with these deficiencies, either partially or fully.

Using conventional powder metallurgy techniques the present work is focused on the development of new iron-copper base antifriction materials with different addition elements such as lead, tin and molybdenum disulphide powders. In sintered iron based materials copper has unique properties as an alloying element. In small amounts copper improves strength and rust resistance, and also has a rapid surface diffusion over solid iron (Runfors, 1987, as cited in Boivie, 2000). Therefore, copper is rapidly dissolved into the iron particles, forming a substitution solid solution. Since copper atoms have a larger diameter compared to iron atoms, this causes a distortion in the crystal structure and thus resulting in a swelling effect that is larger than the original copper volume (German, 1985, as cited in Boivie, 2000). Lead exhibits excellent self lubricating property and plays a role of solid lubricant to prevent seizure. In addition, since Pb forms a soft dispersion phase, it has conformability and allows solid matter to be embedded therein. Tin is a key player in antifriction alloys because it can influence both corrosion resistance and fatigue strength and friction and wear properties of these materials can be also improved. During compaction, some phenomena including particle deformation, cold welding at points of contact and interlocking between particles occur. The reasons for non-uniform density distribution are friction between particles and die, and internal friction between powder particles (Sustarsic, 1998). Because of its lamellar structure, MoS_2 is one of the most popular and usable solid lubricants (Hutchings, 1992, as cited in Sustarsic, 1998). It is very important to understand the parameters that control the densification behaviour and the resulting microstructure (grain size and shape, pore size and shape, phase distribution etc.) because of their effect on the physical and mechanical properties of the final product.

Liquid phase sintering, i.e. sintering where a proportion of the material being sintered is in the liquid state, is a common processing technique for a variety of systems, including metal, cermets and ceramics. The sintered material usually consists of grains of one or more solid phases at the sintering temperature intermixed with phase that is liquid at sintering temperature. Liquid phase sintering is one of the most popular methods to enhance the sintering behaviour of a powder material. During liquid phase sintering a liquid phase coexists with a particulate solid at sintering temperature (German, 1996). If the liquid has good wetting properties over the solid particle and there is solid solubility in the liquid, then an enhanced sintering over all wetted surfaces will occur. This enhancement involves growth along the wetted grain boundaries. Thus, grains situated at a free surface will expand in all lateral directions, while the un-wetted free surface is depleted. This can lead to a smoothing effect on the material surface (Park et al, 1986, as cited in Boivie, 2000).

For the most part, in conventional powder metallurgy, liquid phase sintering exhibits sufficient internal force through liquid capillary action on the particulate solid that external forces are not required for the compaction during sintering (German, 1996).

In this study the formation of the liquid phase during sintering process of Fe-Cu-Sn-Pb system was investigated. The liquid phase produced during sintering led to a considerable accurate process and swelling of the sintered compacts was observed and studied. There are many factors that cause dimensional changes and their combined effect make it more difficult to forecast and control these changes. Consequently, in order to avoid subsequent operations such as sizing and machining it is imperative to improve the final dimensional tolerances obtained after sintering (Takata & Kawai, 1995). Therefore, the evaluation of quantitative effects of the dimensional changes of the sintered compacts was also investigated due to variations of the sintering time, temperature and chemical composition.

The material porosity plays an important role in tribological applications for an adequate functioning of the machines and devices. Self-lubricating bearings accomplish superior performance when the porosity level is high, so sufficient oil is accumulated in the pores during inactivity periods and it leaks when the shaft begins rotate in it. Also, the material density plays an important role for self-lubricating bearings for exhibiting good load bearing capacity. Therefore, self-lubricating bearings need to have a high open porosity level for better performance. The porosity due to closed pores shall be reduced, so that density remains high and hence the load bearing capacity of the bearing (Yusof et al., 2006). Among the most important material characteristics required for self-lubricating bearings are identified to be high porosity and reasonably high density. Thus the effect of sintering parameters (temperature and time) and compacting pressure on porosity of the sintered antifriction materials was systematically examined. The mathematical model of the sintering process of the iron-copper base material for tribological applications was developed by establishing the relationship between density of the material and sintering parameters (temperature and time) and the chemical composition.

2. Liquid phase sintering characterization

2.1 Test materials and experimental procedure

The composite materials are made of elemental powders of iron (DP 200HD), electrolytic copper, tin, lead and MoS_2 as solid lubricant. In this study three compositions are considered and labelled as MAS1, MAS2 and MAS2. Their physical characteristics and chemical composition are presented in table 1. Elemental powders were weighed to selected proportions and mixed in a three-dimensional rotating turbula-type device for two hours. Then the powder mixtures were cold compacted in a single-acting hydraulic press using three pressures (350, 500 and 700 MPa) and cylindrical specimens, 10.0 mm in diameter and 7.0 mm in height, were obtained and used for metallographic examination, measurement of green and sintered density, porosity and dimensional change after sintering. Reference densities for the selected compositions were calculated by the rule of mixtures and green and sintered densities were evaluated gravimetrically using Archimedes principle. Volumetric dimensional change of the sintered compacts was calculated and the total porosity of the specimens was evaluated from the difference between the reference density and the measured density.

Composition of the mixtures [wt %]					
	Copper	Tin	Lead	MoS_2	Iron
Composition 1 – MAS1	5	1	2.5	1	bal.
Composition 2 – MAS2	5	1.5	5	1	bal.
Composition 3 – MAS3	5	2.5	7.5	1	bal.
Characteristics of powders					
Particle size [µm]	< 125	< 100	< 100	1 - 5	< 160
Apparent density [g/cm^3]	2.63	3.7	5.63	-	2.92
Flow rate [s/50g]	28	-	-	-	28.1

Table 1. Physical characteristics of powders and alloys composition.

The compacted samples were placed in a tubular furnace with uniform heating zone and sintered at 800°C, 850°C and 900°C for 20, 35 and 50 minutes. The upper limit of the sintering temperature is chosen considering the high evaporability of the lead at temperatures above 900 and for its sweating phenomenon (Baranov et al., 1990).

The sintering process was performed in dry hydrogen atmosphere (dew point of -15 °C) with a flow rate of 1l/min and samples were cooled in furnace by switching off the power and maintaining the same flow rate of the hydrogen gas.

Material	Theoretical density [g/cm^3]	pressure [MPa]					
		350		500		700	
		green density and total porosity after pressing					
		ρ [g/cm^3]	P [%]	ρ [g/cm^3]	P [%]	ρ [g/cm^3]	P [%]
MAS1	8,14	6,56	16,05	6,8	13,27	7,02	5,75
MAS2	8,05	6,48	16,39	6,71	15,61	6,77	6,13
MAS3	7,98	6,37	17,25	6,63	14,88	6,73	5,63

Table 2. Green density and total porosity of the compacted samples.

Material	pressure [MPa]								
	350			500			700		
	ρ [g/cm^3]	P [%]	ΔV[%]	ρ [g/cm^3]	P [%]	ΔV[%]	ρ [g/cm^3]	P [%]	ΔV[%]
MAS1	6,55-6,85	15,42-19,12	0,4-3,14	6,68-6,98	13,81-16,9	0,65-4,67	7,22-7,58	6,4-10,85	0,21-3,91
MAS2	6,48-6,7	16,51-19,25	0,75-2,9	6,52-6,78	15,51-18,75	1,02-2,91	7,18-7,45	7,16-10,52	1,15-4,47
MAS3	6,38-6,5	18,01-19,52	1,4-2,32	6,5-6,71	15,36-18,01	1,49-3,5	7,44-7,52	5,14-6,15	1,95-4,61

Table 3. Sintered density, total porosity and volumetric dimensional change for sintered compacts.

After cooling, the samples were polished and chemically etched in order to investigate the microstructure and the porosity by optical microscopy. The green density of the samples compacted at specified pressures and the theoretical density of the three compositions are shown in table 2. After sintering at temperatures and maintaining times mentioned above the iron base compacts achieve densities, porosities and volumetric dimensional change as presented in table 3.

2.2 Dimensional changes

Volumetric diffusion and viscous flow of the matter are the dominant processes that govern linear and volumetric dimensional changes during sintering. Fundamentally admitting the sintering process is materialized by the growth and consolidation of the grain contact, the shrinkage results as a sintering law. But this phenomenon does not always accompany the sintering in all cases obligatorily, very often the opposite occurrence being noticed. There are multitudes of explanations given to this event. Because it cannot always occur simultaneously and with the same intensity in all the material mass the initial approach of the grains, one to each other, causes as expected tensile stresses in some regions. Although in as-isolated domains the material undergoes the compaction by grain closeness and porosity reduction, yet globally the material volume can grow. Tensile stresses might originate as a result of a non simultaneous annihilation of the residual stresses caused by the compacting operation or as a result of a non simultaneous thermal dilatation, a non homogeneous distribution of the grain size or a varying repartition of the compacting pressure. Gases enclosed through the pressing process and those resulted from oxides reduction or emanate from protection atmosphere are sometimes collected in closed pores carrying out pressures on the adjacent material and thus majoring the volume of the pores.

Additionally, swelling process can be observed in those systems where reciprocal diffusion markedly exhibits the Kirkendall effect, which is associated with different diffusion rates of the atoms, or where the liquid phase rapidly diffuses in the solid phase or in those systems where evaporation-condensation process plays an important role (Domsa, 1966).

After sintering the specimens were measured in order to investigate linear and volumetric modifications. During sintering both swelling and shrinkage of the samples were detected, but the most predominant occurrence was swelling of the samples sintered at all three temperatures. The characterization of the sintering behaviour in the presence of the liquid phase was performed by analyzing the influence of the sintering temperature and chemical composition (fig. 1) on the dimensional change of the iron-copper based samples compacted at 350 MPa, 500 MPa and 700 MPa and sintered for 50 min.

The following plots are given on the basis of a predictive modelling and data analysis of the sintering process using Statistica software in order to reach the optimum correlation between sintering parameters and chemical composition for an antifriction material. Swelling of the sintered compacts can be observed at all temperatures for all three compacting pressures. The lower expansion corresponds to all samples sintered at 800°C with a minimum value for the sample containing 1 wt% Sn and compacted at 700 MPa

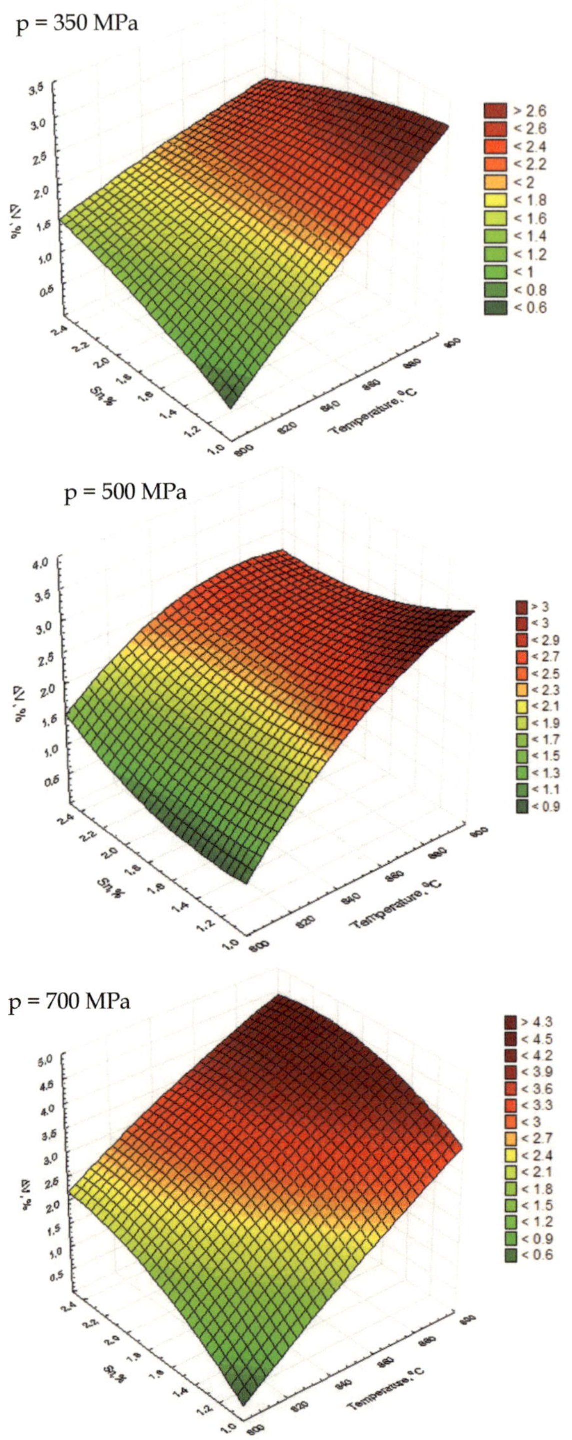

Fig. 1. The effect of tin content and sintering temperature on dimensional changes.

(0,21%). Also, the higher dimensional growth was recorded for all samples sintered at 900°C, the maximum value being detected for the sample compacted at 700 MPa and with a content of 2.5 wt% Sn (4,38%).

A sharper rise of dimensional growth is observed for specimens with a higher starting porosity, particularly at high sintering temperatures. From the plotted graphs almost the same variation of the dimensional changes with tin content is noticed, except for specimens sintered at 900°C and compacted at 700 MPa.

Taking into account that, for copper-tin alloys, the initial melting temperature (solidus) sharply decreases with rising of the tin content, from 1083°C for copper to 798°C for an alloy with 13,2% Sn (Sorokin, 1966), then molten copper-tin will exist in larger quantities and penetrate the solid iron boundaries. Liquid penetration of the grain boundaries causes grain separation and swelling on liquid formation. This penetration action pushes more solid particles apart, increasing the distance between particle centres and consequently leading to the growth of the compacts (Wang, 1999). This behaviour is attributed probably to the formation of Cu-Sn, Fe-Cu and Fe-Sn compounds in larger amounts at higher temperatures as the tin content increases. Diffusion of copper into the iron particles leads to dimensional growth (Sands & Shakespeare, 1966, as cited in Chandrasekaran & Singh, 1996a), and formation of the copper tin alloy along the neck regions in the Fe-Sn intermetallic network also contributes to swelling (Watanabe et al., 1988; Watanabe & Kim, 1984, as cited in Chandrasekaran & Singh, 1996a). The extent of interconnecting pores in the compacts is increased by the formation of the Fe-Sn intermetallic compound, which thus results in compact swelling. Compact swelling due to pore formation at prior particle sites is observed if the liquid particles have substantial solubility in the solid during heating (Lee & German, 1985; Xydas & Salam, 2006, as cited in German et al., 2009). Decreasing of the dimensional changes is attributed to the formation of the Fe-Sn phase at higher temperatures, which has a lower coefficient of thermal expansion than iron (Watanabe & Kim, 1984, 1987, as cited in Chandrasekaran & Singh, 1996b). Another cause of the dimensional growth might be attributed to the trapped hydrogen in the pores as the sintering process was performed in hydrogen atmosphere. Unlike solid coarsening, where volume is conserved, gas-filled pores change volume as they grow since the internal pressure depends on the inverse of the pore size. Thus, as the pores grow the gas pressure decreases and the pore volume increases both due to coalescence and due to the declining pressure, resulting in long-term swelling (German et al,, 2009). The internal pressure in the pore increases with temperature, leading to compact swelling (German & Churn, 1984, as cited in German, 2009). Pore growth occurs in liquid phase sintering due to gas diffusion in the liquid, partially due to annihilation of the smaller pores, but also due to vapour production during sintering (German et al., 2009).

2.3 Total porosity

When the liquid is soluble in the solid swelling occurs, and it is most useful in forming porous structures, such as self-lubricating bearings. Therefore, a great attention should be devoted to the porosity of the compacts for a better description of the sintering process when a liquid phase forms. 3D surface plots from figure 2 show the total porosity of the samples with different tin content compacted at 500 MPa and sintered at 800, 850 and 900°C for 20, 35 and 50 min.

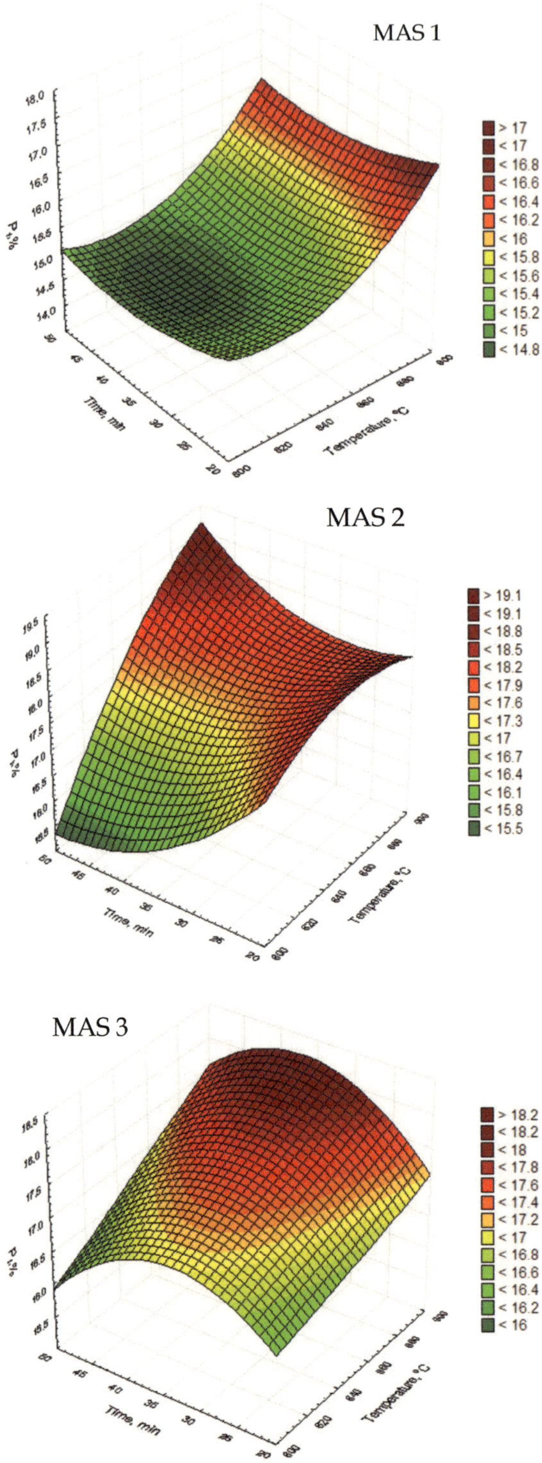

Fig. 2. The effect of sintering time and temperature on total porosity.

From these graphs a non uniform variation of the total porosity with sintering temperature and time can be seen for all three compositions. Samples with 1% tin addition (MAS1) show the same variation of the total porosity at all sintering temperatures. First, a slightly increase of the porosity occurs, up to 35 minutes of maintaining time and then total porosity decreases in the same manner for further sintering time. For samples containing 2,5% Sn (MAS3) a dissimilar tendency can be noticed comparatively with those previously mentioned. As for specimens with 1,5% Sn (MAS2), there is a decreasing in the total porosity as the sintering time is longer for heating at 800°C and 850°C, while sintering up to 50 minutes at 900°C produces a continuous increase of the porosity.

Decreasing of the porosity level is probably due to the formation of the soft phases of Cu-Sn, Fe-Sn-S or Pb-Sn-S, which diffuse into the pores. Capillarity drives the liquid to fill smaller pores prior to greater ones (Shaw, 1993, as cited in German, 2009). As the smaller pores fill, the mean pore size increases while the porosity and the number of pores decrease. Usually, during liquid phase sintering porosity is decreasing, but since smaller pores are annihilated first the mean pore size increases while the grain size is increasing (German et al., 2009).

The formation of a copper-tin alloy at the neck of the iron-tin intermetallic compound generates the initial increase of the porosity and thus restricting diffusion of low melting phases lead, lead-tin or sulphides into the network to fill the pores. Since the diffusion coefficient of copper in tin is higher than that of tin in iron at a temperature of about 850°C, the occurrence of iron-tin intermetallics contribute to the increase in intercommunicating porosity, and in that way increasing total porosity (Watanabe & Kim, 1984, 1987, as cited in Chandrasekaran & Singh, 1996b). Meanwhile, the excess copper surrounding the iron-copper combines to form a copper-tin liquid phase which diffuses into the iron-copper skeleton. Regarding the excess of lead amount, it is possible for this to migrate at the surface of the component forming a soft cover over the hard matrix of iron-copper (Watanabe & Iwatsu, 1980, as cited in Chandrasekaran & Singh, 1996a).

Microscopic investigation of the porosity was performed on the polished and un-etched sample surfaces made from selected compositions compacted at 500 MPa and sintered in the range of 800-900°C for 20 to 50 minutes. Figures 3 through 5 illustrate the optical micrographs for the as-sintered microstructure of MAS1, MAS2 and MAS3 showing the lower and the higher level of porosity for each one.

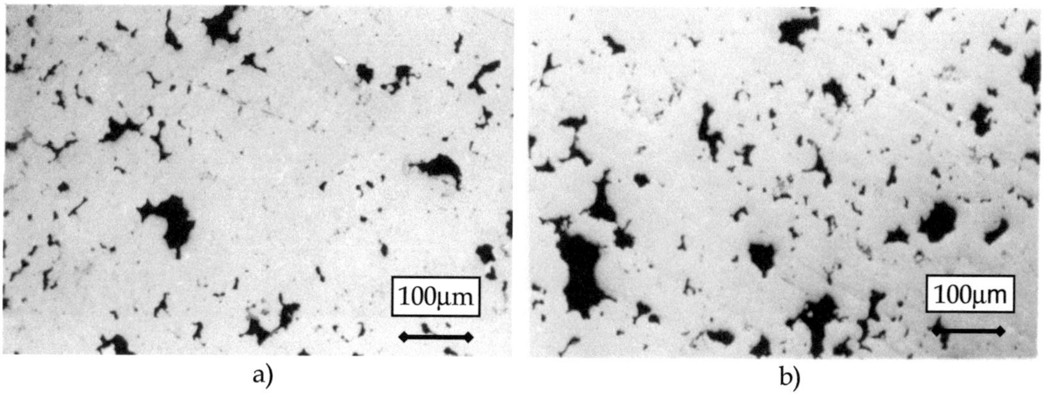

Fig. 3. Micrographs showing the porosity of the composition with 1 wt% Sn (MAS1) sintered at: a) 800°C for 35 minutes (13,8%) and b) 900°C for 20 minutes (17,5%).

Figure 3(a) shows some pores of relatively smaller size and few larger pores but both types with predominant irregular shapes and sharp edges or needle-like shapes, and an overall non-uniform distribution. In figure 3(b) the number of the larger pores increases, the majority having the same irregular shape except for few pores with rounded shape. In the same time the distribution of the pores becomes little more uniform.

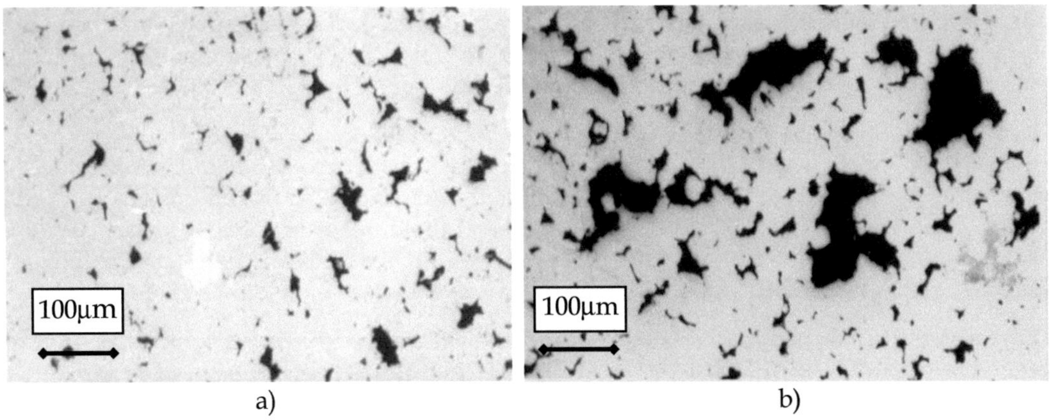

Fig. 4. Micrographs showing the porosity of the composition with 1,5 wt% Sn (MAS2) sintered at: a) 800°C for 50 minutes (15,2%) and b) 900°C for 50 minutes (18,7%).

Figures 4 (a) and 4 (b) present a higher degree of porosity with irregular and sharply edged pores of different sizes. In the left micrograph, figure 4 (a), the number of large pores is decreased, and a more uniform distribution can be noticed. A similar distribution of the pores is also observed in the right micrograph, figure 4 (b), still few much larger pores exist.

Microstructure of MAS3, figures 5 (a) and (b) exhibits the highest level of porosity with finer and smaller pores than previous microstructures. Although some agglomeration of pores still exists, the overall repartition of the pores is more homogeneous. Some large pores can be remarked in the sample sintered at 900°C, with irregular shape, but acicular and spot-like shape, even spherical one, are the most predominant profiles of the pores. From the above micrographs it is obvious that pore coarsening occurs at higher sintering temperatures.

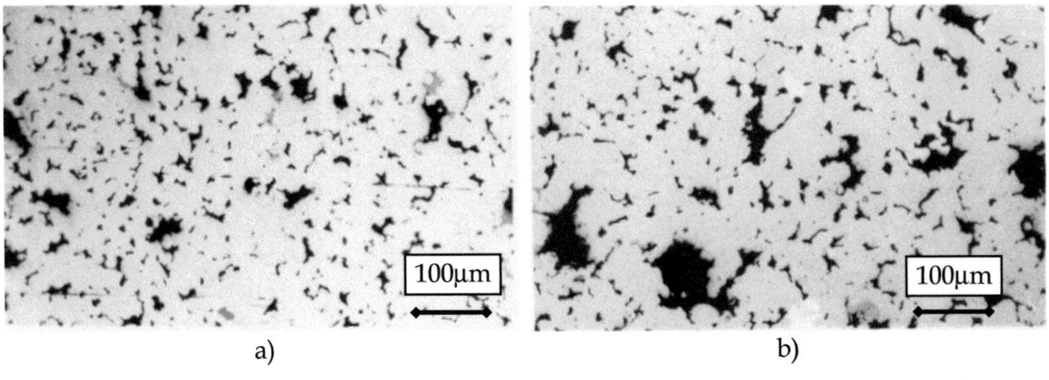

Fig. 5. Micrographs showing the porosity of the composition with 2,5 wt% Sn (MAS3) sintered at: a) 800°C for 50 minutes (15,4%) and b) 900°C for 35 minutes (18,0%).

2.4 Microstructure

Representative microstructures developed during sintering in the presence of the liquid phase for the selected compositions are presented in figures 6, 7 and 8.

The liquid-phase sintering starts from a non equilibrium condition involving mixed powders of differing composition. The compact swelling during sintering with a liquid phase is mainly attributed to the boundary penetration of solid particles by the liquid phase (Wanibe et al., 1990, as cited in Wang, 1999).

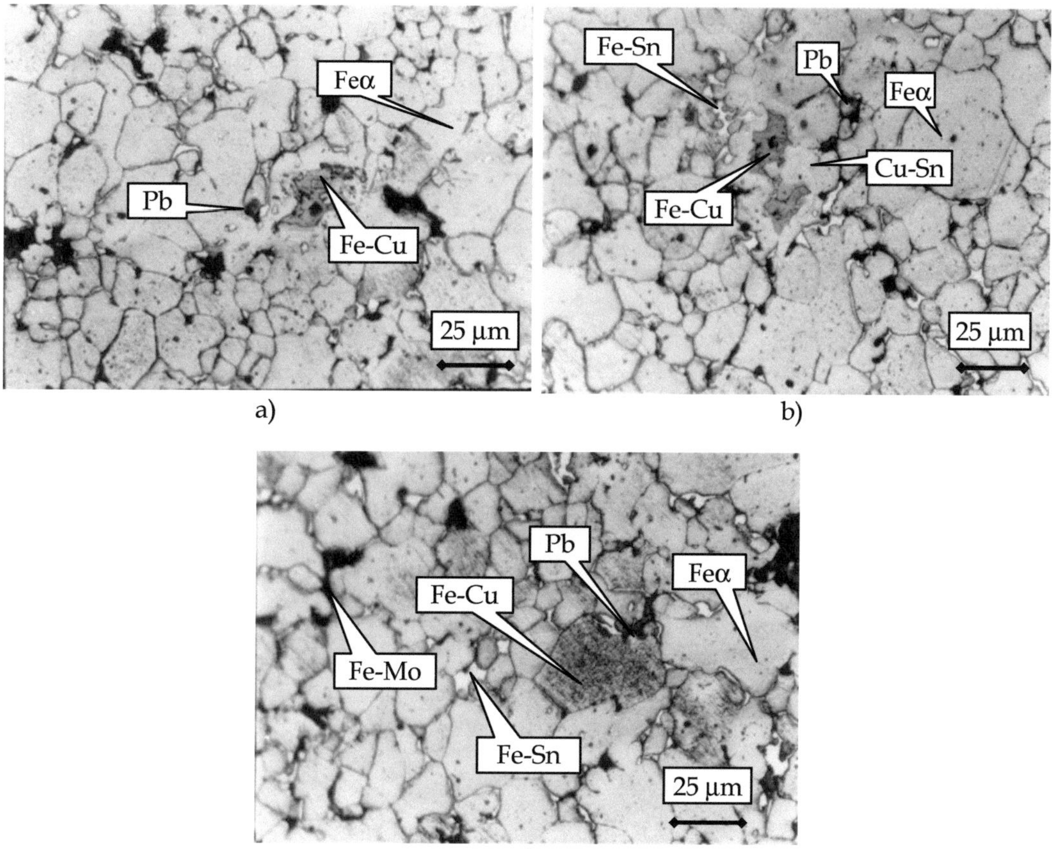

Fig. 6. Microstructure of the MAS1 sintered at: a) 800°C, b) 850°C, c) 900°C for 50 minutes.

Figure 6 shows the microstructure of the iron based samples with 1% Sn, 7,5% Pb and 1% MoS_2 sintered at 800°C, 850°C and 900°C for 50 minutes and emphasize a relatively uniform distribution of the phases with distinctive grain boundaries. Few globular Pb grains and Fe-Cu grains in a ferrite matrix are present in all three microstructures as well as some small pores with irregular shape. In addition, some Fe-Sn and Cu-Sn intermetallic compounds can be seen in the sample sintered at 850°C and few Fe-Mo grains in the sample sintered at 900°C. The microstructure of MAS2 composition (1,5% Sn) sintered at 800°C for 50 minutes (fig.7, a) shows a homogeneous distribution of elongated Fe-Sn grains scattered in the αFe matrix, Fe-Cu grains and few globular Pb grains. In figure 7 (b) the microstructure of MAS2 alloy is sintered at 850°C for 50 minutes and illustrates the presence of Fe-Cu grains, Fe-Sn grains, Fe-Mo and also few globular Pb grains distributed in the ferrite matrix. The microstructure from figure 7 (c) corresponds to MAS2 composition sintered at 900°C for 50 minutes and exhibits a uniform distribution of the Fe-Sn grains around the neck of the ferrite grains. Also, few elongated Pb grains, Cu-Sn and Fe-Cu grains are detected.

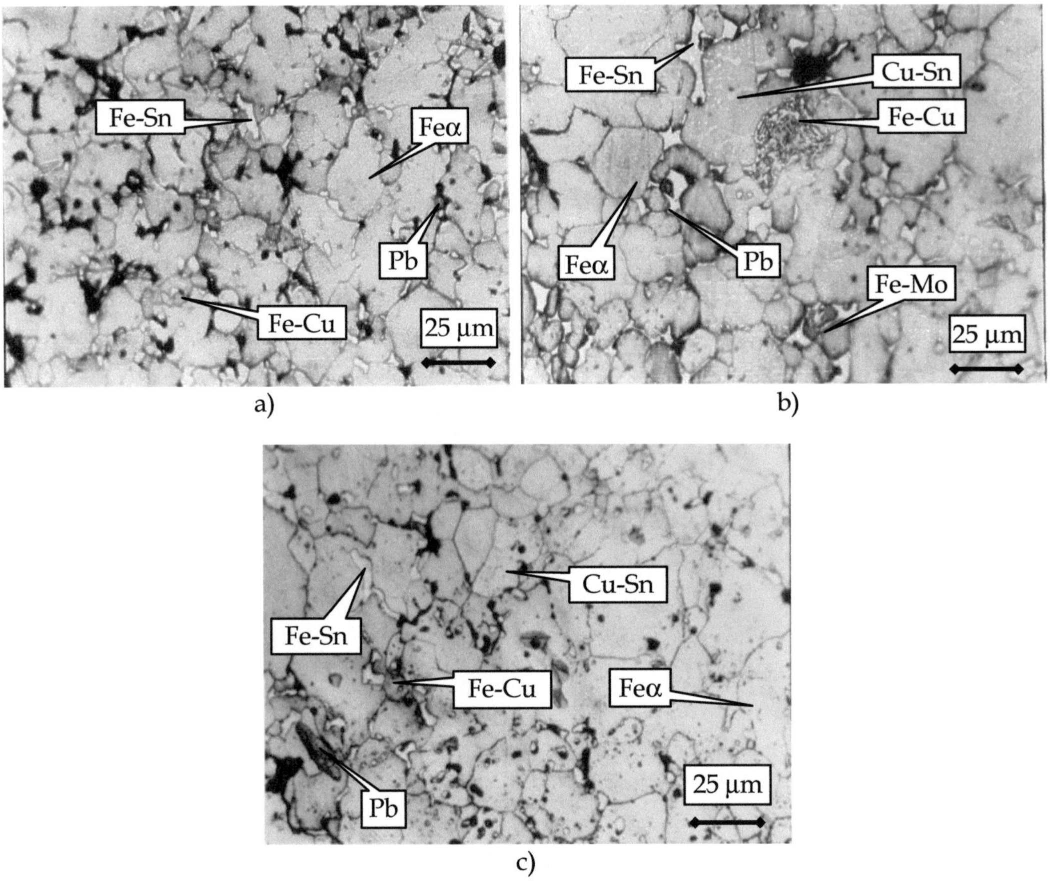

Fig. 7. Microstructure of the MAS2 sintered at: a) 800°C, b) 850°C, c) 900°C for 50 minutes.

Many irregular and small pores can be seen in both samples sintered at 800°C and 900°C placed mainly at the grain boundaries. A large rounded pore can be observed in the sample sintered at 850°C where, prior to its formation, a Cu-Sn grain existed. Large particles generate pores when they form a liquid and when the compact has a low porosity, thus, liquid spreading leads to swelling, but further densification is achieved in longer times (Lu et al., 2001, as cited in German et al., 2009). The sample with 2,5% Sn sintered at 800°C for 50 minutes (fig. 8, a) shows a well-distribution of Fe-Sn grains around the ferrite grains with few spheroidal Pb grains and Fe-Cu grains. Microstructure of the alloy with 2,5% Sn sintered at 850°C contains Fe-Cu and Cu-Sn grains embedded in a Fe-Sn grains network. Also, some globular Pb grains are spotted. The sample with the same composition sintered at 900°C for 50 minutes contains Fe-Sn grains in a uniform distribution over the αFe skeleton and some Pb, Fe-Cu and Fe-Mo grains. Again, a large pore forms where the tin grains were prior to melting.

From these microstructures it can be observed the liquid formation and a progressive growth of the larger grains at the expense of the smaller ones, giving a fewer grains with a larger average size.

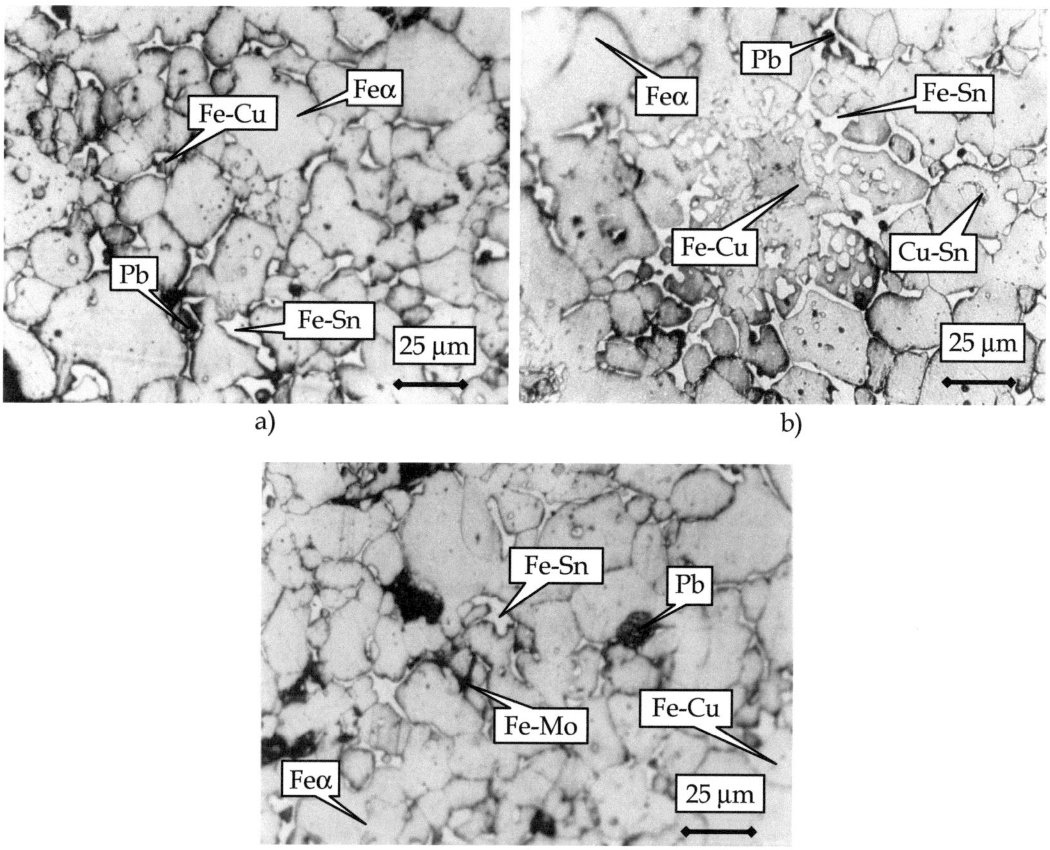

c)

Fig. 8. Microstructure of the MAS3 sintered at: a) 800°C, b) 850°C, c) 900°C for 50 minutes.

Material is transported from the small grains to the large ones by diffusion through the liquid. The liquid spreads and penetrates grain boundaries. After liquid spreading between grains, the film often decomposes into lens-shaped regions forming a necklace microstructure. This occurs because of the increasing surface energy which accompanies completion of a reaction across the solid-liquid interface (German, 1996). This event can be observed on the higher magnification of the optical micrograph shown in figure 9 (a).

The low solubility of the solid copper phase in the liquid lead and tin phase correlated with high solubility of the liquid tin in the solid iron and copper produces the swelling of the system due to decreasing of the liquid volume.

Tin powder particles completely melt and form liquid pools. The liquid phase penetrates the particle boundaries, which possess higher free energy. Figure 9 (b) shows the sample sintered at 800 °C, in which the tin-base liquid phase further penetrates into the grain boundaries of the solid Fe-Cu.

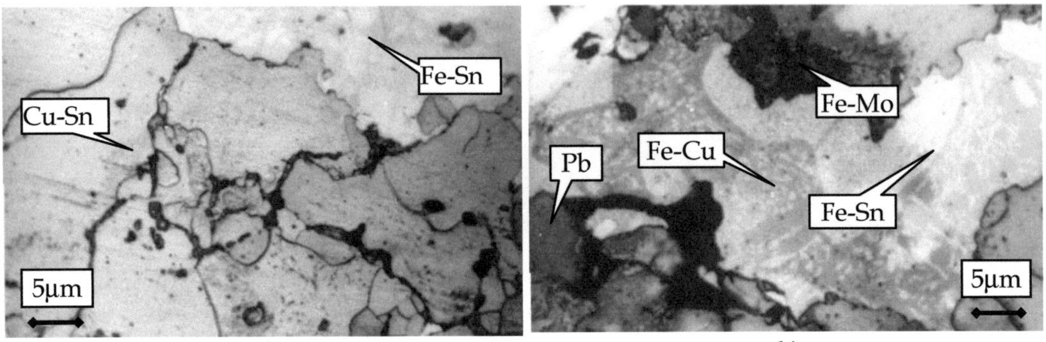

a) b)

Fig. 9. Microstructure of the MAS3 (a) sintered at 850°C/20min and MAS2 (b) sintered at 800°C/50min.

Microstructure coarsening continues and residual pores enlarge if they contain trapped gas, giving compact swelling. Generally, properties of most liquid-phase sintered materials are degraded by prolonged final-stage sintering, thus, short sintering times are preferred in practice (German, 1996). The initial melt induces swelling due to liquid tin and lead penetrating grain boundaries. Pores form at the prior Sn and Pb particle sites. As the Fe particle size increases, swelling goes through a peak. At large sized Fe particles there are fewer interparticle regions for tin base liquid penetration, thus less swelling is observed. Other representative higher magnification microstructures are presented in figure 10.

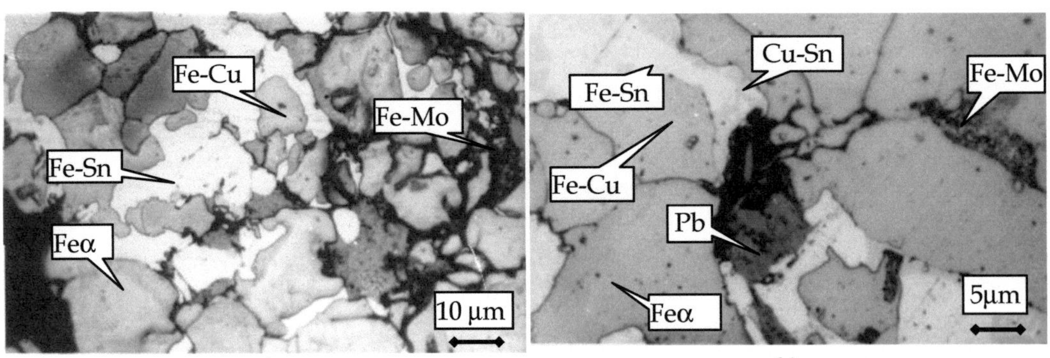

a) b)

Fig. 10. Microstructure of the MAS1 (a) sintered at 800°C/20min and MAS2 (b) sintered at 850°C/50min.

3. Mathematical modelling of the sintering process

A technological process optimization is based on a mathematical model and, generally, it is characterized by two types of parameters or variables: independent parameters (input data) and dependent parameters (output data) (Taloi et al., 1983). The essential target in resolving an optimization problem is to choose the performance or optimization function of the analyzed process. Development of the mathematical models using statistical methods lays emphasis primarily on the concordance between data and the mathematical model, which can be performed by regression analysis (Gheorghe & Ciocardia, 1987). Empirical methods of the mathematical modelling used in regression analysis through passive experiment are based on statistical analysis of the experimental results obtained by varying independent parameters at different levels, intuitively established on previous experiment. The accuracy of the methods increases with the number of the experiments (Micu & Mihoc, 1987). A different situation appears when statistical methods are used in all stages of the experiment: before the experiment by establishing the number of experiments and their performing conditions; during the experiment by processing the results; after the experiment by conclusions referring to development of future experiments. This new way of treating the problem is named "active experiment" and assumes the programming of the experiment (Marusciac, 1973) as follows:

- Establishing the necessary and sufficient number of experiments and their operating conditions;
- Determining the regression equation by statistical methods, which fairly approximate the model of the process;
- Determining the conditions to attain the optimum value of the process performance.

For statistical analysis of the experimental data the sintered density values of MAS1, MAS2 and MAS3 alloys are presented in Table 4. Of a great importance in resolving the optimization problems using the programming of the experiment is to know the influence factors which represent the independent variables.

Temp. [°C]	800°C			850°C			900°C		
Time [min]	20	35	50	20	35	50	20	35	50
Compacting pressure [MPa]	500								
	MAS 1								
ρ_s	6,8	6,98	6,84	6,94	6,73	6,94	6,68	6,83	6,73
	MAS 2								
ρ_s	6,56	6,78	6,8	6,6	6,61	6,56	6,56	6,56	6,52
	MAS 3								
ρ_s	6,59	6,54	6,71	6,64	6,56	6,54	6,54	6,5	6,55

Table 4. Sintered density values for selected materials.

The parameters that influence the sintering process were considered to be sintering temperature, time and lead content. In the regression analysis by active experiment the non-compositional program was used.

3.1 Process performance and variables

The process performance or optimization function proposes to establish the relationship between the density of the powder metallurgy iron-copper based alloys having different additional elements, which refers as a dependent variable, and the sintering process parameters (temperature and time) as well as the lead content, which represent the independent variables.

For the development of the non-linear mathematical model of the sintering process the non-compositional program uses a matrix of experiments. The technological process can be approached by a two degree polynomial function (1), where the dependent parameter (y) is density and the independent parameters (x_1, x_2, x_3) are sintering temperature, maintaining time and lead content.

$$y = b_0 + b_1x_1 + b_2x_2 + b_3x_3 + b_{12}x_1x_2 + b_{23}x_2x_3 + b_{13}x_1x_3 + b_{11}x_1^2 + b_{22}x_2^2 + b_{33}x_3^3 \quad (1)$$

A base level will be established for each variable (independent parameter), which actually represents factorial space coordinates of the starting points. Also, variation ranges (steps) Δz_i will be defined. The superior level is obtained by adding the variation range to the base level and the inferior level is found by extracting the variation range from the base one. If x_i is the codified value of the variable z_i, and is determined from the following relation:

$$x_i = \frac{z_i - z_{0i}}{\Delta z_i} \quad (2)$$

then the superior level will be marked with "+1", the inferior level with "-1" and the base level with "0".

Variable	Sintering temperature		Maintaining time		Lead content	
	natural units, °C	codified values	natural units, min	codified values	natural units, %	codified values
Base level	$z_0 = 850$	$\frac{850-850}{50}=0$	$z_0 = 35$	$\frac{35-35}{15}=0$	$z_0 = 5$	$\frac{5-5}{2,5}=0$
Variation range	$\Delta z = 50$	-	$\Delta z = 15$	-	$\Delta z = 2,5$	-
Superior level	900	$\frac{900-850}{50}=+1$	50	$\frac{50-35}{15}=+1$	7,5	$\frac{7,5-5}{2,5}=+1$
Inferior level	800	$\frac{800-850}{50}=-1$	20	$\frac{20-35}{15}=-1$	2,5	$\frac{2,5-5}{2,5}=-1$
Random value	920	$\frac{920-850}{50}=+1,4$	30	$\frac{30-35}{15}=-0,3$	4,5	$\frac{4,5-5}{2,5}=-0,2$

Table 5. Correspondence between natural and codified units of the variable levels.

Table 5 shows the correspondence between natural and codified units of the independent variable considered in the sintering process optimization: temperature, time, Pb content.

Therefore, the codified notations of the three variables levels are:

x = +1 – Upper level (z_1 = 900°C; z_2 = 50 min.; z_3 = 7, 5%Pb);
x = 0 – Base level (z_1 = 850°C; z_2 = 35 min.; z_3 = 5%Pb);
x = -1 – Lower level (z_1 = 800°C; z_2 = 20 min.; z_3 = 2, 5%Pb).

The variation step for the tin content is different for the three levels and for this reason to establish the optimization function the lead content was considered.

The programming matrix of the experiment is presented in Table 6.

Exp.	Variables													ρ [g/cm³]
	x_0	x_1	x_2	x_3	$x_1 x_2$	$x_1 x_3$	$x_2 x_3$	x_1^2	x_2^2	x_3^2	T[°C]	t[min]	%Pb	
1	+1	+1	+1	0	+1	0	0	+1	+1	0	900	50	5	6,52
2	+1	+1	-1	0	-1	0	0	+1	+1	0	900	20	5	6,56
3	+1	-1	+1	0	-1	0	0	+1	+1	0	800	50	5	6,8
4	+1	-1	-1	0	+1	0	0	+1	+1	0	800	20	5	6,56
5	+1	+1	0	+1	0	+1	0	+1	0	+1	900	35	7,5	6,83
6	+1	+1	0	-1	0	-1	0	+1	0	+1	900	35	2,5	6,5
7	+1	-1	0	+1	0	-1	0	+1	0	+1	800	35	7,5	6,98
8	+1	-1	0	-1	0	+1	0	+1	0	+1	800	35	2,5	6,54
9	+1	0	+1	+1	0	0	+1	0	0	+1	850	50	7,5	6,94
10	+1	0	+1	-1	0	0	-1	0	+1	+1	850	50	2,5	6,54
11	+1	0	-1	+1	0	0	-1	0	+1	+1	850	20	7,5	6,94
12	+1	0	-1	-1	0	0	+1	0	+1	+1	850	20	2,5	6,64
13	+1	0	0	0	0	0	0	0	0	0	850	35	5	6,61
14	+1	0	0	0	0	0	0	0	0	0	850	35	5	6,61
15	+1	0	0	0	0	0	0	0	0	0	850	35	5	6,61

Table 6. The matrix of the second degree non-compositional experiment programming.

3.2 Statistical validation of the regression equation

The experimental error (dispersion of the results reproducibility), s_0^2, is determined with the following formula (3):

$$s_0^2 = \frac{\sum_{u=1}^{n} (y_u - \bar{y})^2}{n-1} \qquad (3)$$

u - experiment number
n - number of the experiments
$\bar{y}$ - the average of the results obtained in n parallel experiments

The coefficients of the regression equation (1) are calculated with relations (4) and (5):

For linear effects

$$b_i = \frac{\sum_{u=1}^{N} x_{iu} \cdot y_u}{\sum_{u=1}^{N} x_{iu}^2} \tag{4}$$

For interacting effects

$$b_{ij} = \frac{\sum_{u=1}^{N} x_{iu} \cdot x_{ju} \cdot y_u}{\sum_{u=1}^{N} (x_{iu} \cdot x_{ju})^2} \tag{5}$$

The values of these coefficients are statistically evaluated using the relation (6):

$$|b_i| \geq |\Delta b_i| \tag{6}$$

$$|\Delta b_i| = t_{\alpha,N} \cdot s_{b_i} \tag{7}$$

$t_{\alpha, N}$ – Student criterion for α significance level and N degrees of freedom;
s_{bi} – mean square deviation for calculation of b_i coefficient:

$$s_{bi} = \pm\sqrt{s_{bi}^2} \tag{8}$$

s_{bi}^2 – dispersion for calculation of b_i coefficient:

$$s_{bi}^2 = \frac{s_0^2}{\sum_{u=1}^{N} x_{iu}^2} \tag{9}$$

The hypothesis of the model concordance is verified using Fischer criterion (10):

$$F_c = \frac{s_{conc}^2}{s_0^2} \tag{10}$$

s^2_{conc} – calculated model dispersion
s_0^2 – results reproducibility dispersion

The calculated model is concordant when

$$F_c < F_{0,05;v1;v2} \cdot$$

v_1 – number of degrees of freedom to calculate s^2_{conc}: $v_1 = N - k'$
v_2 – number of degrees of freedom to calculate s^2_0: $v_2 = n - 1$

The coefficients of the regression equation that established the link between process performances (sintered density) and independent variables (sintering temperature, maintaining time and lead content) and also the statistical verifying of the regression equation were performed using Maple software:

b0 := 18.21431; b1 := -2.27833; b2 := -.04472; b3 := .154167; b11 := -.0033; b22 := -.022; b33 := .00266; b12 := .16; b13 := .0122; b23 := .001

3.3 Sintering process optimization

Using statistical analysis of the experimental data regression equations were obtained for independent variables as functions of two and three input data (dependent variables). For analysis of the mathematical models adequacy a multiple regression analysis was performed in order to investigate the errors introduced by mathematical models comparatively with experimental results. For an accurate description of the investigated process using regression analysis, different scale order of the variables possibly causing difficulties for some algorithms should be considered. If the typical values of the problem variables are known, then the problem can be transformed so all variables will have the same scale order. Therefore, it is imperative to have similar scale order for all variables in the interest region and thus, providing a "weight" compensation of all variables during the optimization process. The mathematical models of the density as a function of sintering temperature, maintaining time or lead content (wt%) are presented in table 7, considering two of these input data as independent variables and the third remaining constant.

Input data			Mathematical models
Dependent Variables		Independent Variables	
Pb [wt%]	2.5	Temperature (T_S) Maintaining time (t_S)	$\rho = 11.382 - 1.073T_S + 0.0287t_S + 0.06T_S^2 - 0.007T_St_S + 0.004t_S^2$
	5		$\rho = 16.489 - 2.56T_S + 0.967t_S + 0.16T_S^2 - 0.093T_St_S - 0.022t_S^2$
	7.5		$\rho = -1.769 + 2.045T_S + 0.192t_S - 0.127T_S^2 + 0.003T_St_S - 0.03t_S^2$

Table 7. Mathematical models of the density as a function of two independent variables.

Also, it is very important for final density analysis of the self-lubricating bearings to study the mathematical model of the dependent variable (density) as a function of all three independent variables (temperature, time, chemical composition). This mathematical model is given by the following equation (11):

$$\rho = 18.21431 - 2.72833T_S - 0.04472t_S + 0.154167C_{Pb} - 0.0033T_St_S - 0.022T_SC_{Pb} + 0.00266t_SC_{Pb} + 0.16T_S^2 + 0.0122t_S^2 + 0.01C_{Pb}^2 \quad (11)$$

For an accurate analysis of the process factors influences on the independent variable the results of the statistical analysis are depicted in figures 11 through 13. The 3D graphic representations show the dependence of the density on two input parameters (sintering temperature and maintaining time) for each of the three compositions (the lead content is considered constant).

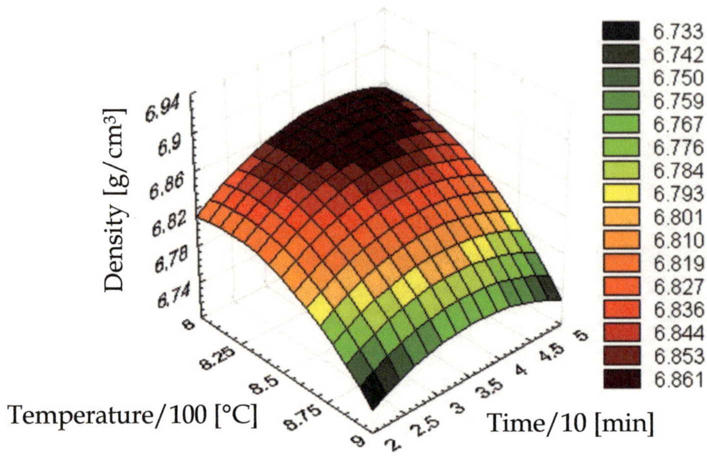

Fig. 11. The dependence of the density on sintering temperature and maintaining time (2.5wt% Pb).

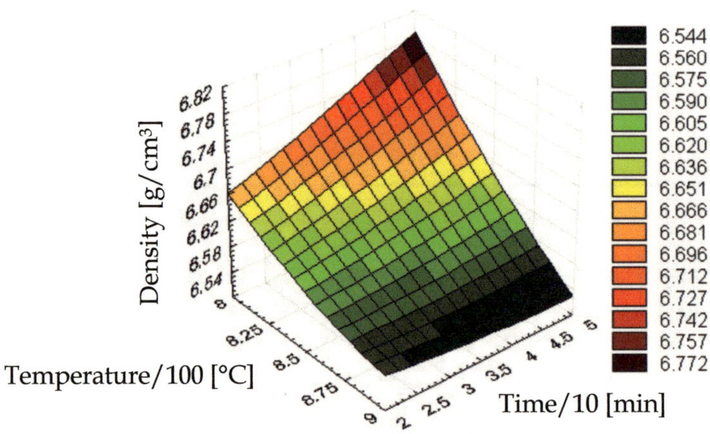

Fig. 12. The dependence of the density on sintering temperature and maintaining time (5wt% Pb).

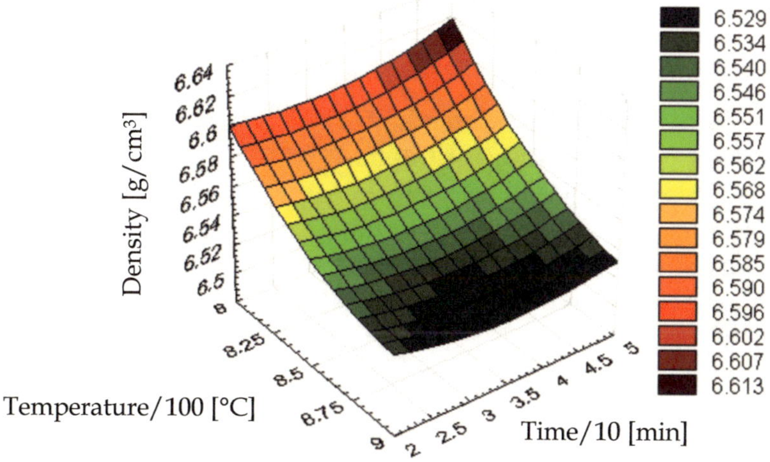

Fig. 13. The dependence of the density on sintering temperature and maintaining time (7.5wt% Pb).

MULTIPLE REGRESSION RESULTS:

Dependent Variable: rho
Multiple R: .96304871
Multiple R-Square: .92746282
Adjusted R-Square: .90767995
Number of cases: 15
F(3, 11) = 46.88212 p < .000001
Standard Error of Estimate: .49053156
Intercept: 7.311416667 Std.Error: .2998758 t(11) = 24.381 p < .000

STAT. MULTIPLE REGRESS.

Regression Summary for Dependent Variable: rho
R= .96304871 R^2= .92746282 Adjusted R^2= .90767995
F(3,11)=46.882 p<.00000 Std.Error of estimate: .04905

N = 15	BETA	St. Err. of BETA	B	St. Err. of B	t(11)	p-level
Intercept			7.311417	.299876	24.38148	.000000
T_s	-.304351	.081205	-.130000	.034686	-3.74793	.003221
t_s	.181440	.081205	.025833	.011562	2.23434	.047166
%Pb	.895496	.081205	.076500	.006937	11.02756	.000000

Table 8. Multiple regression analysis for the dependence of the density on three variables.

The variables estimation coefficients and the residual values were determined by multiple regression analysis of the density dependence on sintering temperature, maintaining time and lead content and the results of these analyses are presented in table 8.

The residual values offer information about errors introduced by using mathematical models comparing with the results obtained by experimental study. These errors represent the difference between the experimental data and analytical data.

The figures 14, 15 and 16 show the relationship between the experimental and calculated values of the density in accordance with the mathematical model, between the residual values and experimental densities and between residuals and calculated densities.

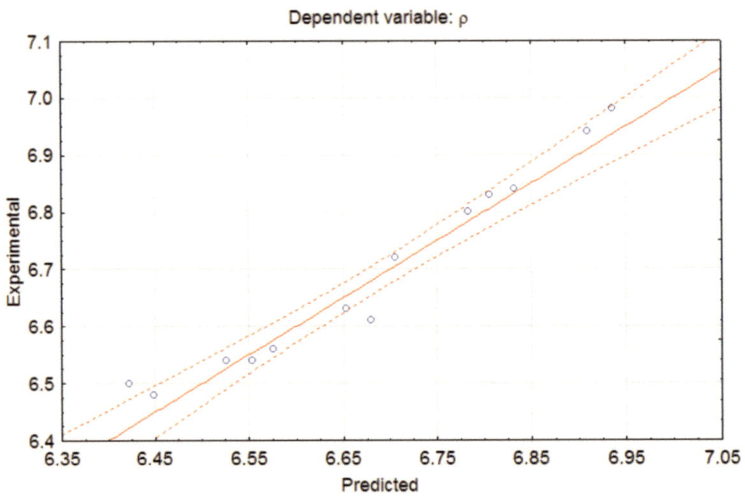

Fig. 14. The dependence between experimental and calculated values of the density.

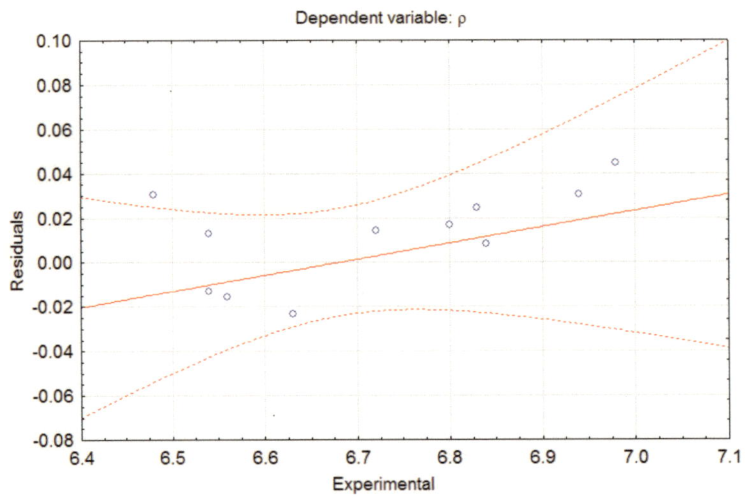

Fig. 15. The dependence between the residuals and experimental values of the density.

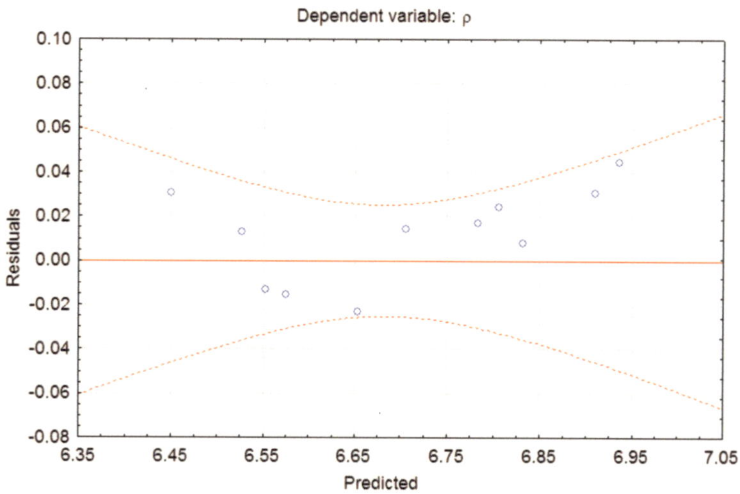

Fig. 16. The dependence between the residuals and calculated values of the density.

4. Conclusions and future research

Using powder metallurgy technologies new Fe-Cu-Sn-Pb materials were developed for tribological applications where the material porosity plays an important role for an adequate functioning of the machines and devices by permitting the retention of a sufficient oil amount in the part pores. Besides the advantages that additions of copper, lead and tin bring to tribological and mechanical properties of self lubricating bearings, these additions direct to the formation of the liquid phase during sintering process causing swelling of the sintered compacts due to pore formation.

From the investigations and observations regarding dimensional changes, total porosity and microstructure development of the iron-copper based alloys with different addition elements during sintering in dry hydrogen atmosphere at 800°C, 850°C and 900°C, the following remarks are drawn:

- Due to high solubility of the liquid in the solid, formation of a transient liquid phase of Sn is observed. Considering the sintering temperature and tin content, swelling of all samples sintered under specified conditions was observed. Lower values of the dimensional change were noticed for the sample with 1 wt% Sn compacted at 700 MPa and sintered at 800^{0}C. This decrease of the dimensional change is attributed to the formation of the Fe-Sn phase at higher temperatures, which has a lower coefficient of thermal expansion than iron. A sharper rise of dimensional growth is detected for specimen with 2.5 wt% Sn (4,38%) sintered at 900°C and compacted at 700 MPa, and is due possibly to the formation of Cu-Sn, Fe-Cu and Fe-Sn compounds in larger amounts. The volume expansion behavior can be ascribed to the boundary penetration of solid particles by the tin-base liquid phase. Cu-Sn liquid phase wets the iron skeleton, diffuses into the iron and enhances growth. Also, the dimensional growth might be attributed to the tapped hydrogen in the pores as the sintering process was performed in hydrogen atmosphere;

- A non uniform variation of the total porosity with sintering temperature and time can be seen for all three compositions. Usually, during liquid phase sintering porosity is decreasing, but since smaller pores are annihilated first the mean pore size increases while the grain size increases. The highest value of total porosity was observed for sample with 1,5% Sn sintered at 900°C for 50 minutes and the lowest one was for sample with 1% Sn sintered at 800°C for 35 minutes. This behaviour is attributed to the formation of the soft phases of Cu-Sn, Fe-Sn-S or Pb-Sn, which diffuse into the pores and hence, decreasing total porosity, and to the formation of iron-tin intermetallics which contribute to the increase in intercommunicating porosity, and in that way increasing total porosity. Microscopic investigation of the selected materials porosity shows pore coarsening at higher sintering temperatures. Also, the number of pores increases as the tin content increases, except for the specimen with 1,5% Sn sintered at 900°C for 50 minutes. Few larger pores with main irregular shapes and a non uniform distribution can be identified in all samples. Smaller size pores present an overall homogeneous distribution with acicular and spot-like shape and even spherical;
- Representative microstructures developed during liquid phase sintering show a ferrite matrix with distinctive grain boundaries, but irregular shape and different size. Cu-Sn and elongated Fe-Sn grains are relatively uniformly distributed at the intergranular sites in the ferrite matrix. Also, globular Pb grains, Fe-Cu, some Fe-Cu sulphide and Fe-Mo grains can be found in the ferrite matrix. Many irregular and small pores can be seen in all samples, placed mainly at the grain boundaries, and few large pores where the tin grains were prior to melting, surrounded by molten tin. From these microstructures the liquid formation can be observed and a progressive growth of the larger grains at the expense of the smaller ones gives a fewer grains with a larger average size. Penetration of the solid Fe-Cu particles boundary by the tin base liquid phase generates swelling of the compact. A necklace microstructure forms when the tin liquid between grains decomposes into lens-shaped regions due to the increasing surface energy which accompanies completion of a reaction across the solid-liquid interface.

Mathematical model or the optimization function of the sintering process was developed using statistical methods which lays emphasis primarily on the concordance between data and the mathematical model. Using regression analysis of the density dependence on two independent variables (sintering temperature and time), the third one (lead content) being considered constant, and on three independent variables (sintering temperature, sintering time and lead content), the following conclusions are drawn:

- Mathematical models are sufficiently precise and the errors introduced are no greater than 5%;
- The adequacy of the mathematical model is confirmed by the admissible values of the parameters and coefficients obtained using regression analysis;
- The lead content has the maximum influence on the sintered parts density and then in lessening order, the sintering temperature and the maintaining time;
- The mathematical model can be applied to the sintering processes in order to obtain an optimal relationship between antifriction properties of the self-lubricating bearings, and sintering process parameters and the material composition.

Considering the purpose for which these materials were developed, at this point of the research the alloy containing 2,5% Sn presents the highest level of total porosity among these three materials taken in this study, but its microstructure homogeneity should be improved for a better distribution of the phases and pores into the ferrite matrix with more uniform size and regular shape of grains. To reach this goal further research will be proceed by employing mechanical alloying of elemental powders and microwave sintering for the achievement of nanostructures with unique properties.

5. References

Baranov, N. G., Ageeva, V. S., Zabolotnyi, L. V., Il'nitskaya, A., I., Mokrovetskaya, V. S. & Sabaldyr, V. Ya. (1990). Structure and Properties of the Powdered Antifriction Material Iron-Copper-Tin-Lead, In: *Powder Metallurgy and Metal Ceramics*, Vol.29, No.7, 08.09.2011, Available from: http://www.springerlink.com/content/j73vk6147mp13545/

Boivie, K. (2000). SLS Application of the Fe-Cu-C System for Liquid Phase Sintering, In: *Proceedings of the Solid Freeform Fabrication Symposium*, 03.09.2011, Available from: http://utwired.utexas.edu/lff/symposium/proceedingsArchive/pubs/Manuscripts/2000/2000-18-Boivie.pdf

Chandrasekaran, M. & Singh, P. (1996a), Effect of Pb Additions on the Friction and Ware of Sintered Fe-Cu-Sn-MoS_2, In: *The International Journal of Powder Metallurgy*, Vol.32, No.1, pp. 51-58, ISSN 08887463

Chandrasekaran, M. & Singh, P. (1996b), Effect of Sn on the Properties of Fe-Cu-Pb-MoS_2 Antifriction Alloy In: *The International Journal of Powder Metallurgy*, Vol.32, No.4, pp. 323-330, ISSN 08887463

Domsa, Al. (1966). *Tehnologia Fabricarii Pieselor din Pulberi Metalice (Manufacturing Technology of the Metallic Powder Parts)*, Ed. Tehnica, Bucuresti, Romania

German, R., M. (1985). *Liquid Phase Sintering*, Plenum Press, ISBN 0-306-42215-8, New York

German, R., M. (1996). *Sintering Theory and Practice*, John Wiley & Sons, ISBN 0-471-05786-x, New York

German, R., M., Suri, P. & Park, S., J. (2009). Review: Liquid Phase Sintering, In: *Journal of Materials Science*, Vol.44, No.1, 03.09.2011, Available from: http://www.springerlink.com/content/eu8804w248232124/

Gheorghe, M., Ciocardia, C. (1987). *Metodologie de determinare a functiilor de regresie, incluzand variabile independente nesimetrizabile, (Methodology for Determination of the Regression Functions, Including non-symmetrical independent variables)*, In: *Bulletin of Polytechnic Institute of Bucharest*, pp. 35-70

Kostornov, A.G. & Fushchich, O. I. (February 2007). Sintered Antifriction Materials, In: *Powder Metallurgy and Metal Ceramics*, Vol.46, Nos.9-10, 02.09.2011, Available from: http://www.springerlink.com/content/9221327246150781/

Kostornov, A.G., Fushchich, O. I. & Chevichelova, T. M. (May 2007). Structurization in Sintering of Antifriction Powder Materials Based on Iron-Copper Alloys, In: *Powder Metallurgy and Metal Ceramics*, Vol.46, Nos.11-12, 02.09.2011, Available from: http://www.springerlink.com/content/7ml73l6r46245576/

Marusciac, I. (1973). *Metode de Rezolvare a Problemelor de Programare Neliniara (Methods for Resolving Non-linear Programming Problems)*, Ed. Dacia, Cluj-Napoca, Romania

Micu, N. & Mihoc, Gh. (1987). *Elemente de Teoria Probabilitatilor si Statistica Matematica (Elements of Probabilities Theory and Mathematical Statistics)*, Editura Didactica si Pedagogica, Bucuresti, Romania

Shahparast, F. & Davies, B. L. (February 2003). A study of the potential of sintered iron-lead and iron-lead-tin alloys as bearing materials, In: *Ware*, Vol. 50, Issue 1, 24.08.2011, Available from: http://www.sciencedirect.com/science/article/pii/0043164878902521

Sorokin, V. K. (1966). Densification during the Sintering of Copper-Tin Powder Alloys, In: *Powder Metallurgy and Metal Ceramics*, Vol.5, No.11, 03.09.2011, Available from: http://www.springerlink.com/content/x24862v5nmugw681/

Sustarsic, B & Kosec, L. (1998). Engineering Properties of Fe-MoS_2 Powder Mixtures, In: *Proceedings of the 1998 Powder Metallurgy World Congress & Exhibition*, pp. 316-321, ISBN 1-899072-09-8, Granada, Spain, October 18-22, 1998

Takata, J. & Kawai, N. (1995). Dimensional Changes During Sintering of Iron Based Powders, In: *Powder Metallurgy*, Vol.38, No.3 (1995), pp. 209-213

Taloi, D., Florian, E., Bratu, C. & Berceanu, E. (1983). *Optimizarea Proceselor Metalurgice (Optimization of the Metallurgical Processes)*, Editura Didactica si Pedagogica, Bucuresti, Romania

Verghese, R & Gopinath, K. (1989). The Influence of Antimony Additions on Sintered Iron-Copper Bearing Materials, In: *Key Engineering Materials*, Vol. 29-31, P. Ramakrishnan, (Ed.), pp. 457-464, Trans. Tech. Publications, ISBN 978-0-87849-577-1, Switzerland

Wang, W.-F. (1999). Effect of Tin Addition on the Microstructure Development and Corrosion Resistance of Sintered 304L Stainless Steel, In: *Journal of Materials Engineering and Performance*, Vol.8, No.6, 31.08.2011, Available from: http://www.springerlink.com/content/w772221x0w030u28/

Yusof, F., Hameedullah, M. & Hamdi, M. (2006). Optimization of Control Parameters for Self-lubricating Characteristics in a Tin Base Composite, In: *Engineering e-Transaction, University of Malaya*, 31.08.2011, Available from: http://ejum.fsktm.um.edu.my/article/346.pdf

8

Photonic Sintering of Silver Nanoparticles: Comparison of Experiment and Theory

Jeff West, Michael Carter, Steve Smith and James Sears
South Dakota School of Mines and Technology
USA

1. Introduction

Photonic sintering is a low thermal exposure sintering method developed to sinter nanoparticle thin films. The process involves using a xenon flash lamp to deliver a high intensity, short duration (< 1 ms), pulse of light to the deposited nanoparticles. Photonic sintering was developed by Nanotechnologies (now NovaCentrix) of Austin, Texas, and was first made public in 2006 (Schroder et al., 2006). As photonic sintering is a new technology it is also known as pulsed thermal processing (PTP) (Camm et al., 2006) and intense pulsed light (IPL) sintering (Kim et al., 2009). Conductive thin films composed of nanoparticle depositions, when exposed to a short pulse of high intensity light, are transformed into functional printed circuits. The printed circuits can be tailored for use as flexible circuit boards, RFID tags, flat panel displays (Carter & Sears, 2007), photovoltaics, or smart packaging (Novacentrix, 2009). One of the primary advantages of the method is that the high intensity pulse of light produces minimal damage on low temperature substrates. This allows the nanoparticles to be deposited and cured on a high variety of low temperature substrates such as cloth, paper, and Mylar (Carter & Sears, 2007; Farnsworth, 2009). Another advantage of using photonic curing is the speed at which nanoparticle depositions can be sintered. Rather than spending hours in an oven or programming a laser to follow the deposition path, the photonic curing process can sinter large areas (~ 200 cm^2 per 10 cm long lamp) in < 2 ms (Novacentrix 2009).

One of the main objectives of the work reported here was to determine the effectiveness of photonic sintering of silver nanoparticle depositions. This was done by measuring the densification of silver nanoparticles films following photonic sintering. The absorption of light emitted by a flash lamp for varying thicknesses of silver nanoparticle layers was also measured. To determine the amount and depth of sintering, SEM images were taken of a cross section of a sintered film. To better understand the process through which the nanoparticles are sintered, we calculate the absorption of the light emitted by the flash lamp by the silver nanoparticle film using the Bruggeman effective medium theory. Using the heat transfer software package Fluent™ to model the temperature profile of the films during and following sintering, we propose a model for the photonic process.

2. Photonic sintering overview

Photonic sintering was first introduced at the 2006 NSTI Nanotechnology Conference and Trade Show (Schroder et al., 2006). It was developed by NovaCentrix for the purpose of rapidly sintering metal nanoparticle based films (Schroder et al., 2006). The technology allows the nanoparticles to sinter without significantly raising the temperature of the substrate. This is accomplished by using a flash lamp. Two main parameters control the degree of sintering: the intensity of the lamp and the duration of the light pulse. The flash lamp is held between 0.5 cm to 20 cm above the deposition and an intense current is run through the flash lamp (Novacentrix, 2007). Due to this intense current, the xenon flash lamp issues a high intensity, broad spectrum pulse of light. This pulse of light is absorbed by the nanoparticles, which heats them to such a degree that they sinter into a single component.

Because photonic sintering has a minimal effect on the substrate, it enables nanoparticle films to be cured on low temperature substrates such as paper, Mylar, and PET. In addition to allowing low temperature substrates to be used, the speed at which the films are sintered allows the use of inks which would oxidize if sintered over long periods of time using traditional methods, such as copper (Novacentrix, 2009; Schroder 2006). Both of these advantages allow printed electronics manufacturers to reduce the cost of production by using lower cost substrates and less expensive inks.

The Novacentrix PCS-1100 Photonic Curing System is a research and development model in use at South Dakota School of Mines & Technology (SDSM&T). The PCS-1100 has a pulse duration that can be set from 35 µs to 1000 µs (Novacentrix, 2007). The voltage of the flash lamp (which controls the intensity and spectral distribution of the lamp) can also be adjusted with a maximum operational voltage of 4000V (Novacentrix, 2007). The PulseForge 3100 is a production model (currently in operation at Oak Ridge National Laboratory) that can be incorporated with a roll-to-roll production system or a conveyor belt to sinter continuous or discrete items (Novacentrix, 2009).

During the development of photonic sintering, three basic assumptions were considered for sintering nanoparticles: (i) nanoparticles are predominantly black, they should absorb light very well (Schroder, 2006), (ii) once light is absorbed by the nanoparticles, due to their high surface area to mass ratio, the nanoparticles would heat easily and sinter quickly, and (iii) as nanoparticle films are very thin, they would not retain heat very well and cool rapidly, minimizing damage to the substrate. Photonic sintering has been shown to sinter conductive nanoparticle metals (*e.g.*, silver, gold, and copper) as well as dielectric nanoparticles made of alumina, zirconia, barium titanate, and mica, as well as the soft magnetic materials cobalt, ferrite and iron-nickel permalloy (Carter & Sears, 2007).

A comparison of the resistivities obtained using furnace, laser, and photonic sintering was reported by Carter and Sears (Carter & Sears, 2007). For the comparison, two different silver inks were used. One ink was based on the AgSt2 Novacentrix 25 nm diameter silver nanoparticles and the other ink was composed of UT Dot 7 nm diameter silver nanoparticles. The inks were deposited into 1 cm^2 pads and sintered using various furnace temperatures, laser fluences, and flash lamp voltages. The resistivities were then measured and the lowest values obtained for each method. The lowest resistivities were found using the following settings for each method. The furnace sintering was done at a temperature of 500°C for two hours. The laser sintering was done using a laser fluence of 2800 J/cm^2. The photonic curing was done using a lamp voltage of 1200V with a pulse length of 900 μs. The lowest surface resistivity for each method is shown in Table 1. The measurements showed that the resistivity of the photonically sintered silver was comparable to the resistivity of the oven and laser sintered silver, and that smaller nanoparticles approach bulk silver resistivities after photonic or oven sintering.

Material	Sintering Method	Resistivity (μΩ-cm)
UT Dot (7 nm silver)	Furnace	2.1 ± 0.9
	Photonic	2.8 ± 0.8
AgSt2 (Novacentrix 25 nm silver)	Furnace	3.8 ± 0.3
	Laser	5.3 ± 0.3
	Photonic	7.9 ± 0.5
Bulk Silver		1.59

Table 1. Table comparing resistivities obtained using different sintering methods for two silver nanoparticle based inks.

3. Experimental results

After observing the comparisons to traditional sintering methods, experiments were run to gain insight into the process by which particles are sintered during photonic curing. The densification as a function of the flash lamp voltage and pulse duration was measured to determine the effects of those parameters. Depositions of varying thickness were tested using a UV-Vis spectrometer to measure the absorption of the depositions in the wavelength region produced by the flash lamp. Finally, SEM images were taken of the cross section of a thick sintered sample to determine the depth and amount of sintering in the deposition.

3.1 Parametric study of densification

We measured the densification of silver nanoparticles as a function of the pulse duration and flash lamp voltage to find the optimal settings to sinter V2 silver ink. The V2 silver ink consists of Novacentrix 25 nm silver nanoparticles suspended in DMA. The measurements also allow a determination of the effect of lamp voltage and pulse duration on the sintering process.

The process of measuring the densification began by finding the volume fraction of nanoparticles in the deposition prior to sintering. This was accomplished by weighing a clean glass slide and then depositing V2 ink in a square pattern on the slide. The thickness of the deposition was then measured using a Zeiss Imager M1M microscope. The deposition thickness was determined by focusing on the surface of the glass slide and then again on the surface of the deposition. The focal distances were then compared to find the thickness of the deposition. The glass slide with the deposition was then weighed again to find the total mass. The area and thickness of the deposition gave the total volume of the deposition and from the weight measurements the mass of the deposition was also known. The mass of the deposition was then divided by the density of silver to find the volume of the deposited silver. The calculated volume of the silver was then divided by the volume of the deposition to find the volume fraction of the silver in the deposition. The volume fraction of silver in the deposition was calculated to be 36% using this method with a density of the deposition being 3.8 g/cm^3.

After finding the volume fraction of the silver in the deposition prior to sintering, the next step was to deposit the material and measure the volume prior to and after sintering to determine the densification of the deposition. This was accomplished by depositing V2 silver ink in a 3 mm by 10 mm pad on a glass slide. The thickness of the deposition was measured using the Zeiss Imager M1M microscope. The deposition was then placed into the photonic curing system and sintered. The thickness of the deposition was then measured again. As the thickness was the only dimension to change during the sintering process the change in the volume was found. As the mass stayed approximately constant after sintering, the density of the deposition and volume fraction of silver in the deposition were then calculated. This process was completed ten times for each setting on the photonic curing system to obtain an adequate sampling size. The results are shown in Figure 1.

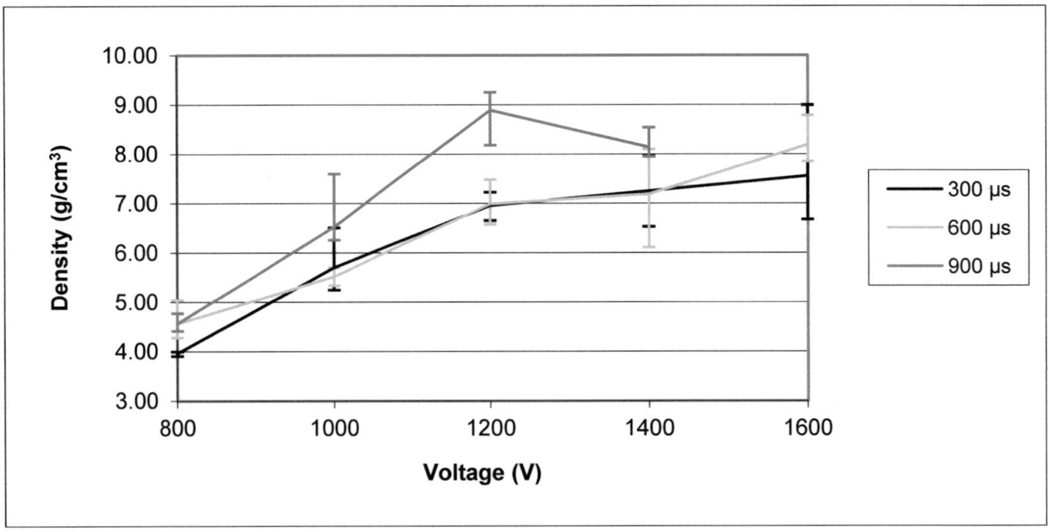

Fig. 1. Graph of Sintered Density vs. Flash Lamp Voltage.

Figure 1 shows that there is a definite optimum voltage setting for the photonic curing of the silver depositions tested. As the density of bulk silver is 10.6 g/cm^3, the average density using settings of 900 µs and 1200V of 8.89 g/cm^3, gives a volume fraction of 83.87% for the sintered silver. Figure 1 also shows that a longer pulse length produces a denser deposition than a shorter pulse. Using flash lamp voltages above 1600V the depositions cracked and experienced blow off due to the high intensity of the light. This was taken as an upper limit to the voltages that can be used to sinter silver films using these pulse lengths.

3.2 Measurement of UV-vis spectrum absorption

To find which wavelengths and how much of the light output by the flash lamp are absorbed by different thicknesses of silver depositions, measurements of the absorption were made by a HP 8452A Diode Array Spectrophotometer. These measurements allowed us to determine the thickness of the silver deposition beyond which the light will no longer penetrate. To make these measurements, depositions were made on a quartz slide to minimize absorption and scattering effects from the substrate. The depositions were made using the GPD aerosol jet system to deposit V2 silver in 1 cm^2 pads. Thicknesses of 1.0, 3.1, and 5.0 μm were measured using the Zeiss Imager M1M microscope and taken to the HP 8452A Diode Array Spectrophotometer for the absorption measurements. The spectrophotometer takes a measurement of the absorption every 2 nm in wavelength, from 190 to 820 nm. The absorption is found by sending a pulse of light through the deposition and substrate to a sensor on the other side. The absorption is measured as the amount of light that does not reach the sensor. The spectrophotometer gives the results as the absorbance. The absorption is found using the equation

$$A = (1 - 10^{-a}) * 100, \quad (1)$$

where A is the absorption and a is the absorbance (Skoog, 1998). The results of the measurements are shown in Figure 2.

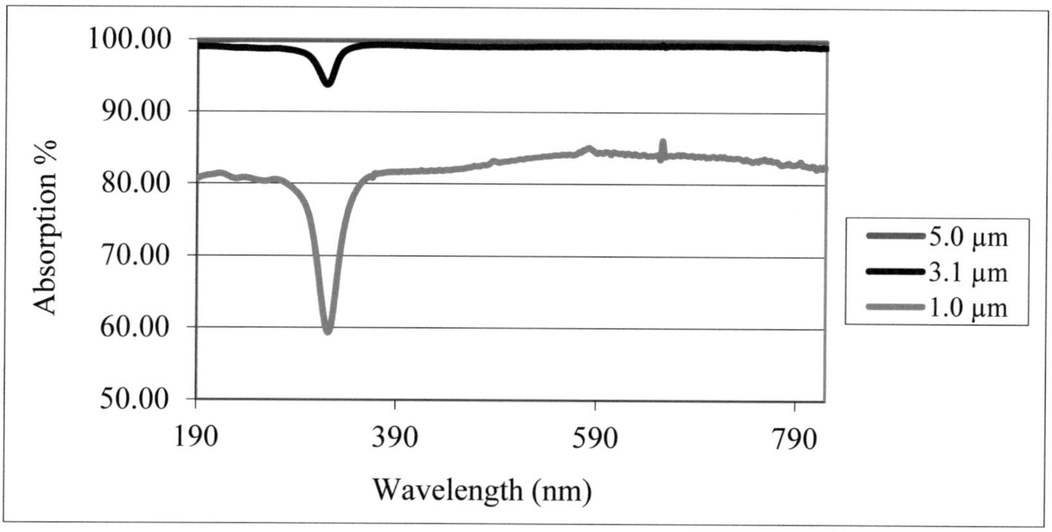

Fig. 2. Absorption vs. Wavelength for different thicknesses of V2 silver depositions.

Figure 2 shows that there is strong absorption of the light by the silver deposition at operational thicknesses. There is a drop in the absorption resulting in a minimum absorption at 322 nm. There is near total absorption of the light using the 3.1 μm thickness with the absorption after the minimum averaging 99%. There is complete absorption of the light for the 5.0 μm thickness. This indicates that if there is any sintering of silver beyond that thickness then it is due to thermal effects as there would be no light to impart energy to the deposition at that depth. The absorption measurements do not show the plasmon resonance peak that has been observed in silver nanoparticles suspended in solution (Chen et al., 2004; Hutter et al., 2001; Lim et al., 2005). This indicates that the absorption for silver nanoparticle films with volume fractions of 30% or higher of silver, more closely mimic the absorption properties of bulk silver than of nanoparticle silver. Figure 2 also shows that the V2 silver absorbs the light strongly in the visible portion of the spectrum which is the area where bulk silver reflects a majority of the light. These measurements show that for the UV and visible spectrums of light the V2 silver consistently absorbs a majority of the light output from the lamp allowing photonic sintering to take place.

3.3 SEM Imaging of microstructure

SEM images were taken of a cross section of photonically cured V2 silver to determine the depth and amount of sintering in a deposition beyond the thickness light will penetrate. Twenty six layers of V2 silver were deposited in a 1 cm^2 area upon a glass slide using the GPD aerosol jet system. The deposition was then cured using a 900 μs, 1200 V pulse of light from the photonic curing system. After etching, the slide was broken in half to provide a cross section of the deposition. One of the slide halves was embedded in an epoxy mold for mounting and polishing followed by examination in the SEM. An image was taken using the Zeiss optical microscope to show the thickness of the deposition in cross section. This image is shown in Figure 3. The thickness of the silver cross section after sintering was measured to be 21.59 μm.

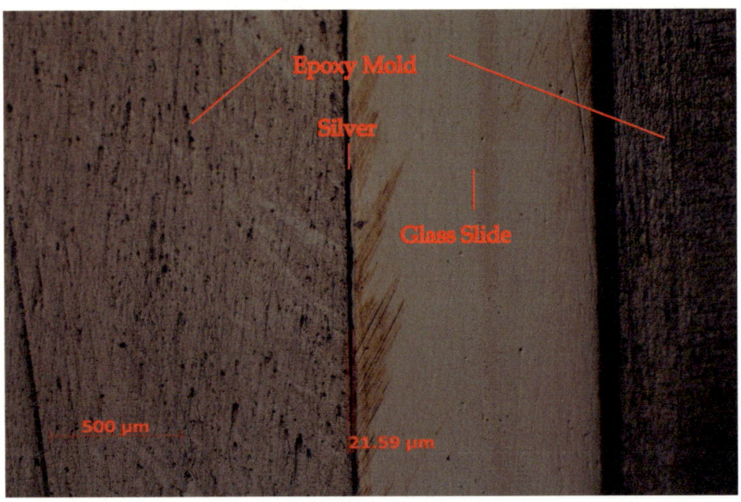

Fig. 3. Optical cross section of the V2 silver on a glass slide in an epoxy mold.

The SEM images that were taken show that slightly over half of the deposition was sintered, while there are substantial voids in the section of deposition closest to the glass. A SEM image of the silver deposition cross section is shown in Figure 4. Figure 4 shows approximately 11 μm of sintered silver which proves that there is a heat transfer effect in photonic curing. This is shown by densification beyond the penetration of the light as the light does not penetrate more than 5 μm into the deposition. The voids in the deposition are a consequence of the polishing procedure and show that that portion of the deposition was not sintered. Figure 4 shows a silver volume fraction of approximately 80%. This is consistent with the densification measurements made in Figure 1.

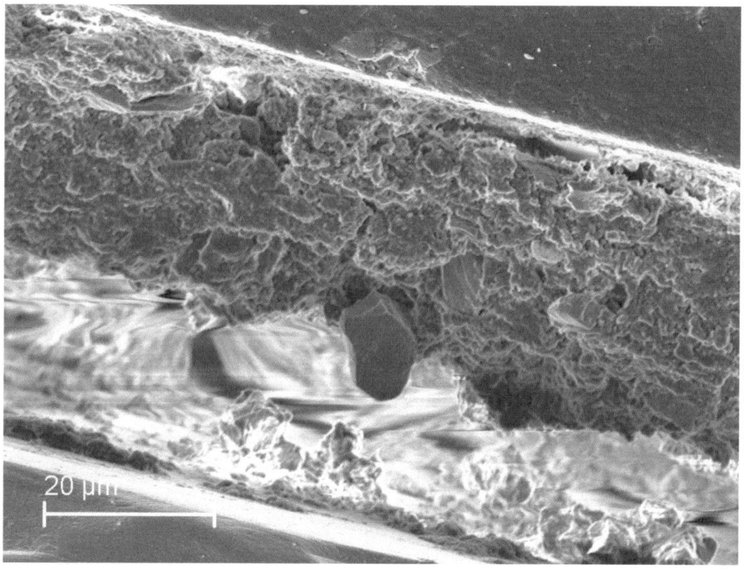

Fig. 4. SEM image of silver deposition cross section.

As thickness measurements were not made prior to sintering the silver, due to problems with the Zeiss microscope software, the actual volume reduction was not found. However, the previous densification measurements can be used to assist in back calculating the thickness of the deposition prior to sintering. Using the observed silver volume fraction of 80% throughout the 11 μm sintered thickness, the 11 μm thickness that was sintered is calculated to have been 24.6 μm thick prior to sintering. 24.6 μm is much thicker than the 5 μm depth at which the light can no longer penetrate. The calculated thickness of the sintered area prior to sintering was combined with the approximately 10.6 μm thickness that was not sintered, which gives the total thickness of the deposition of 35.2 μm.

4. Optical absorption in metallic nanoparticle composites

Effective medium theories have been used to calculate the effective dielectric constant for particles within a medium. The effective medium theories that appear the most often in the literature are the Maxwell-Garnett and Bruggeman effective medium theories. The Maxwell-Garnett effective medium theory was originated in 1904 by J.C. Maxwell Garnett (Maxwell Garnett, 1904). The Maxwell-Garnett theory is only applicable when encountering low volume fractions of material, typically below the percolation threshold or a volume fraction below 30% (Claro & Rojas, 1991; Datta et al., 1993; Davis & Schwartz, 1985; Gibson & Buhrman, 1983; Grechko et al., 2000; Lamb et al., 1980; Li et al., 2006; Lidorikis et al., 2007;

Stroud & Pan, 1978; Wang et al., 2005; Yannopapas, 2006). However, other papers have demonstrated that the Maxwell-Garnett effective medium theory will work for higher volume fractions (Fu et al., 1993; Mallet et al., 2005). The Maxwell-Garnett dielectric function is defined by the expression:

$$\frac{\varepsilon-\varepsilon_b}{\varepsilon+2\varepsilon_b} = f_a \frac{\varepsilon_a-\varepsilon_b}{\varepsilon_a+2\varepsilon_b}, \tag{2}$$

where ε is the dielectric function of the effective medium, ε_a is the dielectric function of the particles in the medium, ε_b is the dielectric function of the medium, and f_a is the volume fraction of the particles contained in the medium (Aspnes, 1982).

The Bruggeman effective medium approximation was formulated in 1935 by D.A.G. Bruggeman (Bruggeman, 1935). Unlike the Maxwell-Garnett model the Bruggeman model does not have a preference toward the embedded particle or the medium, the Bruggeman model treats the inclusions and medium equally (Aspnes, 1982; Landauer, 1978). The Bruggeman model has been used to calculate the dielectric constant of particles embedded in materials and compared with the measured values (Brosseau, 2002; Chen, L.Y. & Lynch, 1987; Mendoza-Galván et al., 1994; Rousselle et al., 1993). The Bruggeman model has also been modified to better model the dielectric constant of nonlinear media and different shaped particles (Goncharenko, 2003; Pellegrini 2001). The Bruggeman dielectric function has been shown to work well with higher volume fractions (Bohren & Huffman, 1983; Bruggeman, 1935). The Bruggeman dielectric function is defined by the expression:

$$f_a \frac{\varepsilon_a-\varepsilon}{\varepsilon_a+2\varepsilon} + f_b \frac{\varepsilon_b-\varepsilon}{\varepsilon_b+2\varepsilon} = 0, \tag{3}$$

where ε is the dielectric function of the effective medium, ε_a is the dielectric function of the particles in the medium, ε_b is the dielectric function of the medium, f_a is the volume fraction of the particles contained in the medium, and f_b is the volume fraction of the particles contained in the medium (Aspnes, 1982). Once the average dielectric function is found, the imaginary portion of the refractive index and the absorption coefficient for the effective medium can be calculated by the expression:

$$\alpha' = \frac{4\pi k}{\lambda}, \tag{4}$$

where k is the imaginary part of the refractive index and λ is the wavelength of light (Skoog et al., 1998). From this, using the Beer-Lambert law (Skoog et al., 1998), the absorption spectra can be calculated.

To test the models, the absorption coefficients for 30% silver in air was calculated using the Maxwell-Garnett and Bruggeman models and compared with the absorption coefficient for bulk silver. The comparison is shown in Figure 5. From Figure 5 it is observed that the absorption coefficients obtained using the effective medium models are lower than for bulk silver. In the UV range of the spectrum the effective medium models match well and have the same dip at the 318 nm wavelength. The absorption coefficient minimum at 318 nm corresponds to the absorption minimum observed in the UV-Vis measured absorption. This indicates the measured absorption dip is a property of bulk silver and not an effect of the plasmon resonance. After the absorption coefficient dip the data from the models begins to diverge. The absorption coefficient from the Maxwell-Garnett model rises sharply, becomes higher than for bulk silver, and peaks at the 380 nm wavelength mark. This peak in the absorption coefficient is near the surface plasmon resonance that has been seen in the literature for silver nanoparticles dispersed in an aqueous solution (Chen, T.C. et al., 2004; Hutter et al., 2001; Lim et al., 2005). The Maxwell-Garnett data then drops rapidly to near zero for the remainder of the modeled wavelength area. After the 318 nm wavelength mark the Bruggeman data gently rises for the remainder of the wavelengths modeled. Thus the Bruggeman model will most closely matches the measured data in our experiments.

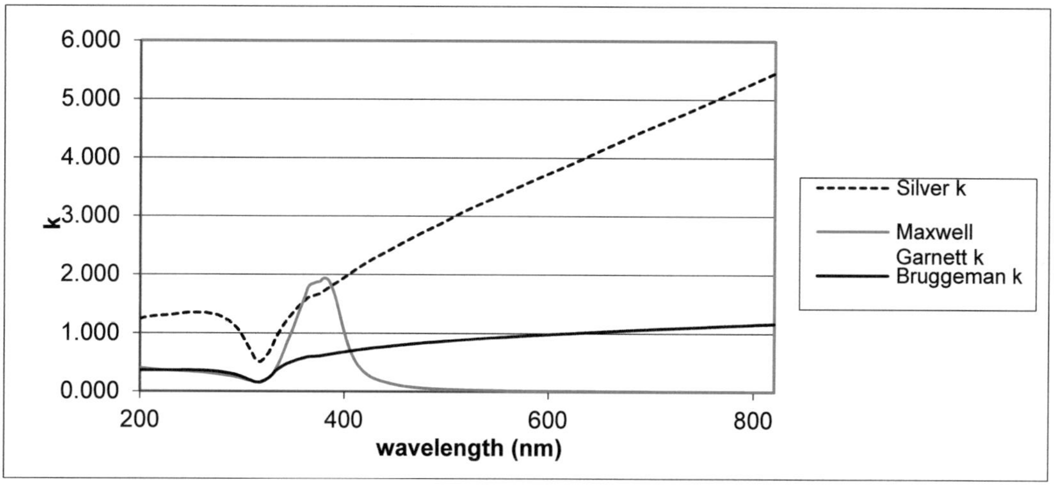

Fig. 5. Comparison of the absorption coefficients for 30% silver in air using Maxwell-Garnett and Bruggeman effective medium models with the values for bulk silver.

To obtain a closer look at how the Bruggeman absorption curve and the measured absorption curve compare, the thickness of the deposition for the Bruggeman model was set to 160 nm and is shown in Figure 6 compared with the 1.3 μm thick measured absorption. The absorption minimums in Figure 6 are 62% for the Bruggeman model at 316 nm and 64% at 322 nm for the measured silver. In the UV and near IR regimes the absorption curves have a similar slope. After the absorption minimum the Bruggeman model absorption is relatively constant through the visible spectrum while the measured data rises constantly before leveling off. Though the absorption as modeled using the Bruggeman effective medium theory is stronger than what was measured, unlike the Maxwell-Garnett model, the Bruggeman model closely follows the measured absorption dip and does not decrease significantly at longer wavelengths. The discrepancy in the theoretical and experimental thicknesses would suggest that most of the absorption takes place near the surface of the film. However, the spectral characteristics of the Bruggeman model reproduce very well the experimental data, and thus were was used to model the absorption of the deposited materials in this work.

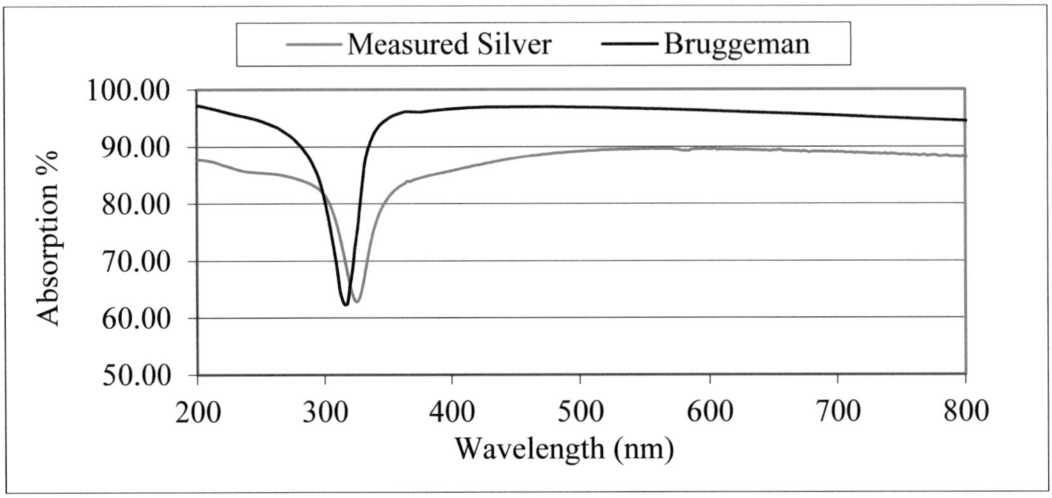

Fig. 6. Absorption vs. wavelength for 30% volume fraction silver, 160 nm thick film using Bruggeman model and a 1.3 µm thick measured silver film.

4.1 Modifying the bruggeman model with the extended drude model

As the calculated absorption using the Bruggeman model does not perfectly match the measured absorption, the complex dielectric function was modified using the extended Drude model in an attempt to bring the calculated and measured absorption closer together (Haiss et al., 2007; Kreibig & Vonfrags, 1969). The extended Drude model was used to modify the dielectric function of silver for particle diameters of 25 nm and 5 nm. The absorption was then calculated using a volume fraction of 30% and a film thickness of 150 nm and compared with the measured absorption and the unmodified Bruggeman model. This is shown in Figure 7. The thickness of the film was thinned to 150 nm to bring the modified Bruggeman model with the 25 nm diameter silver particles in line with the measured silver absorption. Figure 7 shows that in the UV portion of the spectrum the particle size did not affect the absorption of the silver film. The particle size did affect the absorption minimum slightly with the minimum occurring at 314 nm for the 5 nm diameter silver with a minimum absorption of 68.5%. The 5 nm diameter silver also had a calculated absorption in the visible range slightly below the unmodified Bruggeman and Drude modified Bruggeman absorptions. The 25 nm diameter calculated absorption using the extended Drude and Bruggeman models agreed well with the unmodified Bruggeman calculated absorption in the UV and visible ranges. The main difference was that the absorption minimum at 316 nm was 3.5% stronger for the Drude modified Bruggeman model than for the unmodified Bruggeman absorption. These comparisons indicate that the particle size does not play a large role in the absorption by nanoparticle films with volume fractions of 30% or greater. The nanoparticle silver films appear to act more like bulk silver than like individual silver nanoparticles. This is shown through a surface plasmon resonance peak not appearing in the measured absorptions. The reason for the absorption appearing to be so strong in the calculated absorptions as opposed to the measured data could be due to the majority of the absorption occurring within the first few hundred nanometers of the films, with the absorption only slightly increasing beyond that depth.

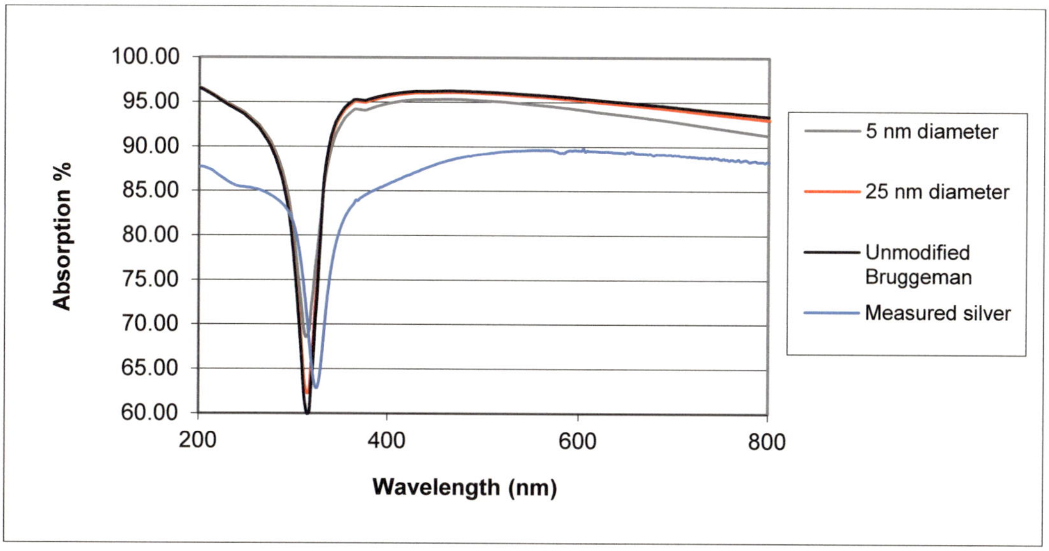

Fig. 7. Absorption vs. Wavelength for a 30% volume fraction of silver, 150 nm thick film calculated using the unmodified Bruggeman model, the Bruggeman model modified by the extended Drude model for particle diameters of 5 nm and 25 nm, and for a 1.3 μm thick measured silver film.

5. Modeling of sintering processes using fluent

The Gambit modeling software and Fluent heat transfer software were used to create a slab of material and model the heat transfer through the slab. Fluent and Gambit are two software programs that model fluid flow and heat transfer through solids and fluids. In Gambit the volume and boundaries are created and then the created volume is transferred to Fluent. In Fluent the material and boundary conditions are input, the energy equation and other thermal models are activated, and the simulation is run. Using the Gambit and Fluent software, a 10 μm thick, 100 x 100 μm wide and long slab of silver was created that would absorb radiative energy along its top surface and was surrounded by air and had a substrate of glass. In Fluent the temperature of the silver and the amount of silver melted as a result of the energy absorption were simulated.

The Fluent simulation used boundary conditions of convective cooling for the sides and top of the silver slab, conductive cooling through the glass substrate, and radiative heating on the top surface of the slab. The volume fraction of silver in the slab was set to 0.33 with the remainder of the volume taken up with air. The melting point of the silver was set to the observed value of 600°C. The ambient temperature surrounding the slab was set to 21°C and the glass substrate thickness was set to 18 μm for computational efficiency. The simulated silver slab was heated using a flash lamp voltage of 1200V and a pulse duration of 900 μs. The heat transfer coefficient between the silver and the air was set to 3500 W/m^2K. Following heating, the slab was allowed to cool for 900 μs. The temperature and the amount of melting was measured every 100 μs. The temperature profile of the silver slab from the simulation is shown in Figure 8.

From Figure 8 it is shown that the simulated temperature reaches the melting point of the nanosilver by 200 μs. After spending 500 μs at 600°C the temperature of the slab begins to rise again. Once the flash lamp pulse is turned off the silver slab cools rapidly. The silver slab solidifies within 100 μs and cools to a temperature below 100°C within 800 μs of the pulse ceasing which matches observations of the slab being room temperature upon immediate removal from the photonic sintering system.

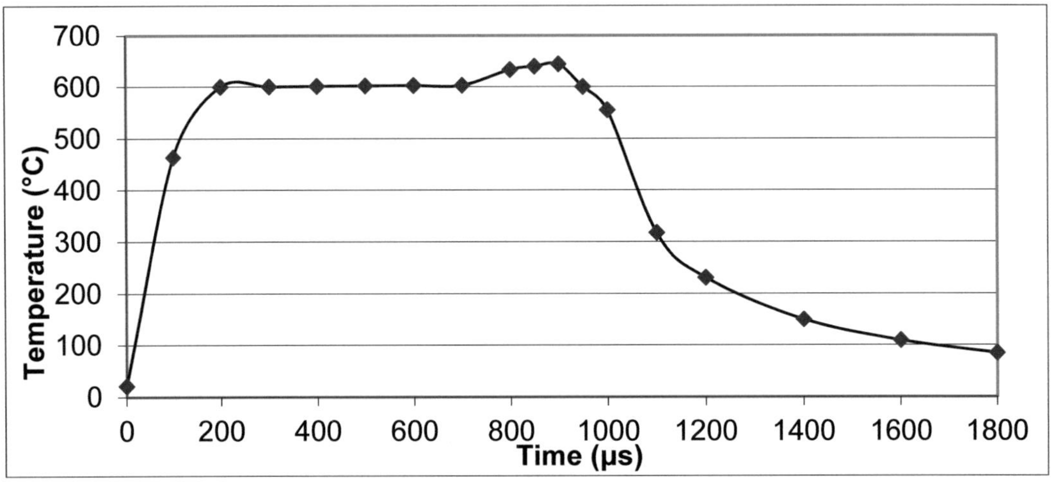

Fig. 8. Temperature vs. Time for a 10 μm thick silver slab on a glass substrate with a flash lamp voltage of 1200V for 900 μs.

To complement the temperature profile, the liquid fraction at the top, center, and bottom of the slab was also modeled. The liquid fraction of the slab is shown in Figure 9. The silver has slightly melted at 200 μs and by 700 μs the top surface of the silver has completely melted.

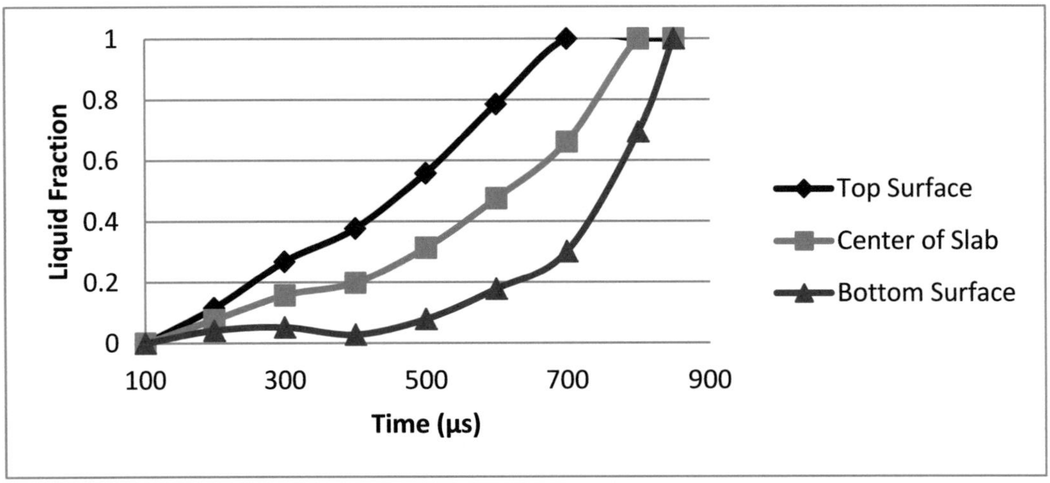

Fig. 9. Liquid Fraction vs. Time for three positions on the silver slab.

The bottom surface of the silver takes until 850 μs to completely melt. After the top surface of the silver has completely melted the temperature begins to rise again as seen in Figure 8. This matches the premise based on the densification measurements and simulation data that at 300 μs the silver has just begun to melt, the melting is still in progress at 600 μs, and the material has completely melted by 900 μs. This gives a relatively slow melting time allowing the material to sinter at a rate that will not crack or ablate the material off the substrate.

After simulating the temperature of the 10 μm thick silver slab on glass the silver slab was thinned to 3 μm, as 3 μm is the typical deposition thickness. For the thinner slab some conditions were modified to make them more accurate. The porosity of the silver slab remained at 33%. The heat transfer coefficient was set to 3100 W/m²K as the simulations on glass showed the silver remaining near 600°C. The emissivity was set to 0.99 to allow the appropriate amount of the pulse to pass through the deposition. The melting temperature remained at 600°C and the ambient temperature was set to 21°C. The flash lamp voltage was set to 1200V and the pulse was set to a length of 900 μs, after which 900 μs of cooling was simulated. The thickness of the glass substrate was increased to 20 μm. The resultant graph of temperature versus time is shown in Figure 10. From observing the liquid fraction the simulation indicates that for the thinner slab the deposition has melted 93% after the pulse has been active for 100 μs. The temperature reaches 701°C by 200 μs and then decreases to 644°C at 300 μs and proceeds to increase at a constant rate for the duration of the pulse. Once the pulse ends the temperature rapidly decreases. One difference from using the 3 μm slab instead of the 10 μm slab is the early peak of slab temperature at 200 μs. Another difference is that the thinner slab melts quicker, which is expected as the thin slab is absorbing nearly the same amount of energy and has less mass to heat.

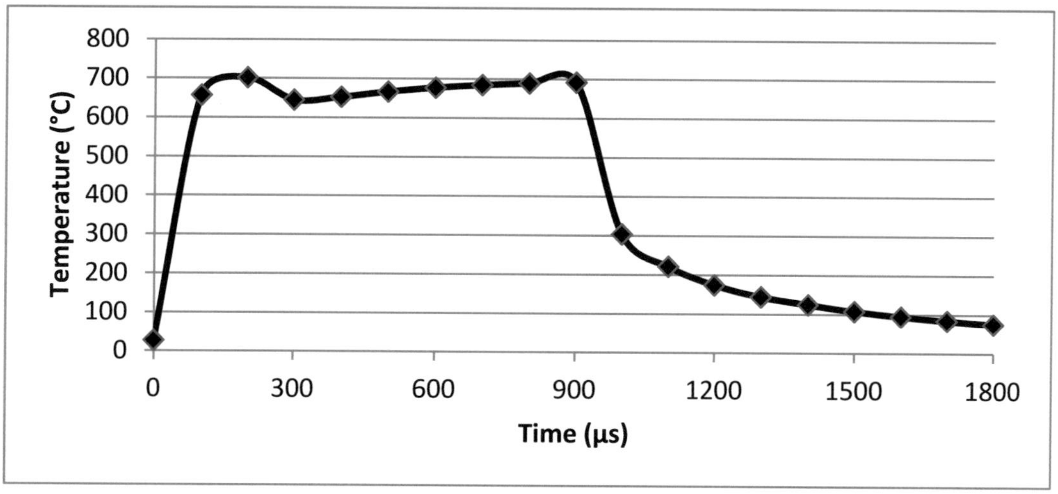

Fig. 10. Temperature vs. time for a 3 μm thick silver slab using a 1200V flash lamp voltage and a 900 μs pulse length using Fluent.

6. Conclusions

In examining the photonic sintering process a number of interesting results were found. It was shown that there is significant densification of the photonically sintered silver, with the photonically sintered silver reaching 84% of the density of bulk silver. It was observed that there is an optimum flash lamp voltage, beyond which the silver nanoparticle film becomes damaged through cracking and blow off. UV-Vis spectroscopy shows that the silver nanoparticle films absorb in excess of 80% of the light emitted by the flash lamp at operational thicknesses, with total absorption occurring at a thickness of 5 μm. The UV-Vis spectroscopy also showed that the light absorption by the uncured silver films is broadband, and that the plasmon resonance is not the dominating feature of the absorption. This indicates the absorption of the silver films acts like that of bulk silver rather than that of individual silver nanoparticles. Sintering a 35 μm thick silver deposition with the optimal pulse of 1200 V and pulse length of 900 μs, the deposition decreased in thickness to 21.6 μm. Of the sintered thickness of 21.6 μm, 11 μm was sintered to a volume fraction of 80% dense, while 10.6 μm of the sintered thickness did not sinter. This proves that photonic sintering involves heat transfer between nanoparticles and that the photonic sintering process does not solely depend on the amount of light absorbed by each nanoparticle.

After deriving the equations and calculating the light absorption by silver nanoparticle films based on the Maxwell-Garnett effective medium theory and the Bruggeman effective medium approximation, the Bruggeman model matched the measured absorption the best. This result shows that to obtain the best match with the measured data, the nanoparticles should be considered as an effective medium, and that the effective medium theory used should be valid above the percolation threshold. Modifying the Bruggeman model with the extended Drude model showed that the absorption does not have a strong dependence on the particle size. The absorption models indicate that the majority of the absorption takes place within a few hundred nanometers of the surface of the film.

The temperature simulations using the heat transfer software Fluent indicate that the nanoparticle films melt within 300 μs from the beginning of the pulse during the photonic sintering process. This result indicates that traditional solid state sintering models do not apply to photonic sintering. The Fluent simulations also showed that the temperature throughout the deposition does not vary to a wide extent. However, the liquid fraction of silver at various locations within the film can vary greatly depending on the speed of the melting process and the thickness of the silver layer.

7. References

Aspnes, D.E. (1982). Local-field effects and effective-medium theory: A microscopic perspective. *Am. J. Phys.* Vol. 50, No. 8, pp. 704-709.

Bohren, C. F. and Huffman, D. R. (1983). *Absorption and Scattering of Light by Small Particles.* New York: John Wiley & Sons, Inc.

Brosseau, C. (2002). Generalized effective medium theory and dielectric relaxation in particle-filled polymeric resins. *Journal of Applied Physics.* Vol. 91, No. 5, pp. 3197-3204.

Bruggeman, D.A.G. (1935). Berechnung verschiedener physikalischer Konstanten von heterogenen Substanzen. *Annalen der Physik (Leipzig).* Vol. 24, pp. 636-679.

Camm, D.M., Gelpey, J.C., Thrum, T., Stuart, G.C., and McCoy, S. (2006). Flash-Assist RTP for Ultra-Shallow Junctions. *JOM,* Vol. 58, No. 6, pp. 32-34.

Carter, M. and Sears, J. (2007). Photonic Curing for Sintering of Nano-Particulate Material. *Advances in Powder Metallurgy & Particulate Materials – 2007: Proceedings of the 2007 International Conference on Powder Metallurgy & Particulate Materials, May 13-16, Denver, Colorado.*

Chen, L.Y., Lynch, D.W. (1987). Effect of liquids on the Drude dielectric function of Ag and Au films. *Physical Review B*. Vol. 36, No. 3, pp. 1425-1431.

Chen, T.C., Su, W.K., and Lin, Y.L. (2004). A Surface Plasmon Resonance Study of Ag Nanoparticles in an Aqueous Solution. *Japanese Journal of Applied Physics*, Vol. 43, pp. L119-L122.

Chen, T.C., Su, W.K., and Lin, Y.L. (2004). A Surface Plasmon Resonance Study of Ag Nanoparticles in an Aqueous Solution. *Japanese Journal of Applied Physics*. Vol. 43, pp. L119-L122.

Claro, F., Rojas, R. (1991). Correlation and multipolar effects in the dielectric response of particulate matter: An iterative mean-field theory. *Physical Review B*. Vol. 42, No. 8, pp. 6369-6375.

Datta, S., Chan, C.T., Ho, K.M., Soukoulis, C.M. (1993). Effective dielectric constant of periodic composite structures. *Physical Review B*. Vol. 48, No. 20, pp. 14936-14943.

Davis, V.A., Schwartz, L. (1985). Electromagnetic propagation in close-packed disordered suspensions. *Physical Review B*. Vol. 31, No. 8, pp. 5155-5165.

Farnsworth, Stan. Curing Copper and other Thin-Film Materials at Production Speeds. *SEMICON West*. Moscone Center, San Francisco. 16 July 2009.

Fu, L., Macedo, P.B., Resca, L. (1993). Analytic approach to the interfacial polarization of heterogeneous systems. *Physical Review B*. Vol. 47, No. 20, pp. 13818-13829.

Gibson, U.J. and Buhrman, R.A. (1983). Optical response of Cermet composite films in the microstructural transition region. *Physical Review B*. Vol. 27, No. 8, pp. 5046-5051.

Goncharenko, A.V. (2003). Generalizations of the Bruggeman equation and a concept of shape-distributed particle composites. *Physical Review E*. Vol. 68, 041108.

Grechko, L.G., Whites, K.W., Pustovit, V.N., Lysenko, V.S. (2000). Macroscopic dielectric response of the metallic particles embedded in host dielectric medium. *Microelectronics Reliability*. Vol. 40, pp. 893-895.

Haiss, W., Thanh, N., Aveyard, J., and Fernig, D. (2007). Determination of Size and Concentration of Gold Nanoparticles from UV-Vis Spectra. *Analytical Chemistry*. Vol. 79, pp. 4215-4221.

Hutter, E., Fendler, J.H., and Roy, D. (2001). Surface Plasmon Resonance Studies of Gold and Silver Nanoparticles Linked to Gold and Silver Substrates by 2-Aminoethanethiol and 1,6-Hexanedithiol. *J. Phys. Chem. B.*, Vol. 105, pp. 11159-11168.

Hutter, E., Fendler, J.H., and Roy, D. (2001). Surface Plasmon Resonance Studies of Gold and Silver Nanoparticles Linked to Gold and Silver Substrates by 2-Aminoethanethiol and 1,6-Hexanedithiol. *J. Phys. Chem. B.* Vol 105, pp. 11159-11168.

Kim, H.S., Dhage, S.R., Shim, D.E., and Hahn, H.T. (2009). Intense pulse light sintering of copper nanoink for printed electronics. *Applied Physics A*, Vol. 97, pp. 791-798.

Kreibig, U. and Vonfrags. C. (1969). The Limitation of Electron Mean Free Path in Small Silver Particles. *Z. Physik*. Vol. 224, pp. 307-323.

Lamb, W., Wood, D.M., Ashcroft, N.W. (1980). Long-wavelength electromagnetic propagation in heterogeneous media. *Physical Review B*. Vol. 21, No. 6, pp. 2248-2266.

Landauer, R. (1978). Electrical conductivity in inhomogeneous media. *AIP Conference Proceedings.* Vol. 40, No. 1, pp. 2-45.

Li, J., Sun, G., Chan, C.T. (2006). Optical properties of photonic crystals composed of metal-coated spheres. *Physical Review B*. Vol. 73, 075117.

Lidorikis, E., Egusa, S., Joannopoulos, J.D. (2007). Effective medium properties and photonic crystal superstructures of metallic nanoparticle arrays. *Journal of Applied Physics.* Vol. 101, 054304.

Lim, S.K., Chung, K.J., Kim, C.K., Shin, D.W., Kim, Y.H., and Yoon, C.S. (2005). Surface-plasmon resonance of Ag nanoparticles in polyimide. *Journal of Applied Physics*, Vol. 98, No. 8, 084309.

Lim, S.K., Chung, K.J., Kim, C.K., Shin, D.W., Kim, Y.H., and Yoon, C.S. (2005). Surface-plasmon resonance of Ag nanoparticles in polyimide. *Journal of Applied Physics*. Vol. 98, No. 8, 084309.

Mallet, P., Guérin, C.A., Sentenac, A. (2005). Maxwell-Garnett mixing rule in the presence of multiple scattering: Derivation and accuracy. *Physical Review B*. Vol. 72, 014205.

Maxwell Garnett, J. C. (1904). Colors in metal glasses and in metallic films. *Philos. Trans. R. Soc.*, Vol. A203, pp. 385-420.

Mendoza-Galván, A., Martínez, G., Martínez, J.L. (1994). Effective dielectric function modeling of inhomogeneous and anisotropic silver films. *Physica A*. Vol. 207, pp. 365-371.

Novacentrix. (2007). *PCS-1100 Photonic Curing System Operations and Safety Manual*. Austin: Novacentrix Corp.

Novacentrix. (2009). *PulseForge 3100 Development and Production: Metals Processing*. Available from <http://www.novacentrix.com/product/pf3100.php>.

Novacentrix. (2009). *PulseForge 3300 Development and Production: Semiconductor and Photovoltaic Materials Processing*. Available from <http://www.novacentrix.com /product/pf3300.php>.

Pellegrini, Y.P. (2001). Self-consistent effective-medium approximation for strongly nonlinear media. *Physical Review B*. Vol. 64, 134211.

Rousselle, D., Berthault, A., Acher, O., Bouchaud, J.P., Zérah, P.G. (1993). Effective medium at finite frequency: Theory and experiment. *Journal of Applied Physics*. Vol. 74, No. 1, pp. 475-479.

Schroder, K.A., McCool, S.C., and Furlan, W.R. (2006). Broadcast Photonic Curing of Metallic Nanoparticle Films. *Technical Proceedings of the 2006 NSTI Nanotechnology Conference and Trade Show, Volume 3,* pp. 198-201.

Skoog, D., Holler, F. J., Nieman, T. (1998). *Principles of Instrumental Analysis*. Thompson Learning.

Stroud, D., Pan, F.P. (1978). Self-consistent approach to electromagnetic wave propagation in composite media: Application to model granular metals. *Physical Review B*. Vol. 17, No. 4, pp. 1602-1610.

Wang, J., Lau, W.M., Li, Q. (2005). Effects of particle size and spacing on the optical properties of gold nanocrystals in alumina. *Journal of Applied Physics*. Vol. 97, 114303.

Yannopapas, V. (2006). Effective-medium description of disordered photonic alloys. *J. Opt. Soc. Am. B*. Vol. 23, No. 7, pp. 1414-1419.

9

Microwave-Induced Combustion Synthesis of Luminescent Aluminate Powders

A. Potdevin[1,2], N. Pradal[1,3], M.-L. François[1,3],
G. Chadeyron[1,2], D. Boyer[1,2] and R. Mahiou[1,3]
[1]*Clermont Université, Université Blaise Pascal,*
Laboratoire des Matériaux Inorganiques, Clermont-Ferrand,
[2]*Clermont Université, ENSCCF, Laboratoire des Matériaux Inorganiques,*
Clermont-Ferrand,
[3]*CNRS, UMR 6002, LMI, F-63177 Aubière,*
France

1. Introduction

Due to the current environmental directives concerning the use of hazardous mercury in lighting devices and the necessity to save global energy, new economical and eco-friendly light sources are being developed. In this context, phosphor-converted light emitting diodes (LEDs) appear as the most interesting candidates to replace energy-greedy incandescent light bulbs and fluorescent lamps. Indeed, it is now a mature technology that can compete with the former ones, especially thanks to their long lifetime, low-power consumption, high energy efficiency and robustness (Höppe 2009; Huang et al. 2009; Smet et al. 2011). However, some drawbacks remain, such as the lack of high colour rendering devices.

Until recently, white LEDs were almost only based on the association of blue GaN LED with a yellow phosphor, $Y_3Al_5O_{12}:Ce^{3+}$ (YAG:Ce). These devices are characterized by a high colour temperature due to a lack of red contribution. Another approach consists in combining (Ga,Al)N near ultraviolet (UV) pumping LEDs with red, green and blue phosphors. This gives access to a wide range of colours and controlled colour rendering. The optical performances of the LEDs-based light sources rely on the physical characteristics of the semi-conductor as well as the features of the phosphors. In particular, the luminescence properties of phosphors are known to be dependent on their morphology, their size and their crystallinity. Defects have also to be avoided because they act as luminescence shifters or quenchers. Furthermore, it has been proved in several recent studies that white LEDs using smaller particles size, with narrower grain size distribution, required a lower amount of phosphor to get similar efficiency than LEDs using bigger particles (Huang et al. 2009; Jia et al. 2007). It has been related to the reduction of internal light scattering when these nanoparticles are directly coated onto the LEDs surface at the front of this.

Based on this behaviour, it seems necessary to develop highly luminescent nanosized or nanostructured phosphors, easy to shape as luminescent films. Among the phosphors suitable for white LEDs, we have chosen to work on YAG:Ce and $BaMgAl_{10}O_{17}:Eu^{2+}$ (BAM:Eu) for blue and UV LEDs-based devices respectively. Moreover, together with its possible application in UV-LEDs based lighting devices, BAM:Eu represents the most interesting blue phosphor for Hg-free lamps based on Xe-Ne plasma and for plasma display panels(Chen & Yan 2007; Shen et al. 2010; Smet et al. 2011).

Traditionally, these phosphors are obtained via energy-greedy solid-state reactions involving very high synthesis temperatures, associated with grinding and milling (Shionoya 1998). These synthesis routes lead to microsized particles exhibiting surface defects and a large particle size distribution, which is prejudicial to optical performances. The necessity to develop nanostructured phosphors has entailed the emergence of numerous chemical synthesis techniques for these aluminates: solvothermal, sol-gel, co-precipitation, spray pyrolysis...(Kang et al. 1998; Lee et al. 2009; Lu et al. 2006; Pan et al. 2004; Potdevin et al. 2007; Zhang et al. 2010). These methods involve a homogeneous mixing of metal ions by using precursors in solution. Next to them, the combustion process appears as an attractive alternative thanks to its simplicity and its low cost; in particular, solution combustion synthesis (SCS) methods allow to prepare various nanosized oxides of high purity, using different organic fuels (Mukasyan et al. 2007).

Recently, a new kind of SCS routes has been used for the synthesis of oxides: microwave-induced SCS (MISCS). The use of microwave in combustion synthesis presents several advantages (shortened synthesis times, enhanced reaction kinetics...) thanks to dissimilar heating mechanisms compared to the conventional combustion synthesis. For example, microwave radiations are immediately absorbed by the entire sample and interact from bulk to the surface leading to volumetric heating in contrast with conventional sintering in which heating takes place by radiation or conduction of the heat energy from the surface to the core. This process has been recently used for the preparation of various oxides as well as luminescent aluminates such as YAG:Ce and BAM:Eu (Chen et al. 2009; Fu 2006; Jung & Kang 2010).

In this chapter, we report the elaboration of undoped and Ce^{3+}-doped YAG powders as well as BAM:Eu ones by MISCS. Different amounts of urea, employed as fuel, were used and a comparative investigation has been carried out on phosphors properties regarding this content. Structural, morphological and optical features of powders have been studied by means of X-ray diffraction (XRD), infrared (IR) spectroscopy, scanning and transmission electron microscopies (TEM/SEM) and photoluminescence. Morphological and optical properties have also been compared with those of commercial phosphors.

2. Experimental section

2.1 Synthesis procedures

The two synthesis procedures used for YAG and BAM phases were very close to each other. They are based on the works of Y.-P. Fu and Z. Chen *et al.*, respectively (Chen et al. 2009; Fu 2006). For sake of clarity, the differences between the two procedures will be detailed in the following paragraphs.

For both processes, employed precursors are metal nitrates and urea, used as oxidizers and fuel, respectively. Metal nitrates were weighed stoichiometrically, dissolved in distilled water in a beaker and stirred in order to obtain a clear solution. Then, urea was added with different fuel to oxidizer molar ratios (f/o). The theoretical (stoichiometric) molar ratio of urea to nitrate $(f/o)_{th}$ was determined for each process, based on the total oxidizing and reducing valencies of the oxidizer and the fuel (Jain et al. 1981). Then, the solution was either poured into a porcelain crucible or maintained in the beaker and placed into a domestic microwave oven always operating at its maximum power setting (900 W). The solution boiled, underwent a dehydration followed by nitrates and urea decomposition, leading to the formation of a large amount of gases (N_2, CO_2 and H_2O). Then, the medium reached the point of spontaneous combustion and ignited with significant swelling. The whole process, very exothermic, lasted a few minutes and resulted into foamy and friable powders.

2.1.1 $Y_3Al_5O_{12}$ and $Y_3Al_5O_{12}$:Ce^{3+}

$Y_3Al_5O_{12}$ and $Y_{2.97}Ce_{0.03}Al_5O_{12}$ powders were prepared by MISCS using $CO(NH_2)_2$ (Prolabo, purity 99.5%) as fuel and $Y(NO_3)_3{\cdot}6H_2O$ (Sigma-Aldrich, purity 99.9%), $Al(NO_3)_3{\cdot}9H_2O$ (Sigma-Aldrich, purity 99.997%) and $Ce(NO_3)_3{\cdot}5H_2O$ (Sigma-Aldrich, purity 99,9%) as metallic precursors. The $(f/o)_{th}$ in this case was determined to be 2.55/1 from the following equation (Fu 2006):

$$3Y(NO_3)_3 + 5Al(NO_3)_3 + 20CO(NH_2)_2 \rightarrow Y_3Al_5O_{12} + 20CO_2 + 40H_2O + 32N_2 \quad (1)$$

The optimal quantity of fuel has been optimized from the undoped matrix by investigating four molar ratios (f/o) [$(f/o)_{th}$, $2(f/o)_{th}$ $2.5(f/o)_{th}$ and $3(f/o)_{th}$], using the same batch of nitrate solution, divided into four beakers. Depending on the urea amount, the synthesis does not present the same characteristics. Thus, when $(f/o)=(f/o)_{th}$, no flame was observed during the microwave induced heating and the obtained powder is white. For the other molar ratios, an orangey flame is produced, as shown in Fig. 1. The obtained powders are characterized by a grey colour (mixture of grey and white powders for $2(f/o)_{th}$) (for undoped YAG).

Fig. 1. Picture of a porcelain crucible containing nitrates/urea solution, ignited after few minutes of microwave irradiation.

Thanks to the XRD patterns of undoped YAG (detailed subsequently in this chapter), the amount of urea used to prepare YAG:Ce powder wad fixed to $3(f/o)_{th}$. The powder issued from this process presents a grey and a white part. When exposed to UV radiation (λ_{exc}=365 nm), only the grey part exhibits a yellow fluorescence, characteristic of Ce^{3+} ions embedded in the YAG matrix (see Fig. 2).

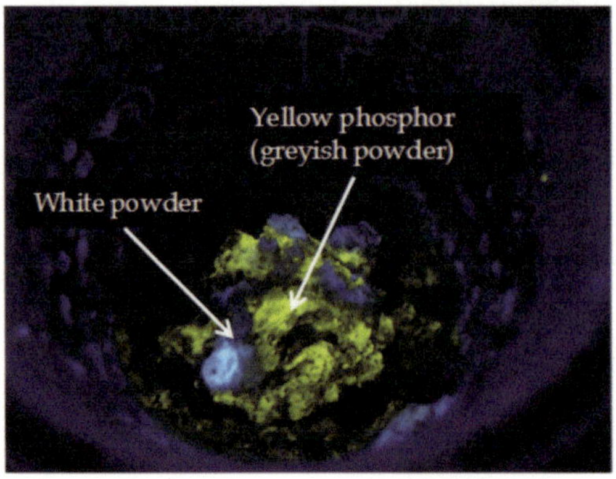

Fig. 2. Picture of as-synthesized YAG:Ce powder upon excitation at 365nm for $3(f/o)_{th}$.

This grey colour is usually characteristic of organic residues. Unfortunately, these organic residues are generally a nuisance for the optical properties since they usually act as quenching sites and result in a strong decrease of the luminescence intensity. Consequently, the YAG:Ce powder was submitted to either an acid washing or a post-calcination in air. The wash does not provoke any modification of the powder but a calcination stage of two hours at 1000°C allows obtaining a yellow powder as evidenced in the Fig. 3.

Fig. 3. Picture of as-synthesized YAG:Ce powder before(a) and after(b) post-calcination at 1000°C for 2 hours.

Hence, optical performances were assessed for the two YAG:Ce powders as well as their structural and morphological properties, in order to evidence the influence of the post-heating treatment.

2.1.2 $BaMgAl_{10}O_{17}:Eu^{2+}$

$Ba_{0.9}Eu_{0.1}MgAl_{10}O_{17}$ powders were prepared by MISCS using $CO(NH_2)_2$ (Prolabo, purity 99.5%) as fuel and $Ba(NO_3)_2$ (Sigma-Aldrich, purity 99+%), $Mg(NO_3)_2{\cdot}6H_2O$ (Acros organics, purity 99+%), $Al(NO_3)_3{\cdot}9H_2O$ (Sigma-Aldrich, purity 99.997%) and $Eu(NO_3)_3 \cdot 5H_2O$ (Sigma-

Aldrich, purity 99.9%) as metallic precursors. The $(f/o)_{th}$ is in this case 2.36/1. Two molar ratios were investigated: $(f/o)_{th}$ and $3(f/o)_{th}$. Contrary to the YAG synthesis, the combustion is characterized by a white flame and obtained powders are white and, when exposed to UV radiation (λ_{exc}=254 nm), they are composed of blue and red parts, as shown in Fig. 4.

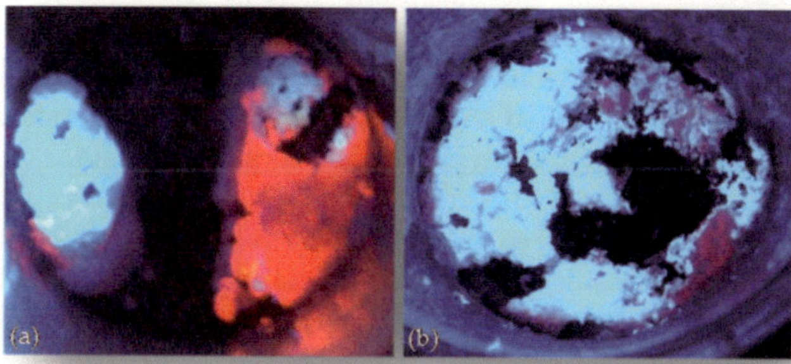

Fig. 4. Pictures of as-synthesized powders upon excitation at 254nm for (a) $(f/o)_{th}$ and (b) $3(f/o)_{th}$.

Since the part intended as phosphors for lighting devices is the blue one, this latter has been isolated before being characterized. This as-synthesized blue part contains a little fraction of $BaAl_2O_4$ (as seen later in the XRD study), consequently an additional washing step has been carried out to purify BAM phase and theoretically improve its optical performances. Thus, this powder emitting in the blue range was washed several times with acidic distilled water before being themselves characterized.

2.2 Characterization techniques

XRD patterns concerning undoped YAG and YAG:Ce phases were performed on a Siemens D500 and a Siemens D5000 diffractometers, respectively, whereas those of BAM:Eu were recorded on a Philips Xpert Pro diffractometer. All the diffractometers operated with the Cu-Kα radiation (λ=1,5406Å). CaRIne Crystallography v3.1 software was used to simulate XRD pattern of BAM from the crystallographic data (Kim et al. 2002). An attenuated total reflection (ATR) accessory installed on a Nicolet FTIR spectrometer was used to record ATR spectra for all powders.

Micrographs were mainly recorded by means of a ZEISS Supra 55VP scanning electron microscope operating in high vacuum between 3 and 5 kV using secondary electron detector (Everhart-Thornley detector). Specimens were prepared by sticking powder onto the surface of an adhesive carbon film. BAM powders morphology was also investigated by means of a transmission electron microscope (Hitachi H7650 120kV) using a 80 kV acceleration voltage and combined with a Hamamatsu AMT HR 1Kx1K CCD camera placed in a side position. Samples were coated into a resin, cut with a microtome and then placed in the microscope.

Emission and excitation spectra of BAM:Eu powders were performed on a Jobin-Yvon set-up consisting of a Xe lamp operating at 400 W and two monochromators (Triax 550 and Triax 180) combined with a cryogenically cooled charge-coupled device (CCD) camera (Jobin-Yvon Symphony LN2 series) for emission spectra and with a Hamamatsu 980 photomultiplicator for excitation ones. Emission and excitation features of YAG:Ce powders as well as absolute photo-luminescence quantum yields values of all powders were measured using C9920-02G PL-QY measurement system from Hamamatsu. The setup comprises a 150W monochromatized Xe lamp, an integrating sphere (Spectralon Coating, Ø = 3.3 inch) and a high sensitivity CCD spectrometer for detecting the whole spectral luminescence. The automatically controlled excitation wavelength range spread from 250 nm to 950 nm with a resolution bandwidth better than 5 nm. This device was also employed to determine the trichromatic coordinates of BAM powders. Thanks to the software Color Calculator from Radiant Imaging, these coordinates have been placed in the CIE colour diagram. These data allow assessing if the obtained phosphors are suitable for replacing commercial phosphors in lighting devices. It must be specified that all luminescence performances have been recorded at room temperature.

3. A yellow phosphor: $Y_3Al_5O_{12}:Ce^{3+}$

3.1 Structural properties

Structural properties of undoped YAG and YAG:Ce powders have been analysed by means of XRD and IR spectroscopy.

3.1.1 Optimization of the urea quantity from undoped YAG powders

First, the influence of the amount of urea, used as fuel, on the crystallization of the YAG phase was scrutinized. The corresponding XRD patterns are gathered in the Fig. 5.

The pattern corresponding to $(f/o)_{th}$ (Fig. 5a) is characteristic of an amorphous phase with no discernable diffraction peak. This can be explained by the fact that no flame was observed during this synthesis; hence, no combustion occurs and the temperature of the medium did not reach a sufficient value to entail any crystallization. For $2(f/o)_{th}$ (Fig. 5b), for which a flame has occurred during the synthesis, XRD pattern is characterized by diffraction peaks mainly ascribable to the YAG structure (JCPDS-file 33-0040) but several weak peaks, labeled ◆, are assigned to the intermediate hexagonal $YAlO_3$ (YAH) phase. When the quantity of urea increases, this phase progressively disappears and, for $3(f/o)_{th}$ (Fig. 5d), only pure YAG phase is obtained, with a good crystallinity. This can be explained on the basis of the thermodynamics of the combustion reaction, investigated by Mukasyan *et al.* (Mukasyan et al. 2007): the maximum combustion temperature (T_{max}) following the ignition is greatly dependent on the fuel amount. It can be assumed that the increase in (f/o) leads to the rise of T_{max}, which entails a better crystallinity and the absence of the YAH phase (directly converted into YAG phase) when (f/o) reaches $3(f/o)_{th}$.

The ATR spectra recorded for the same samples are gathered in Fig. 6. The wavenumber domain has been limited to 2000-400 cm^{-1} because this region allows observing at once vibrations relative to M-O bonds and those associated to the organic groups. As it could be expected, the sample synthesized with a quantity of urea corresponding to $(f/o)_{th}$ (Fig. 6a) is characterized by a broad IR signal below 800 cm^{-1}, ascribed to M-O bonds in an amorphous sample, together with several bands lying from 1600 to 900 cm^{-1} related to organic precursors; indeed, as evidenced in Fig. 7, the vibration bands of the amorphous powder mainly correspond to a mixture of nitrates and urea. The shift and differences observed for these bands between the heated powder (Fig. 7A) and precursors (Fig. 7B and 7C) can be due to the fact that these latters have not undergone a microwave heating-treatment.

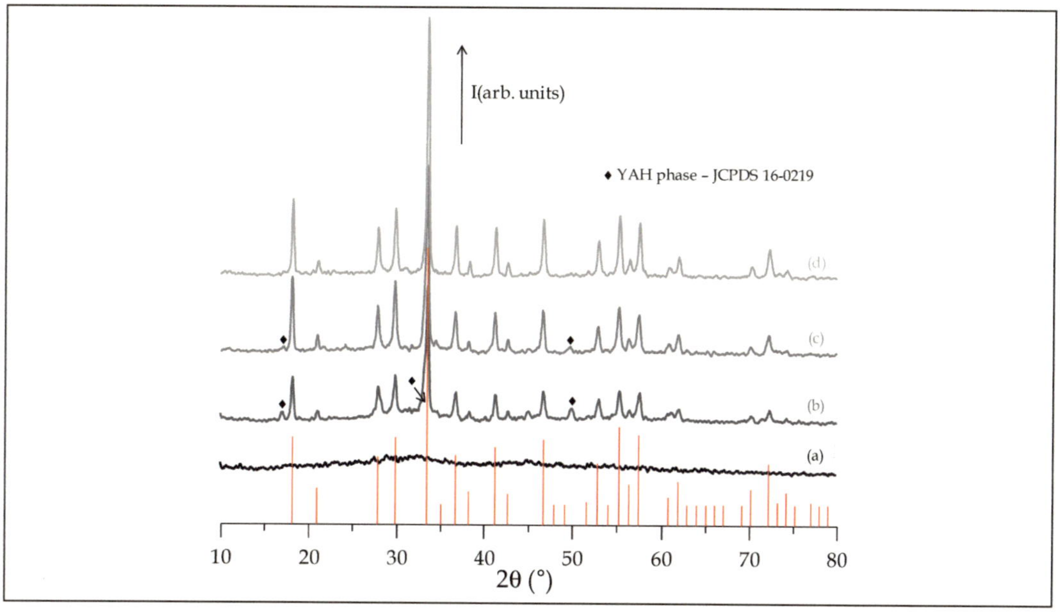

Fig. 5. XRD patterns of undoped YAG powders synthesized with (a) $(f/o)_{th}$, (b) $2(f/o)_{th}$, (c) $2.5(f/o)_{th}$ and (d)$3(f/o)_{th}$, compared to the JCPDS file 33-0040 of YAG.

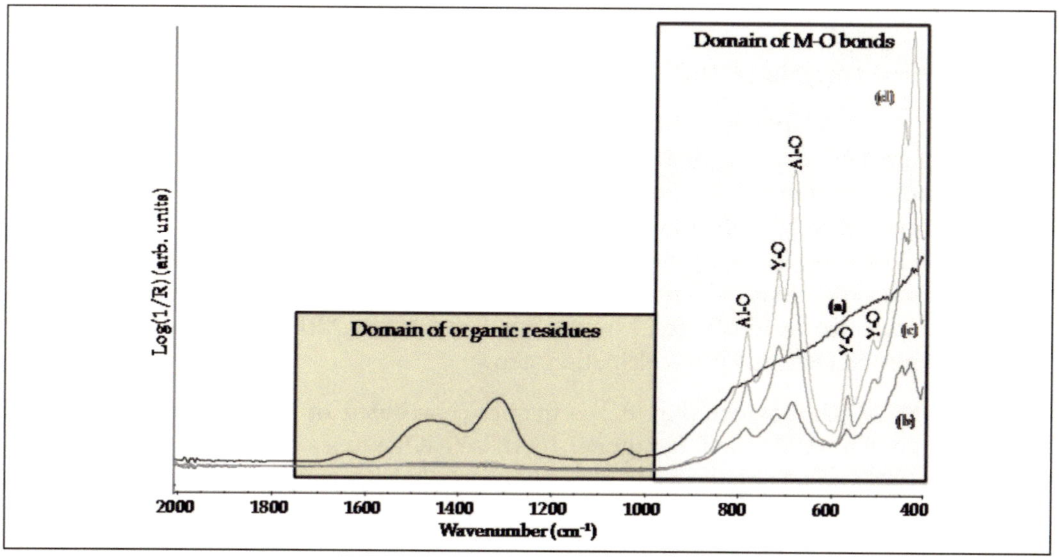

Fig. 6. ATR spectra of undoped YAG powders synthesized with (a) $(f/o)_{th}$, (b) $2(f/o)_{th}$, (c) $2.5(f/o)_{th}$ and (d) $3(f/o)_{th}$.

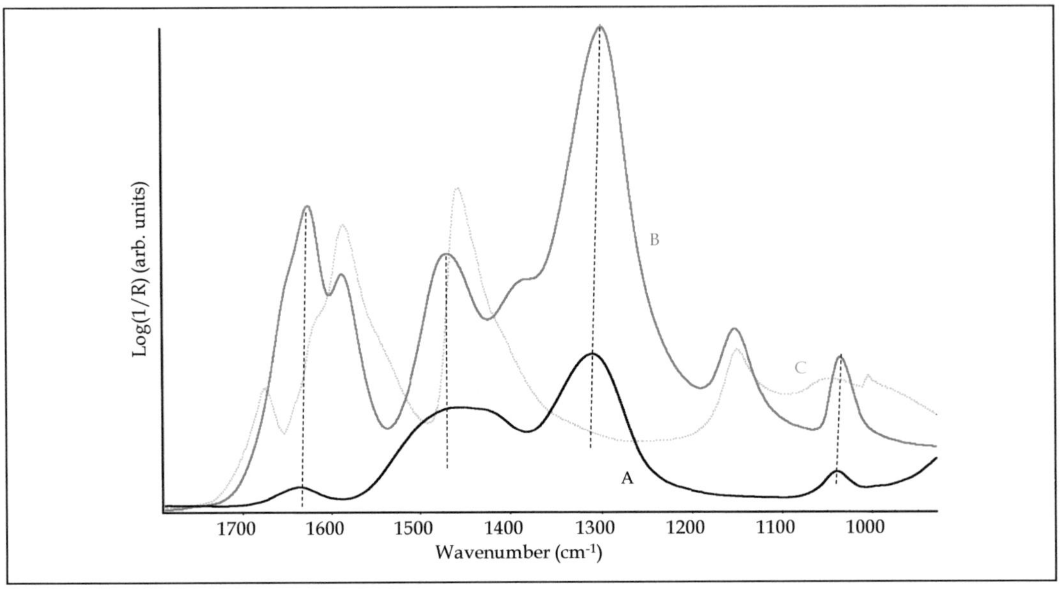

Fig. 7. ATR spectrum of undoped YAG powders synthesized with (a) $(f/o)_{th}$ compared to those of (b)yttrium nitrate and (c)urea.

When (f/o) increases, the intensity of the vibration bands between 1600 and 900 cm^{-1} significantly diminishes and these bands become negligible (Fig. 6b to 6d). The greyish colour of the powder then suggests only the presence of disordered carbonaceous impurities, in all probability resulting from the urea pyrolysis (Guo et al. 2010). Besides, when (f/o) $\geq 2(f/o)_{th}$, several vibration bands appear between 800 and 400 cm^{-1}. They are relative to Y-O and Al-O bonds in the YAG matrix and testifies that the crystallization has occurred (Hofmeister & Campbell 1992; Potdevin et al. 2010).

3.1.2 Structural analysis of YAG:Ce powders

Considering the XRD patterns obtained for the undoped YAG, the optimal ratio (f/o), leading to pure YAG phase with a good crystallinity, corresponds to $3(f/o)_{th}$ (see Fig. 5). As a consequence, Ce^{3+}-doped YAG powders have been elaborated with $3(f/o)_{th}$. As mentioned in the experimental section, the powder obtained was greyish, which is mainly related to the presence of graphite. In order to improve the optical performances of the elaborated phosphors, this carbon has to be removed from the samples. Thus, the as-prepared YAG:Ce powder was calcined at 1000°C for 2 hours. Fig. 8 represents XRD patterns of the YAG:Ce powder before and after this post-calcination step.

As-prepared YAG:Ce powder (Fig. 8a) is mainly constituted of YAG and $YAlO_3$ phases; both orthorhombic (YAP) and hexagonal (YAH) $YAlO_3$ phases coexist with YAG. After sintering at 1000°C (Fig. 8b), the YAH phase have totally been converted into YAG whereas YAP phase is still present. This behaviour can appear as surprising when compared to undoped YAG (Fig. 5) but can be explained by the difference between Y^{3+} and Ce^{3+} ionic radii. ($r_{Y^{3+}}$=1.02 Å and $r_{Ce^{3+}}$=1.14 Å in coordination VIII) (Shannon 1976). Since Ce^{3+} ions are significantly bigger, their substitution to the Y^{3+} can easily lead to the formation of supplementary phases. The presence of YAP phase has obviously to be avoided but the powder obtained after post-calcination presents remarkable yellow luminescence upon UV and blue excitations as it will be detailed in the paragraph 3.3. The crystallinity of the post-heated powder does not seem better than that of the as-prepared one, since the diffraction intensity and the signal-to-noise ratio remain similar.

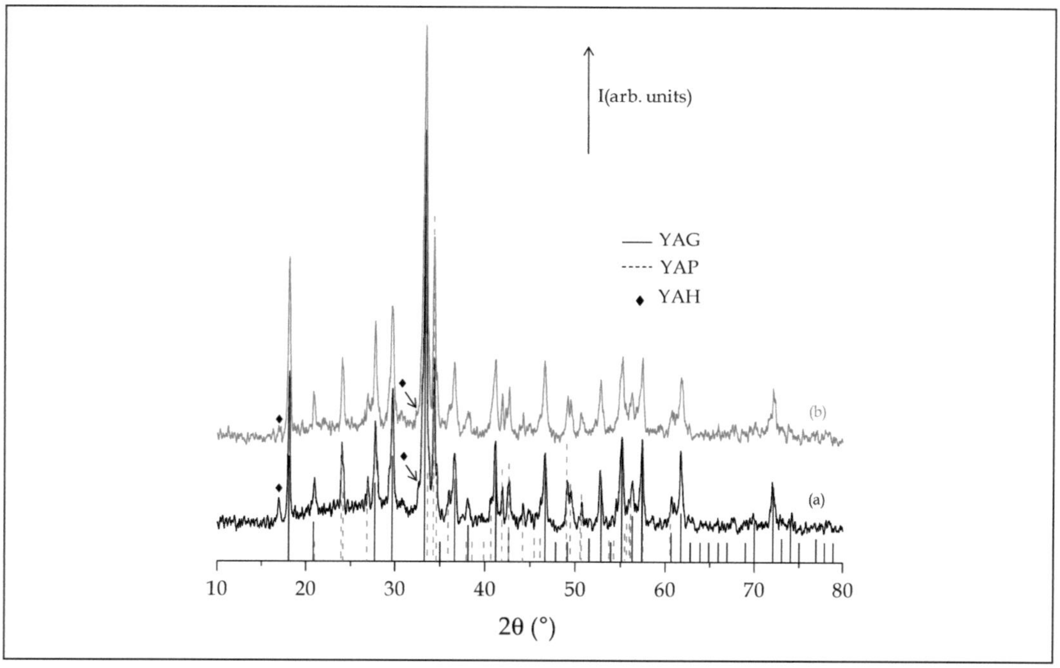

Fig. 8. XRD patterns of YAG:Ce powder synthesized with $3(f/o)_{th}$ (a)before and (b) after a post-heating treatment at 1000°C for 2 hours, compared to the JCPDS files of YAG (33-0040) and YAP (33-0041).

By modifying some synthesis parameters such as the nature of the fuel and its quantity, as well as the amount of water involved in the solution combustion synthesis, it should be possible to improve the phase purity (Mukasyan et al. 2007). The study presented in this chapter consists in a preliminary investigation, intended for evidencing the possibility to synthesize efficient phosphors by a simple and cheap method. A detailed investigation concerning the synthesis parameters of YAG:Ce powders by MISCS has not yet been carried out.

The influence of the heating-treatment at 1000°C on the structural properties of YAG:Ce powder has also been studied by IR spectroscopy. The ATR spectra presented in Fig. 9 confirm the presence of the characteristic Al-O and Y-O bonds of the YAG matrix (Hofmeister & Campbell 1992). Besides, it can be noticed that the domain of organic residues does not exhibit any specific vibration band. Only slight differences are observed between the two kinds of powders. However, the change in colour observed in Fig.3 denotes the removal of carbonaceous impurities, which has already been proved to be an important factor influencing the optical properties (Potdevin et al. 2010).

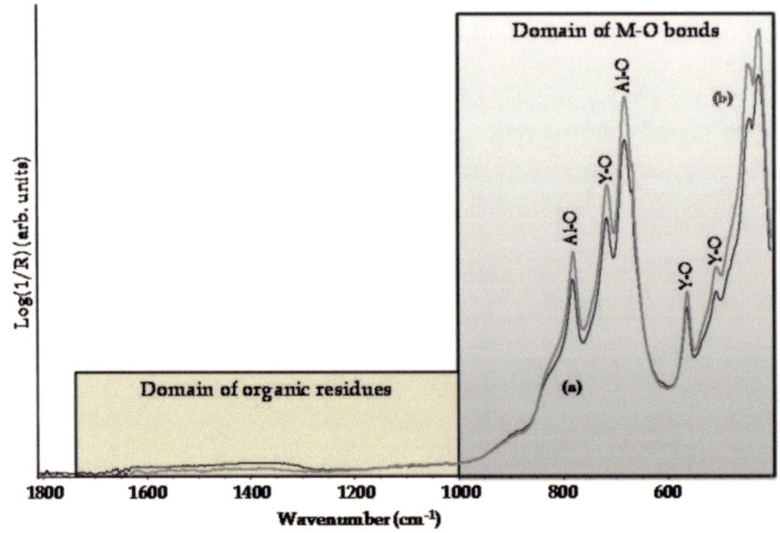

Fig. 9. ATR spectra of YAG:Ce powder synthesized with $3(f/o)_{th}$ (a)before and (b) after a post-heating treatment at 1000°C for 2 hours.

3.2 Morphological study

Micrographs corresponding to undoped YAG powders synthesized with the different amounts of urea are gathered in Fig. 10 and Fig. 11.

The pictures corresponding to the non-ignited powder (Fig. 10a and 10b) reveal dense micro-sized particles of irregular shape and very smooth surface. If compared with the sample synthesized with $(f/o)=2(f/o)_{th}$ (Fig. 10c), it can be seen that combustion reaction gives rise to a nanoporosity, correlated to the large amount of emitted gases during the ignition.

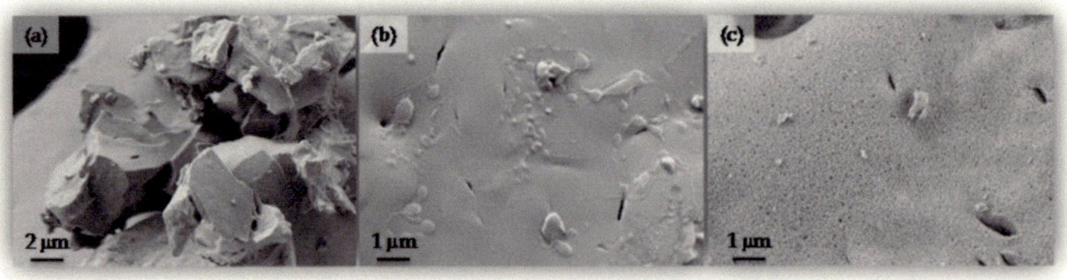

Fig. 10. SEM images of undoped YAG powders synthesized with (a,b) $(f/o)_{th}$ and (c) $2(f/o)_{th}$.

Fig. 11 exhibits SEM images of the particles elaborated with $2(f/o)_{th}$ (Fig. 11a and 11c) and $3(f/o)_{th}$ (Fig. 11b and 11d) : highly agglomerated particles with rough surface are obtained for all specimens. Nevertheless, an increase in the fuel amount entails a decrease in agglomerates size. This can be related to the augmentation of the pressure due to extended volume of emitted gases, which gives rise to the break of interconnected structures. At higher magnifications (Fig. 11c and 11d), voids and nanopores can be well observed; they are due to the explosive character of the ignition and to the homogeneous emission of gases during the combustion, respectively.

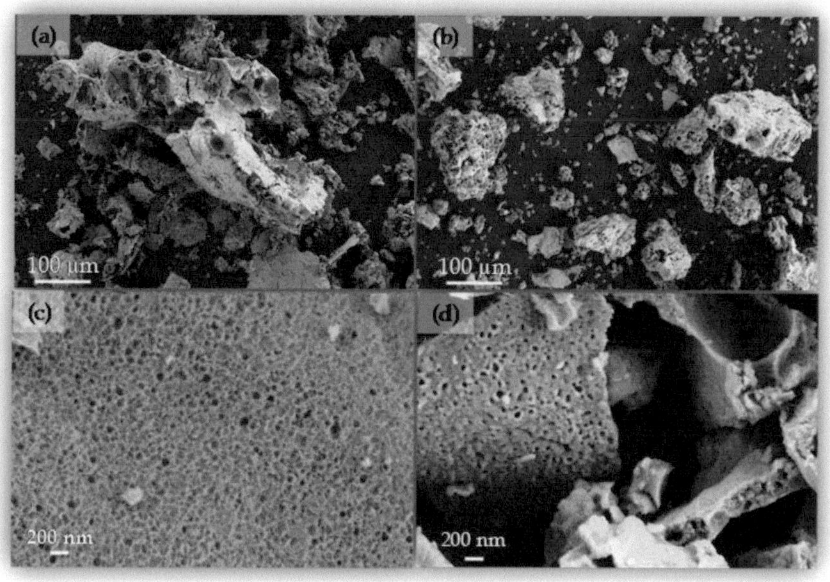

Fig. 11. SEM images of undoped YAG powders synthesized with (a,c) $2(f/o)_{th}$ and (b,d)$3(f/o)_{th}$.

The morphology of the as-synthesized YAG:Ce powder was studied by SEM and correlated to that obtained for the sintered one at 1000°C. Corresponding micrographs are presented in Fig. 12. The as-prepared YAG:Ce powder (Fig. 12a to 12c) is characterized by the same morphology than undoped YAG. No specific shape was observed. Only a highly friable powder, exhibiting large voids and nanoporosity is achieved. After the post-calcination (Fig. 12d), voids and pores remain and similar morphology is obtained. No densification has occurred which is a key point in order to use these phosphors to elaborate films. Indeed, these porous powders can be easily grinded to attain fine particles suitable for aqueous dispersions. These suspensions can then be used to prepare thick or thin luminescent films, as mentioned for BAM:Eu phosphors at the end of this chapter.

This morphology can be tailored by using different fuels, such as glycine, citric acid or carbohydrazide (Mangalaraja et al. 2009).

For comparison, SEM micrographs have been recorded from a commercial YAG:Ce powder (Phosphor Tech QMK58/F-U1). This phosphor has probably been obtained by solid-state reaction, after several hours of sintering at a temperature higher than 1300°C. As a result and as shown in Fig. 13, this phosphor is characterized by highly crystalline and well facetted micro-sized particles. These particles are very dense and a large size distribution is highlighted. This kind of powder cannot be reasonably dispersed in a suspension in order to design homogeneous luminescent films.

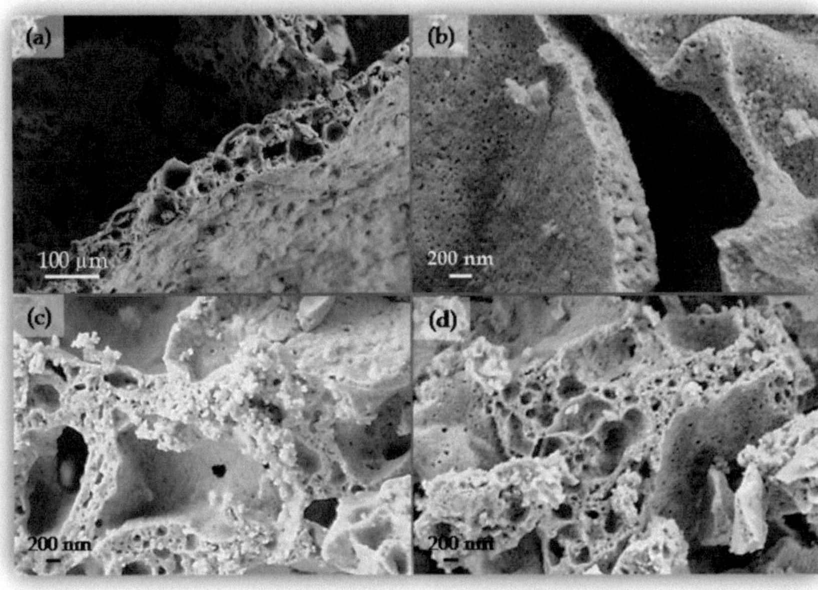

Fig. 12. SEM images of YAG:Ce powder synthesized with $3(f/o)_{th}$ (a,b,c)before and (d) after a post-heating treatment at 1000°C for 2 hours.

Fig. 13. SEM image of a commercial YAG:Ce powder (Phosphor Tech QMK58/F-U1.).

Furthermore, as discussed in the introduction, a large particle size distribution is generally prejudicial to the efficiency of LED-based lighting devices (Huang et al. 2009).

3.3 Photoluminescence features

Excitation spectrum obtained for post-calcined YAG:Ce powders are presented in Fig. 14. Two main excitation bands centred at 342 nm and 460 nm are observed. These excitation bands are due to the electron transitions from the ground-state $^2F_{5/2}$ of Ce^{3+} ions to the different crystal field splitting components of their 5d excited state (Blasse & Grabmaier 1994).

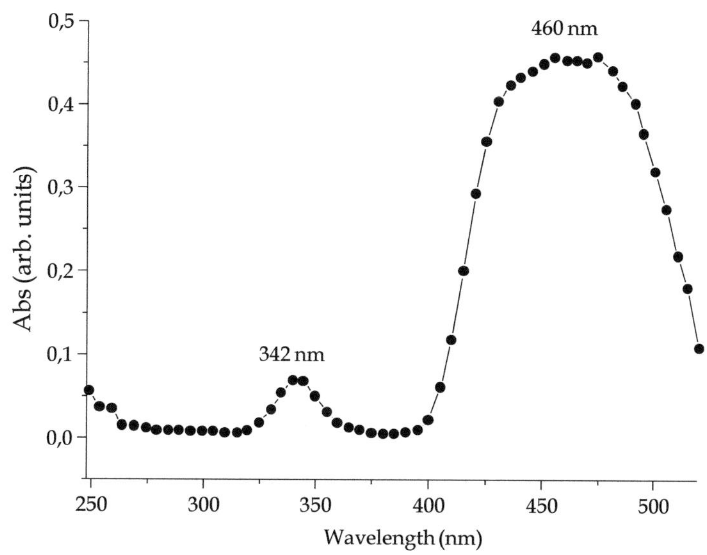

Fig. 14. Room-temperature excitation spectrum of YAG:Ce powder synthesized with $3(f/o)_{th}$ after a post-heating treatment at 1000°C for 2 hours.

The stronger band located at 460 nm matches very well the blue emission of GaN-based LED. Consequently, crystallized YAG:Ce^{3+} powders can absorb efficiently the blue emission of GaN-based LED and convert it efficiently into visible light at longer wavelength range.

Fig. 15 shows the emission spectra of YAG:Ce (1 mol%) powders before and after sintering at 1000°C recorded upon a 460 nm excitation, corresponding to a GaN-based LED. They both consist of a broad emission band due to the overlapped emission transitions relative to Ce^{3+} $^2D_{3/2}$-^7F_J transitions. For the two powders, the maximum emission intensity corresponds to a 540 nm wavelength, which is a yellowish green fluorescence. This specific feature for Ce^{3+} emission is related to strong crystal-field effects in the garnet structure (Blasse & Grabmaier 1994).

When the powder has been further calcined (Fig. 15b) the emission intensity is much more important but no red shifting is observed. Since the 5d orbit is strongly influenced by the strength and symmetry of crystal-field undergone by the Ce^{3+} ions, this testifies that the post-calcination has not entailed any crystal-field modification(Blasse & Grabmaier 1994).

Upon UV-radiation, yellow-green radiation is also observed with a significant enhancement after post-heating treatment, as evidenced in the Fig. 16.

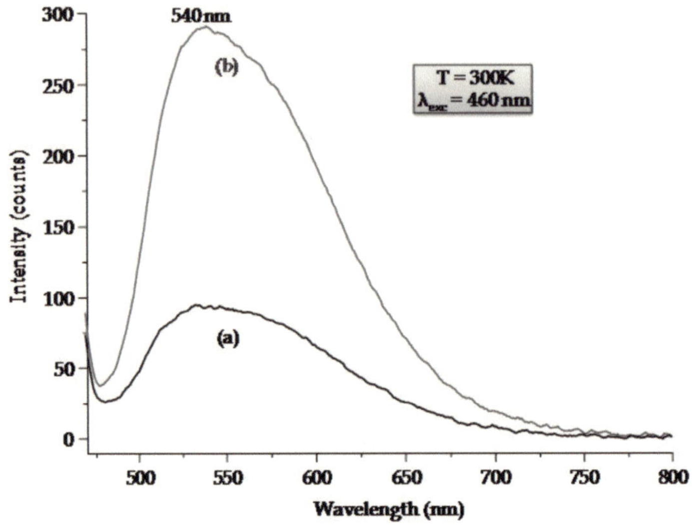

Fig. 15. Room-temperature emission spectra of YAG:Ce powder synthesized with $3(f/o)_{th}$ (a) before and (b) after a post-heating treatment at 1000°C for 2 hours.

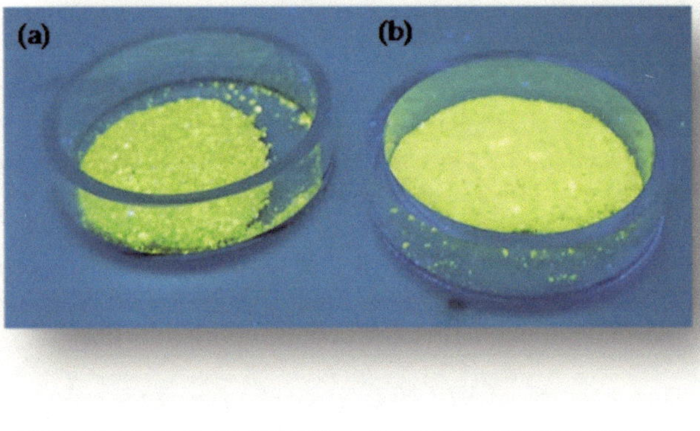

Fig. 16. Picture of as-synthesized YAG:Ce powder before(a) and after(b) post-calcination at 1000°C for 2 hours upon a 365 nm-excitation.

Finally, photoluminescence quantum yields of these YAG:Ce powders have been determined, in the range 475-800 nm, upon a 460 nm excitation, corresponding to the emission wavelength of commercialized blue LEDs. Results are gathered in Table 1.

As foreseen according to the Fig. 15, the post-heating treatment leads to a significant enhancement of the QY of the MISCS-derived YAG:Ce powder to reach 50%. This efficiency concerns a phosphor for which the synthesis parameters have not been optimized. It is better than values obtained from other combustion processes (Purwanto et al. 2008) and becomes relevant to consider the use of these phosphors in lighting devices. The QY of the commercial phosphor is obviously largely superior particularly due to its high crystallinity (very intense diffraction peaks - results not shown). But, thanks to the features detailed before, the YAG phosphors emanating from MISCS are very promising candidates for lighting devices.

	QY(%)
Commercial	87
Before post-calcination	15
After post-calcination	50

Table 1. Absolute quantum yields (QY) for commercial and MISCS-derived YAG:Ce^{3+} powders for $3(f/o)_{th}$.

The second part of this chapter will deal with the meticulous characterization of the blue phosphor $BaMgAl_{10}O_{17}$:Eu^{2+} (BAM:Eu) obtained by MISCS.

4. A blue phosphor: $BaMgAl_{10}O_{17}$:Eu^{2+}

As seen in Fig. 4, a white foamy powder with blue and red parts upon a 254nm excitation was obtained for the two molar ratio (f/o) employed. The best blue/red ratio corresponds to the sample synthesized with the $3(f/o)_{th}$ molar ratio, for which the blue part is more important.

All the characterizations presented in this section have been performed on the powder emitting in the blue range. Characterizations concerning the red part have been published elsewhere (Pradal et al. 2011).

4.1 Structural properties

BAM:Eu^{2+} crystallization was analyzed by means of XRD. Results achieved for the two fuel to oxidizer molar ratios are presented: the stoichiometric molar ratio $(f/o)_{th}$ = 2.36/1 and a fuel-rich one $3(f/o)_{th}$ = 7.08/1.

Fig. 17 shows the XRD patterns of blue phosphors for the two studied molar ratios. Both blue parts correspond to nearly pure $BaMgAl_{10}O_{17}$ phase with a very little fraction of $BaAl_2O_4$, more important for the $(f/o)_{th}$ molar ratio. Since there is a $BaAl_2O_4$ impurity, we have decided to realize an acid-wash. Only washing on the blue powder for the $3(f/o)_{th}$ molar ratio is presented here. Resulting patterns obtained before (Fig. 18a) and after (Fig. 18b) acid-wash are shown in Fig. 18. The experimental XRD pattern after acid-wash reveals that all the diffraction peaks are assigned to the BAM structure. XRD figures also presents the simulated XRD pattern for BAM:Eu^{2+} plotted from the crystallographic data (Kim et al. 2002).

FT-IR spectra of $BaMgAl_{10}O_{17}$:Eu^{2+} blue phosphors for the two studied molar ratios before acid-wash are presented in Fig. 19. Acid-washing does not affect the IR results. Bands ranging from 1000 to 400 cm^{-1} arise from the metal-oxygen (M-O) groups. No band appeared over 1000 cm^{-1}.

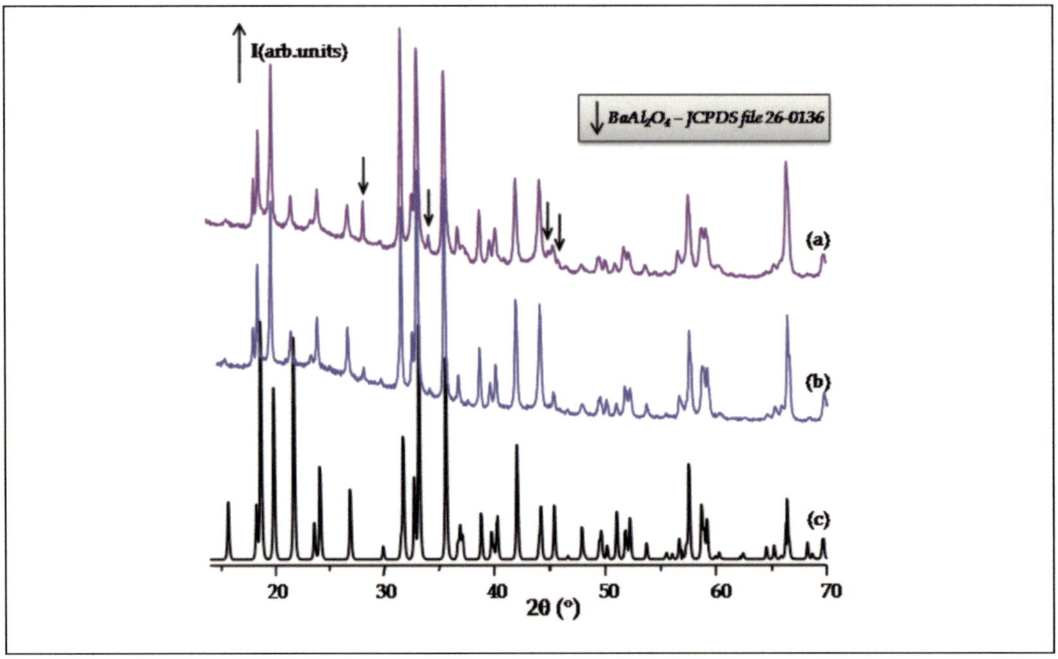

Fig. 17. XRD patterns of blue parts for (a) $(f/o)_{th}$, (b) $3(f/o)_{th}$ and (c) the simulated XRD pattern for BAM:Eu^{2+}.

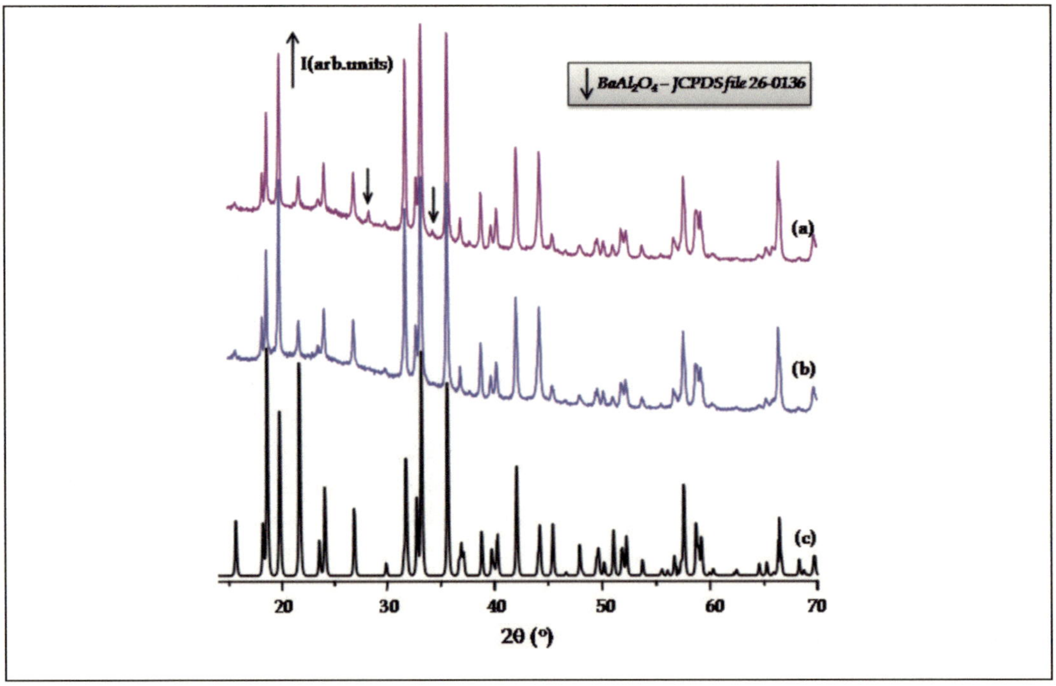

Fig. 18. XRD patterns of blue powder for $3(f/o)_{th}$ (a) before, (b) after acid-wash and (c) the simulated XRD pattern for BAM:Eu^{2+}.

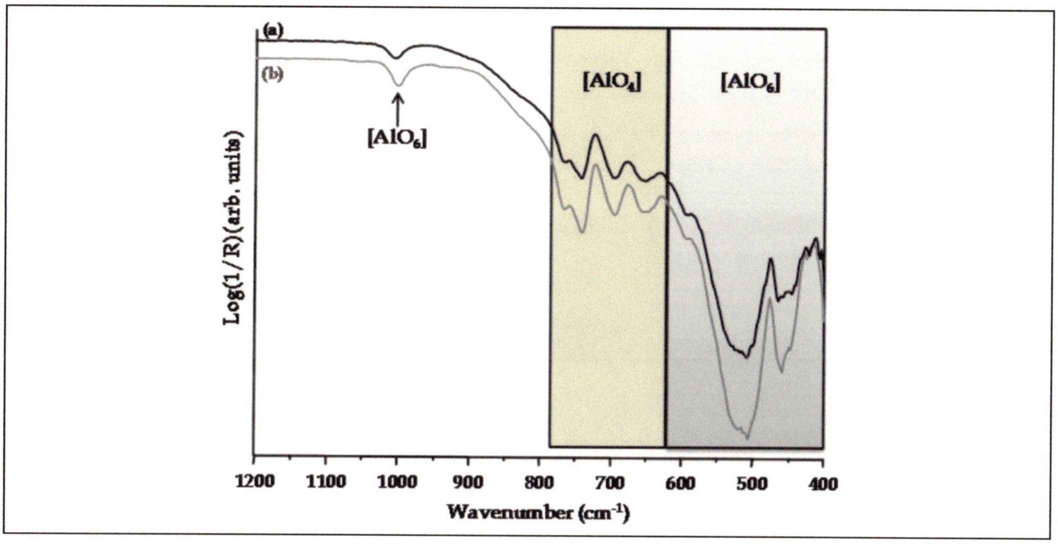

Fig. 19. FT-IR spectra of blue parts for (a) $(f/o)_{th}$, and (b) $3(f/o)_{th}$ before acid-wash.

BAM:Eu^{2+} has a β-alumina structure and belongs to the *P6₃/mmc* space group. This structure consists of a spinel block composed of aluminium, magnesium and oxygen ($MgAl_{10}O_{16}$) organized in AlO_4 tetrahedrons and AlO_6 octahedrons, so there are two sites for Al^{3+} ions. Mg^{2+} ions occupy one Al site. Two spinel layers are connected by a conduction layer (BaO) where Ba^{2+} ions are substituted by Eu^{2+} (Liu et al. 2009).

Peaks located at 1000 cm^{-1} and from 595 to 445 cm^{-1} are attributed to the absorption of [AlO_6] octahedrons. The absorption peaks, located from 765 to 650 cm^{-1}, can be ascribed to [AlO_4] tetrahedrons (Tian et al. 2006; Zhang et al. 2002).

4.2 Morphological properties

Figs. 20 and 21 show SEM images of blue powders for the two molar ratios studied. For both blue powders, they show a very porous opened microstructure (Figs. 20a and 21a). As for the YAG matrix, this porosity can be correlated with the exothermic character of the combustion and especially with the emission of a large amount of gases during the process. For the theoretical molar ratio $(f/o)_{th}$ (Fig. 20), the particles exhibit two morphologies: rods

Fig. 20. SEM images of blue powder for $(f/o)_{th}$ (a) x 5000 and (b) x 30000.

and platelets. The width of individual rods lies from 30 to 80 nm whereas their length is contained between 400 and 700 nm. When $(f/o)_{th}$ is increased by a factor of three (Fig. 21), the particles have only a plate-like shape. The platelets thickness varies from 40 to 80 nm and their diagonal from 200 to 500 nm.

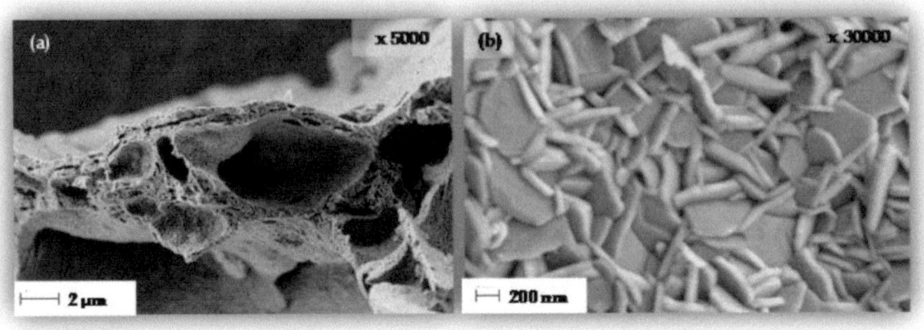

Fig. 21. SEM images of blue powder for $3(f/o)_{th}$ (a) x 5000 and (b) x 30000.

For this ratio, this specific morphology is confirmed with TEM image shown in Fig. 22. Once again acid-washing does not affect the morphological properties.

Fig. 22. TEM image of blue powder for 3(f/o).

Conventionally, BAM:Eu^{2+} phosphor is obtained via an energy greedy solid-state reaction process requiring a several hour-heating treatment at high temperatures (above 1300 °C). It leads to large size and irregular shapes, inappropriate for the devices foreseen.

SEM image of a commercial BAM:Eu^{2+} powder is shown in Fig. 23. As seen, it presents large and heterogeneous grains. It is very different from powders prepared by a MISCS which are nanostructured and homogeneous. This specific feature, associated with a high crystallinity has several consequences on both the optical performances of this phosphor (see Table 2) and the capacity for these particles of being dispersed in an aqueous solution to elaborate films. Indeed, particles with a very dense and large size distribution cannot be well-dispersed in a medium and hence cannot be shaped as coatings.

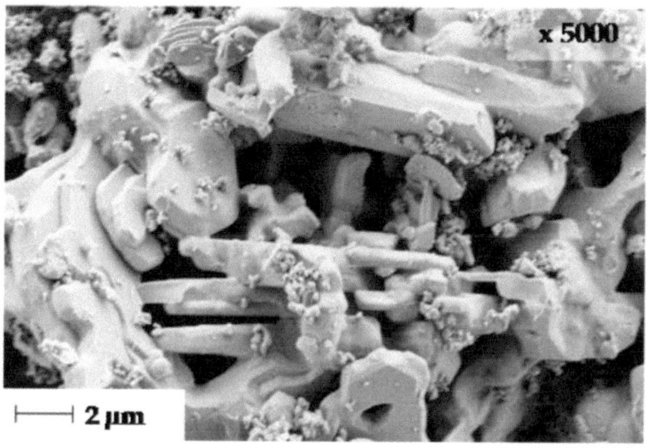

Fig. 23. SEM image of a commercial BAM:Eu^{2+} powder (Rhône Poulenc CRT013).

4.3 Optical properties

Fig. 24 shows emission spectrum of the blue powder for $3(f/o)_{th}$ after acid-wash upon a 382 nm excitation. It consists of a wide band centred at 466 nm. This broad band located in the blue region is due to Eu^{2+} ions transitions from the $4f^65d$ first excited configuration state to the $^8S_{7/2}$ $4f^7$ ground state (Chen & Yan 2007).

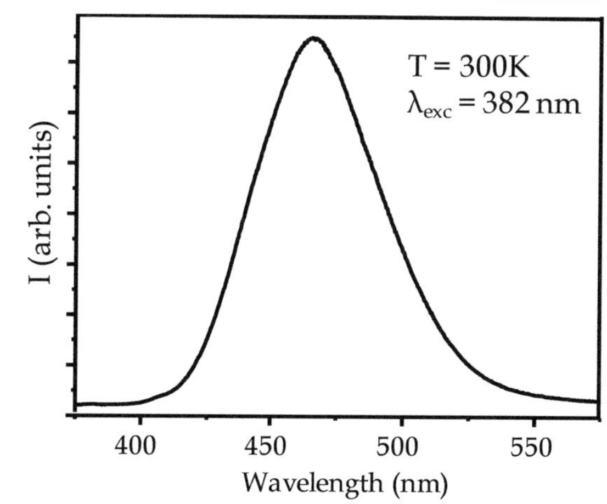

Fig. 24. Room-temperature emission spectrum of blue powder for $3(f/o)_{th}$ upon 382 nm excitation.

The associated excitation spectrum recorded by monitoring the blue emission at 466 nm is shown in Fig. 25. It displays a wide band centred around 382 nm. This absorption band corresponds to the characteristic Eu^{2+} electronic transition from the ground state to the crystal field split 5d levels (Chen & Yan 2007). This broad excitation band allows considering the combination of BAM:Eu with commercial near-UV LED emitting at 365 nm together with red and green phosphors, in order to produce white light (Smet et al. 2011).

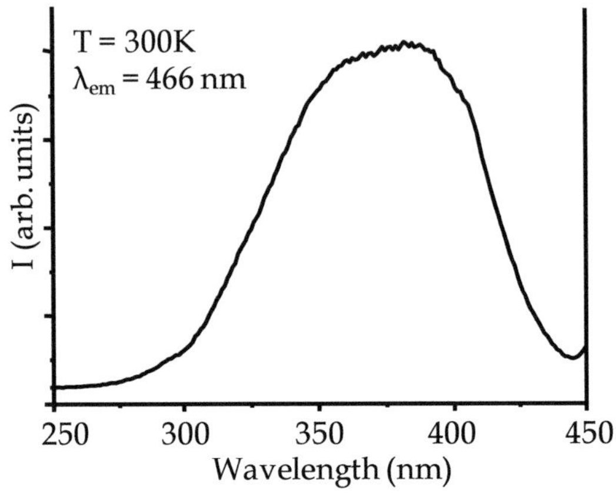

Fig. 25. Room-temperature excitation spectrum of blue powder for $3(f/o)_{th}$ with the emission monitoring at 466 nm.

Finally, photoluminescence quantum yields (in the range 400-650nm) and trichromatic coordinates were recorded by exciting the samples at 365nm, using the equipment detailed in the experimental section. Results are summarized in Table 2.

	QY(%)	x	y
Commercial	88	0.146	0.228
Before washing	79	0.149	0.066
After washing	75	0.149	0.064

Table 2. Absolute quantum yield (QY) and CIE coordinates for commercial BAM:Eu^{2+}powder and MISCS-derived blue powder for $3(f/o)_{th}$ before and after an acid wash, recorded upon a 365 nm excitation.

We can notice that commercial BAM:Eu^{2+} powder presents quantum yield higher than the blue powder resulted from the MISCS for $3(f/o)_{th}$. Furthermore, according to Table 2, acid-washing leads to a decrease of the QY. The main difference between powder before and after acid-wash is that as-prepared powder presents a $BaAl_2O_4$ impurity (see Fig. 18). The presence of $BaAl_2O_4$:Eu^{2+} in our sample does not seem to have a negative effect on BAM:Eu^{2+} luminescence since QY after the acid-wash is worse than that before. Indeed, as it has already been observed (Ravichandran et al. 1999; Singh et al. 2007), MISCS results in the blue shifting of $BaAl_2O_4$:Eu^{2+} emission from 500 nm for a solid-state reaction to about 440 nm. It has to be noticed that the colour coordinates are not modified by washing and that our powders are closer to the ideal blue than the commercial one (see Fig. 26).

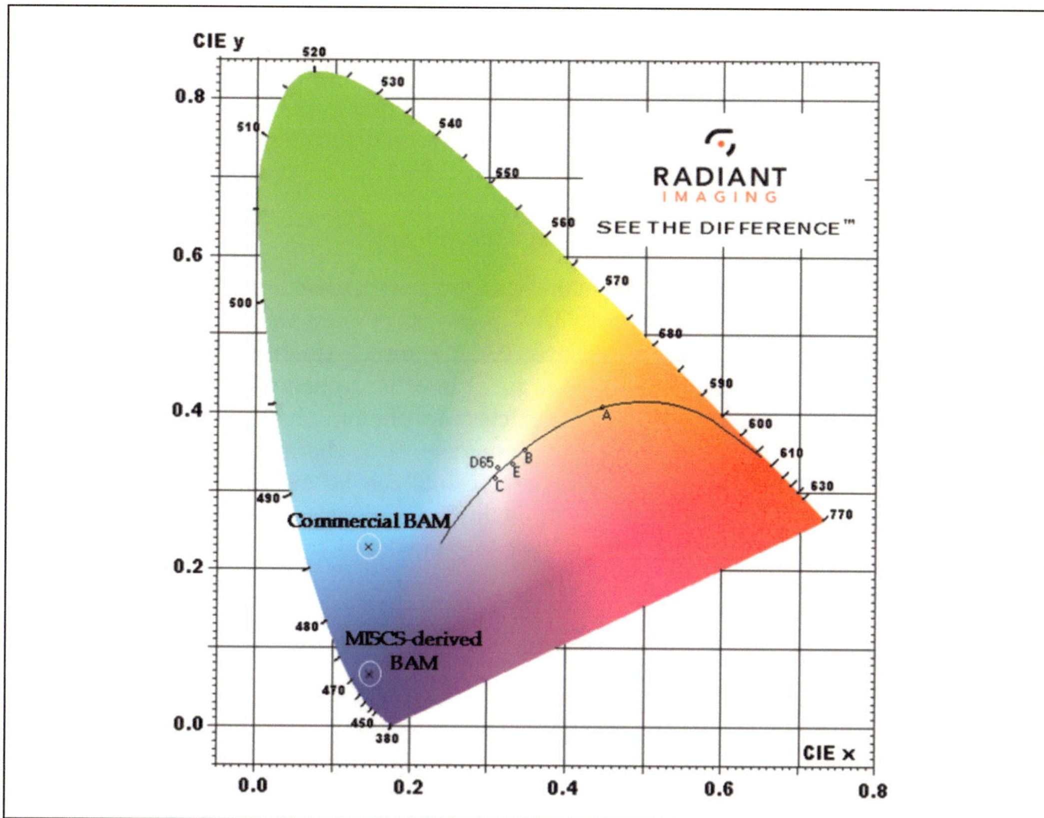

Fig. 26. CIE 1931 colour diagram containing the colour coordinates of BAM:Eu powder synthesized with $3(f/o)_{th}$ before and after an acid wash as well as those of a commercial BAM.

4.4 Shaping of homogeneous films

As mentioned during the study of YAG powders, MISCS-derived phosphors are highly friable and easily grinded into fine particles. These ones can then be homogeneously dispersed into an aqueous or alcoholic solution to give rise to a more or less viscous suspension. This kind of suspension has been used to elaborate a luminescent BAM:Eu film, as shown in Fig. 27.

Fig. 27. Picture of a $BAM:Eu^{2+}$ film under UV excitation.

5. Conclusion

Simple and rapid microwave-assisted combustion procedures have been used to elaborate luminescent YAG:Ce and BAM:Eu powders with suitable quantum efficiencies. The optimal quantity of fuel, leading to the purest aluminates, has been determined to be $3(f/o)_{th}$ in both cases. Furthermore, synthesized powders have revealed to be promising candidates as phosphors for LEDs based lighting devices and can be easily used to make luminescent suspensions, thanks to a uniform and spongious morphology.

Even if the procedures are identical, the features of the powders obtained present significant dissimilarities. For example, as-prepared YAG powders contain carbonaceous impurities, evidenced by their greyish colour, whereas BAM samples do not. As a consequence, YAG:Ce powders need a post-calcination treatment in order to have a respectable quantum efficiency. On the other hand, considering particles morphology, both kinds of samples are very porous and friable but only BAM powders are characterized by a specific nanomorphology (platelets for example). These results can be explained by the fact that microwave radiations absorption depends on the intrinsic properties of ions; hence, the final features of the matrix are closely related to its component and can be tailored by changing the nature of the oxidizers and fuel.

6. Acknowledgment

The authors would like to thank Joël Cellier (LMI), Anne-Marie Gélinaud (Casimir, Aubière, France) and Christelle Blavignac (CICS, Université d'Auvergne, Clermont-Ferrand, France) for their help in acquiring the XRD data, SEM pictures and TEM micrographs, respectively.

7. References

Blasse, G. & Grabmaier, B. C. (1994). *Luminescent Materials* Springer-Verlag, ISBN 978-038-7580-19-7, Berlin

Chen, Z. & Yan, Y. (2007). Morphology control and VUV photoluminescence characteristics of $BaMgAl_{10}O_{17}:Eu^{2+}$ phosphors. *Physica B: Condensed Matter*, 392, 1-2, pp.1-6, ISSN 0921-4526

Chen, Z., Yan, Y., Liu, J., Yin, Y., Wen, H., Zao, J., Liu, D., Tian, H., Zhang, C. & Li, S. (2009). Microwave induced solution combustion synthesis of nano-sized phosphors. *Journal of Alloys and Compounds*, 473, 1-2, pp.L13-L16, ISSN 0925-8388

Fu, Y.-P. (2006). Preparation of $Y_3Al_5O_{12}$:Ce powders by microwave-induced combustion process and their luminescent properties. *Journal of Alloys and Compounds*, 414, 1-2, pp.181-185, ISSN 0925-8388

Guo, K., Zhang, X.-M., Chen, H.-H., Yang, X.-X., Guo, X. & Zhao, J.-T. (2010). Influence of fuels on the morphology of undoped $Y_3Al_5O_{12}$ and photoluminescence of $Y_3Al_5O_{12}:Eu^{3+}$ prepared by a combustion method. *Materials Research Bulletin*, 45, 9, pp.1157-1161, ISSN 0025-5408

Hofmeister, A. M. & Campbell, K. R. (1992). Infrared spectroscopy of yttrium aluminum, yttrium gallium, and yttrium iron garnets. *Journal of Applied Physics*, 72, 2, pp.638-646, ISSN

Höppe, H. A. (2009). Recent Developments in the Field of Inorganic Phosphors. *Angewandte Chemie International Edition*, 48, 20, pp.3572-3582, ISSN 1521-3773

Huang, S. C., Wu, J. K., Hsu, W.-J., Chang, H. H., Hung, H. Y., Lin, C. L., Su, H.-Y., Bagkar, N., Ke, W.-C., Kuo, H. T. & Liu, R.-S. (2009). Particle Size Effect on the Packaging Performance of YAG:Ce Phosphors in White LEDs. *International Journal of Applied Ceramic Technology*, 6, 4, pp.465-469, ISSN 1744-7402

Jain, S. R., Adiga, K. C. & Pai Verneker, V. R. (1981). A new approach to thermochemical calculations of condensed fuel-oxidizer mixtures. *Combustion and Flame*, 40, 0, pp.71-79, ISSN 0010-2180

Jia, D., Wang, Y., Guo, X., Li, K., Zou, Y. K. & Jia, W. (2007). Synthesis and Characterization of YAG:Ce^{3+} LED Nanophosphors. *Journal of the Electrochemical Society*, 154, 1, pp.J1-J4, ISSN 0013-4651

Jung, K. Y. & Kang, Y. C. (2010). Luminescence comparison of YAG:Ce phosphors prepared by microwave heating and precipitation methods. *Physica B: Condensed Matter*, 405, 6, pp.1615-1618, ISSN 0921-4526

Kang, Y. C., Park, S. B., Lenggoro, I. W. & Okuyama, K. (1998). Preparation of non-aggregation YAG-Ce phosphor particles by spray pyrolysis. *Journal of Aerosol Science*, 29, Supplement 2, pp.S911-S912, ISSN 0021-8502

Kim, K.-B., Kim, Y.-I., Chun, H.-G., Cho, T.-Y., Jung, J.-S. & Kang, J.-G. (2002). Structural and Optical Properties of $BaMgAl_{10}O_{17}:Eu^{2+}$ Phosphor. *Chemistry of Materials*, 14, 12, pp.5045-5052, ISSN 0897-4756

Lee, S. H., Koo, H. Y., Jung, D. S., Yi, J. H. & Kang, Y. C. (2009). Fine-sized $BaMgAl_{10}O_{17}:Eu^{2+}$ phosphor powders prepared by spray pyrolysis from the spray solution with BaF_2 flux. *Ceramics International*, In Press, Corrected Proof, pp.-, ISSN 0272-8842

Liu, B., Wang, Y., Zhou, J., Zhang, F. & Wang, Z. (2009). The reduction of Eu^{3+} to Eu^{2+} in $BaMgAl_{10}O_{17}$:Eu and the photoluminescence properties of $BaMgAl_{10}O_{17}:Eu^{2+}$ phosphor. *Journal of Applied Physics*, 106, 5, pp.053102 - 053102-053105 ISSN 0021-8979

Lu, C.-H., Chen, C.-T. & Bhattacharjee, B. (2006). Sol-Gel Preparation and Luminescence Properties of $BaMgAl_{10}O_{17}:Eu^{2+}$ Phosphors. *Journal of Rare Earths*, 24, 6, pp.706-711, ISSN 1002-0721

Mangalaraja, R. V., Mouzon, J., Hedström, P., Camurri, C. P., Ananthakumar, S. & Odèn, M. (2009). Microwave assisted combustion synthesis of nanocrystalline yttria and its powder characteristics. *Powder Technology*, 191, 3, pp.309-314, ISSN 0032-5910

Mukasyan, A. S., Epstein, P. & Dinka, P. (2007). Solution combustion synthesis of nanomaterials. *Proceedings of the Combustion Institute*, 31, 2, pp.1789-1795, ISSN 1540-7489

Pan, Y., Wu, M. & Su, Q. (2004). Comparative investigation on synthesis and photoluminescence of YAG:Ce phosphor. *Materials Science and Engineering B*, 106, 3, pp.251-256, ISSN 0921-5107

Potdevin, A., Chadeyron, G., Boyer, D. & Mahiou, R. (2007). Sol-gel based YAG:Ce^{3+} powders for applications in LED devices. *Physica Status Solidi C: Current Topics in Solid State Physics*, 4, 1, pp.65-69, ISSN 1610-1642

Potdevin, A., Chadeyron, G., Briois, V., Leroux, F. & Mahiou, R. (2010). Modifications involved by acetylacetone in properties of sol-gel derived $Y_3Al_5O_{12}:Tb^{3+}$ - II: optical features. *Dalton Transactions*, 39, 37, pp.8718-8724, ISSN 1477-9226

Potdevin, A., Chadeyron, G., Briois, V., Leroux, F., Santilli, C. V., Dubois, M., Boyer, D. & Mahiou, R. (2010). Modifications induced by acetylacetone in properties of sol-gel derived $Y_3Al_5O_{12}$: Tb^{3+} - I: structural and morphological organizations. *Dalton Transactions*, 39, 37, pp.8706-8717, ISSN 1477-9226

Pradal, N., Potdevin, A., Chadeyron, G. & Mahiou, R. (2011). Structural, morphological and optical investigations on $BaMgAl_{10}O_{17}:Eu^{2+}$ elaborated by a microwave induced solution combustion synthesis. *Materials Research Bulletin*, 46, 4, pp.563-568, ISSN 0025-5408

Purwanto, A., Wang, W.-N., Ogi, T., Lenggoro, I. W., Tanabe, E. & Okuyama, K. (2008). High luminance YAG:Ce nanoparticles fabricated from urea added aqueous precursor by flame process. *Journal of Alloys and Compounds*, 463, 1-2, pp.350-357, ISSN 0925-8388

Ravichandran, D., Johnson, S. T., Erdei, S., Roy, R. & White, W. B. (1999). Crystal chemistry and luminescence of the Eu^{2+}-activated alkaline earth aluminate phosphors. *Displays*, 19, 4, pp.197-203, ISSN 0141-9382

Shannon, R. D. (1976). Revised effective ionic-radii and systematic sutdies of interatomic distances in halides and chalcogenides. *Acta Crystallographica, Section A: Foundations of Crystallography*, 32, SEP1, pp.751-767, ISSN 0108-7673

Shen, C., Yang, Y., Jin, S. & Ming, J. (2010). Luminous characteristics and thermal stability of $BaMgAl_{10}O_{17}:Eu^{2+}$ phosphor for white light-emitting diodes. *Physica B: Condensed Matter*, 405, 4, pp.1045-1049, ISSN 0921-4526

Shionoya, S. (1998). *Phosphor Handbook* CRC Press, ISBN 978-084-9335-64-8, Boca Raton

Singh, V., Natarajan, V. & Zhu, J.-J. (2007). Studies on Eu doped Ba and Zn aluminate phosphors prepared by combustion synthesis. *Optical Materials*, 29, 11, pp.1447-1451, ISSN 0925-3467

Smet, P. F., Parmentier, A. B. & Poelman, D. (2011). Selecting Conversion Phosphors for White Light-Emitting Diodes. *Journal of the Electrochemical Society*, 158, 6, pp.R37-R54, ISSN

Tian, X., Weidong, Z., Xiangzhong, C., Chunlei, Z., Xiaoming, T. & Xiaowei, H. (2006). Low Temperature Luminescence Properties of Tm^{3+} Doped Aluminate Phosphor. *Journal of Rare Earths*, 24, 1, Supplement 1, pp.141-144, ISSN 1002-0721

Zhang, J., Zhang, Z., Tang, Z., Zheng, Z. & Lin, Y. (2002). Synthesis and characterization of BaMgAl10O17:Eu phosphors derived by sol-gel processing. *Powder Technology*, 126, 2, pp.161-165, ISSN 0032-5910

Zhang, Z., Feng, J. & Huang, Z. (2010). Synthesis and characterization of $BaMgAl_{10}O_{17}:Eu^{2+}$ phosphor prepared by homogeneous precipitation. *Particuology*, 8, 5, pp.473-476, ISSN 1674-2001

Part 2

Characterisation and Properties of Sintered Materials

10

On the Application of 3D X-Ray Microtomography for Studies in the Field of Iron Ore Sintering Technology

Volodymyr Shatokha[1], Iurii Korobeynikov[1] and Eric Maire[2]
[1] *National Metallurgical Academy of Ukraine*
[2]*Université de Lyon, INSA-Lyon, MATEIS CNRS*
[1]*Ukraine*
[2]*France*

1. Introduction

Iron ore sinter is obtained by complex processes involving partial smelting of sinter mixture components under conditions of solid fuel combustion in the sintered bed. A simplified scheme of the iron ore sintering process is represented by Fig. 1.

Sinter mixture contains such components as iron ore, solid fuel (usually coke breeze), flux and some additives (e.g. solid wastes like steelmaking slag, gas cleaning dust and sludge etc). The mixture is granulated with the addition of moisture and then loaded to the travelling grate of the sintering plant.

Size distribution and durability of the granules influence both the green bed structure and permeability, thus affecting the sintering productivity and the quality of the sinter (Ellis et al, 2007; Khosa & Manuel, 2007; Naakano et al, 1998; Haga et al, 2005). Granules should not only endure mechanical stresses. They must also withstand destruction under conditions of overmoisturising in the moisture condensation zone of sinter plant. Therefore, study of the structure of granules produced from the sinter mixtures of different compositions is important. Traditional techniques (i.e. optical microscopy) give very limited information providing only random pictures of the structure at the surface of a single cross-section. Moreover, being formed without additives of a binder, granules are very fragile and prone to destruction when dried, which makes study of their internal structure difficult.

On the grate of sintering plant suction is imposed to the bed by the exhauster fan via windboxes (conic pots collecting gas below the travelling grate), gas header and gas cleaning system. After ignition of the solid fuel combustion zone propagates downwards, finally reaching hearth bed layer formed with the sinter return (under screen product <10 mm). Complex processes involving partial smelting of raw materials are resulted in formation of the sintered product.

The product of sintering is a porous lumpy material formed of particles with an irregular shape coalesced due to a binding effect of the solidified melt. The accumulated porosity of a piece of sinter consists of inter-particle voids and intra-particle pores (Debrincat et al, 2004;

Higuchi et al, 2004; Kokubu et al, 1986). The morphological parameters of the porosity such as total porosity pore size distribution, proportion of open and closed pores etc., correlate with specific surface area thus affecting both mechanical and physicochemical properties of sinter.

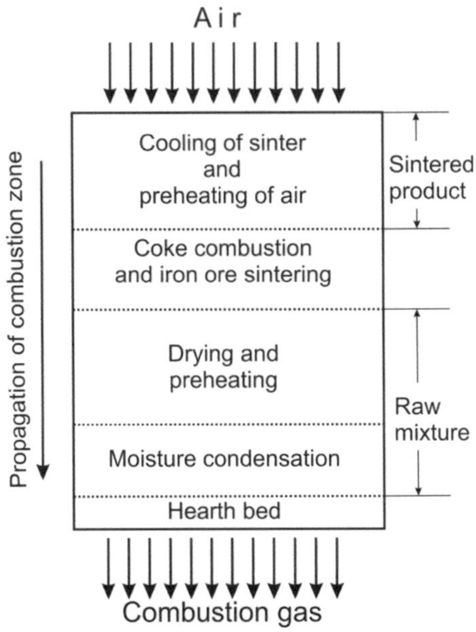

Fig. 1. Scheme of the iron ore sintering process.

The most widely used method for estimating sinter porosity is the mercury porosimetry. However, this method is destructive - it is usually impossible to re-use exactly the same samples for further investigations (e.g. reducibility test) as they are polluted with mercury. Another disadvantage is that the method characterises the size of the access window to the volume of the pore and not really its equivalent diameter; moreover, the material may exhibit so called "closed" pores with no connection to the outer surface, which cannot be measured by mercury porosimetry.

In our previous studies (Shatokha et al, 2009, 2010) a possibility to study the structure of sintering mixture granules and morphology of pores in the sintered product using 3D imaging technique, X-ray tomography, enabling non-destructive investigation of these complex heterogeneous materials, was demonstrated. In this research X-ray tomography technique was applied along with some other experimental methods to study microstructures of sinter granules and of sinter, and their relationship with the materials' properties.

2. Experimental

2.1 Sintering techniques

Sintering experiments were conducted at the National Metallurgical Academy of Ukraine. Characteristics of the sinter mixtures are given at the Table 1. The quantity of return in the mixture was 20% with coke breeze share at 6.5-7.0 % - depending upon consumption of the limestone. Ukrainian raw materials were used such as hematite ore and magnetite concentrate of Kryvyi Rih deposit (Table 2) and limestone of Yelenovskoye deposit. Mixtures were granulated in a laboratory granulation drum with a moisture addition of 7.5-8.0 %. Sintering was conducted in the laboratory pots (steel cylinders) with the grate area of 0.005 m^2 and a height of 0.25 m.

	1	2	3	4	5	6	7	8	9	10
Basicity, CaO/SiO_2	0.8			1.4			2.0			
Concentrate/ore mass ratio	20/80	50/50	80/20	20/80	50/50	80/20	20/80	50/50	80/20	100

Table 1. Characteristics of the sinter mixtures.

	Fe	Mn	P	S	FeO	Fe_2O_3	SiO_2	Al_2O_3	CaO	MgO	TiO_2	K_2O+Na_2O	LOI
Ore	55.12	0.062	0.089	0.025	0.95	77.7	10.3	5.8	0.14	0.34	0.157	0.71	3.4
Concentrate	63.75	0.039	0.017	0.033	27.98	59.97	9.69	0.41	0.35	0.61	0.04	0.27	0.49

Table 2. Chemical composition of the iron ores, used for sintering (mass %).

2.2 The principle of X-ray tomography and its application to study the iron ore material

The principle of X-ray tomography and its application in the field of materials science are explained in details by Maire et al (2001). This non-destructive technique is analogous to medical scanner and allows, from X-ray radiography acquired around a single axis of rotation, to generate a 3D image of the internal structure of an opaque material. The reconstruction involves a computed step and the final image is a 3D map of the local X-ray attenuation coefficient.

X-ray scanning of the samples was performed using the tomograph of the MATEIS laboratory at INSA de Lyon (Lyon, France), a Vtomex system, manufactured by Phoenix X-ray.

The beam was operated at 90 keV and 170 μA with no filtering for the observations made in this study. The highest resolution achieved was 2.85 μm; however because of the trade-off between the resolution and the maximum specimen dimension (e.g. whole pieces of sinter sized from 9 to 25 mm were investigated) the average resolution used in this study was 20 μm. Under given beam parameters and achieved resolution it appeared impossible to make precise phase analysis of the sinter although the images obtained were quite suitable for the analysis of the porosity because the solid phase absorbs the X-rays much better than air.

The tomography setup used is equipped with a standard laboratory X-ray tube, the source is then not monochromatic. Therefore, the values of the attenuation coefficients measured are not absolute values as the exact energy distribution in the incident beam is not well known. The measured apparent value of the attenuation constituent of different phases imaged in this study is shown in Fig. 2. Standard samples made of pure Cr, Fe, Zn were also scanned and Fig.3 shows the measurements performed with these standards: the measured coefficients are in a reasonable qualitative agreement with what could be expected for these materials.

As the preparatory step, several cross-sections were prepared and initially studied using optical microscopy and afterwards - using tomography method. Based on this experience authors were able to master recognition of the sinter mixture components (coarse and fine particles of ore, sinter return, limestone and coke breeze) in undistorted granules. At the same time, the resolution used for acquisition constrained the quality of the images. Therefore, it was not possible to recognise the nature of particles smaller than 0.1 mm in this study. A possibility of distinguishing various phases in the sinter structure (e.g. calcium ferrites, olivines, hematite, magnetite and so on) was also limited due to the same reason.

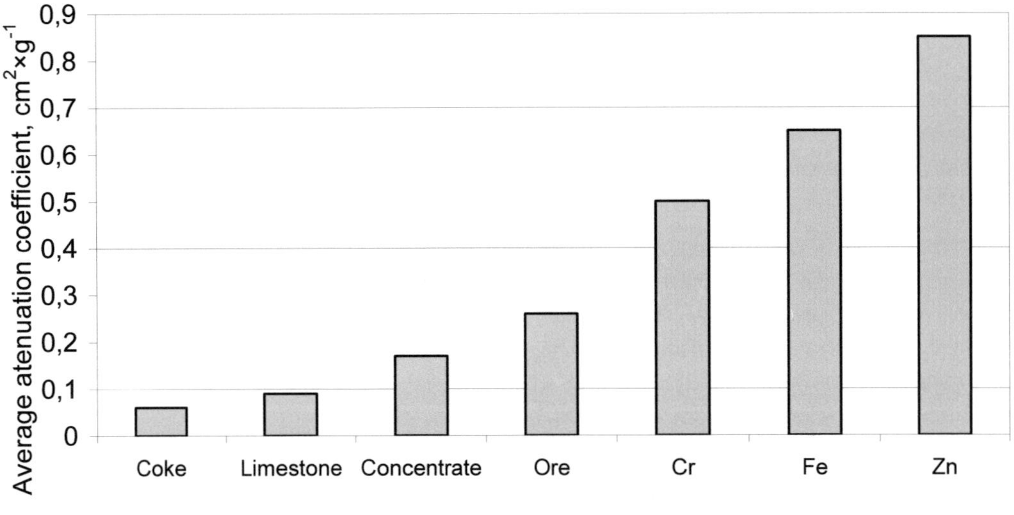

Fig. 2. Average attenuation coefficient measured in the reconstructed images for the studied materials and for three pure standards using an acceleration voltage of 100 kV.

2.3 Methodology of the porosity estimation using 3D X-ray tomography

In the present paper all empty space inside the sample surface i.e. inter-particle voids and intra-particle pores is referred to as pores. After X-ray scanning of the sample following procedures were applied to analyse the porosity parameters:

- Reconstruction from the obtained projections of the 3D images using the commercial DatosX_rec software (from Phoenix X-ray)
- Binarisation of the image using the ImageJ freeware (http:/rsb.info.nih.gov/ij) with a classical threshold. The solid phase of the sample was attributed the maximum grey

level (255) and pores plus the surrounding air were attributed the minimum level (0). An example of a typical slice of 3D image before and after binarisation is shown in Fig.3.

- Calculation of the volume of solid material. A homemade plug-in was used to calculate the fraction of white pixels for each 2D slice. Total amount of white pixels gives the total volume of material (V_m).
- Separation of open and closed porosity. A region growing procedure was used to select the open pores plus the outside air. Fig.4 represents an example of a typical slice of 3D image before (a) and after (b) region growing procedure. Then it was possible to calculate volume of the material plus volume of closed pores ($V_m+V_{c.p}$). Finally the volume of closed pores is obtained by the difference of these two quantities

$$(V_m+V_{c.p}) - V_m = V_{c.p}$$

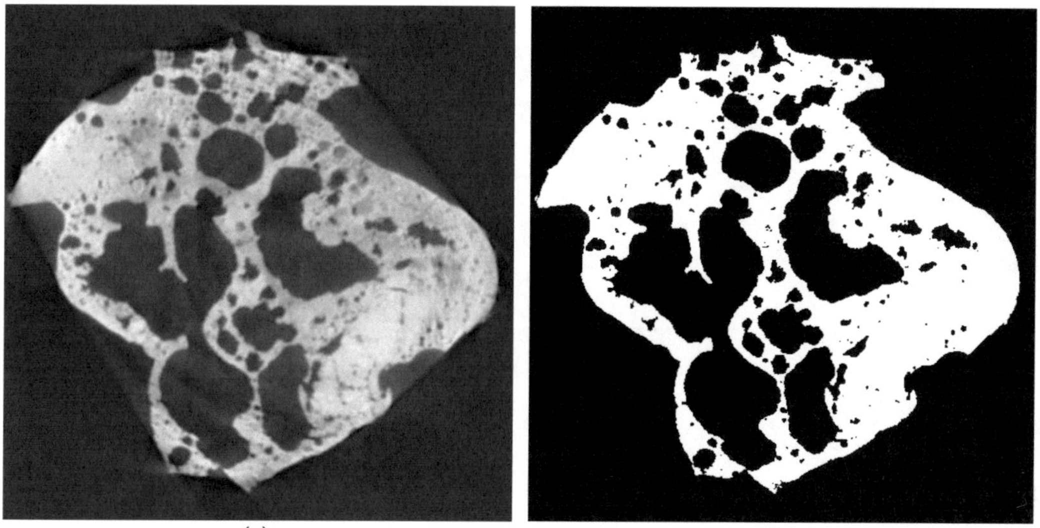

(a) (b)

Fig. 3. Typical slice of 3D image before (a) and after (b) binarisation (white - solid phase, black - gaseous phase).

After completion the procedures, different pores were separated via a 3D watershed type technique. A homemade plug-in system was then used to measure the morphological properties of each open pore (e.g. total volume and equivalent spherical diameter) and to indicate if the pore touches the frame of the 3D reconstruction scan. The pores touching the frame of the scan were assumed to belong to the outside air and they were then excluded from the analysis.

The closed porosity was then separated by making the difference between the open porosity (selected by the region growing procedure as described before) and the total porosity (selected by a simple grey level threshold based on the same value of the grey level as the one used for the region growing). The Fig.4 *c* shows an extracted slice with separated closed porosity. The morphology of each closed pore was then calculated in the same way as for the open porosity.

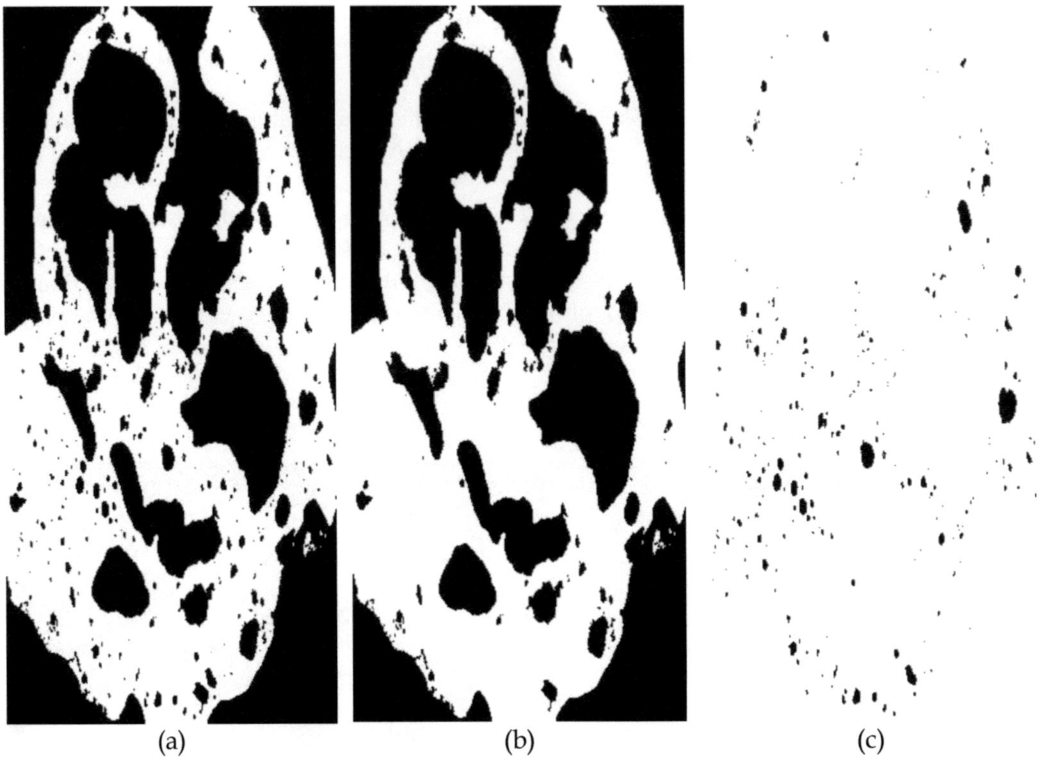

Fig. 4. Reconstructed slice showing the total porosity after binarisation (a), open porosity after region growing (b) and closed porosity (c).

3. Results and discussion

3.1 Morphology of the sinter mixture granules

The granules are formed in sinter mixture due to molecular and capillary forces caused by moistening. Following mechanism is generally accepted:

- particles sized over 1.0-1.5 mm act as nuclei (granule centres);
- superfines, sized below 0.05 mm, adhere to the surface of nuclei;
- particles of 0.05-0.75 mm may adhere to the layer of superfines.

Particles of intermediate size (from 0.75 to 1.0 mm) are sometimes considered as being problematic: they are not large enough to serve as a nucleus for granulation and they are too big to adhere to the surface of existing nuclei. Sinter producers usually take this concept into account while designing requirements for the sizing of the sinter mixture components (Ellis et al, 2007).

Optimal sinter mixtures design in the modern and future conditions is constrained by the following factors (Hunt, 2004, Makkonen et al, 2002, Shatokha et al, 2011):

- the growing deficit of rich coarse iron ores and the corresponding increase of the fineness of ores used for sintering;

- in many cases the only opportunity to keep iron content in the sinter product at the level required by customers is to increase significantly the amount of deeply enriched iron ore concentrate with over 90% of superfines sized below 0.04 mm in the sinter mixture;
- sinter plant is usually considered as the facility needed to recycle fine ferrous wastes (gas cleaning dust and sludge, steelmaking slag, rolling mills scale etc), which sometimes makes unavoidable introduction to the sinter mixture essential quantity (mass share may reach as much as 20 % and more) of such - not perfectly suited for the sintering technology - materials.

Mentioned above may cause deviation of the real granulation behaviour in the sinter mixture from the generally accepted mechanism. These factors are connected with the increase of the share of various superfines (not only iron ore concentrate, but also wastes such as e.g. wet gas cleaning sludge), the role of which in the sintering mixture towards the granulation behaviour, process parameters and quality of final product remains in the focus of many researchers (Ellis et al, 2007, Khosa &Manuel, 2007, Naakano et al, 1998, Haga et al, 2005).

Structure of granules depends on adhering fine and nuclei ratio, moisture content, lime and wastes addition, water absorption capacity of the mixture components etc. (Mou et al, 2007). Bonding of particles in the granule is very weak which makes problematic preparation of section for microscopy analysis. Moreover, irregular form of nuclei and chaotic adhesion of fines decrease representativeness of study based on analysis of a random slice. Therefore application of new, non-destructive experimental methods for studies of the granules' structure is very promising.

Fig. 5-7 demonstrate morphology of the granules obtained from mixtures with the same basicity of 1.4 and with various share of concentrate corresponding to samples 4-6 in Table 1.

Structure of granule, obtained with the lowest share of concentrate (Fig. 5) generally corresponds to the "classic" understanding of sinter mixture granulation mechanism described above. This granule is formed around the large porous particle of a sinter return with an adhered layer of fines (dark). On the outer surface of the fines' layer it is possible to recognise the particles of an intermediate size (well below 0.5 mm).

In granule with the medium share of concentrate the volume of the adhered fines layer becomes generally larger. However, the layer develops unevenly around the nucleus, in a way that the granule has a "front" with thin layer of fines and a "tail" with thick one, which perhaps owes to the granulation dynamics (Fig. 6).

Granule, obtained from the mixture with largest share of concentrate (Fig. 7) has much more regular shape, quite similar to those of pellets (another type of iron ore material, produced from fine concentrate with additives of binder). This granule has a coarse particle of ore inside which could be assumed as nucleus. However, it contains also a distinct spherical clod of fines, which might be formed by concentrate around a drop of water and could be considered as pseudo-particle, potentially able to serve as the nucleus itself.

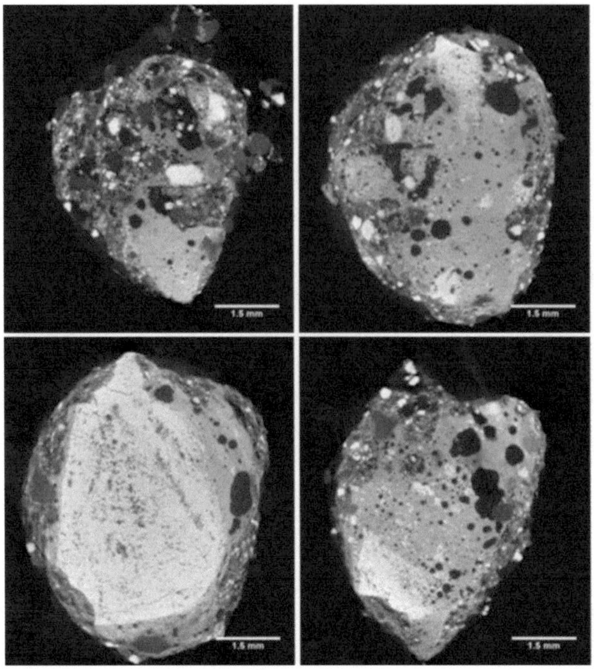

Fig. 5. Reconstructed slices of a granule (sample #4 in Table 1, four parallel sections). Almost entire granule's volume is occupied by sinter return particle (light grey matter of irregular shape with black pores) sized over 5 mm.

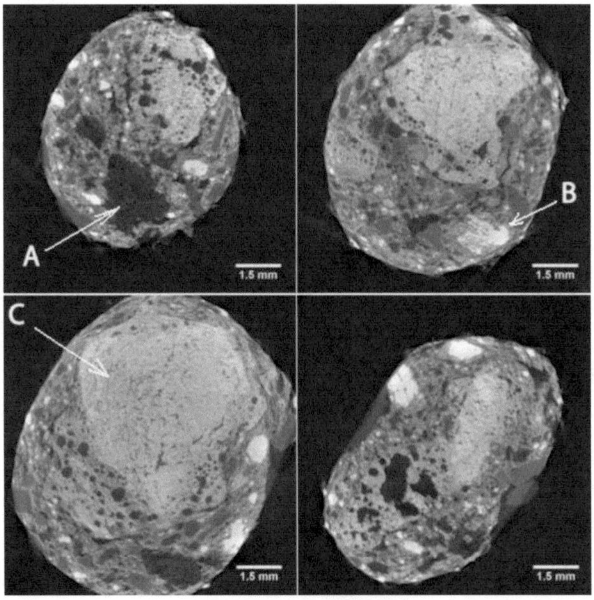

Fig. 6. Reconstructed slices of a granule (sample #5 in Table 1, four parallel sections): A - coke breeze, B - ore, C - sinter return, served as nucleus.

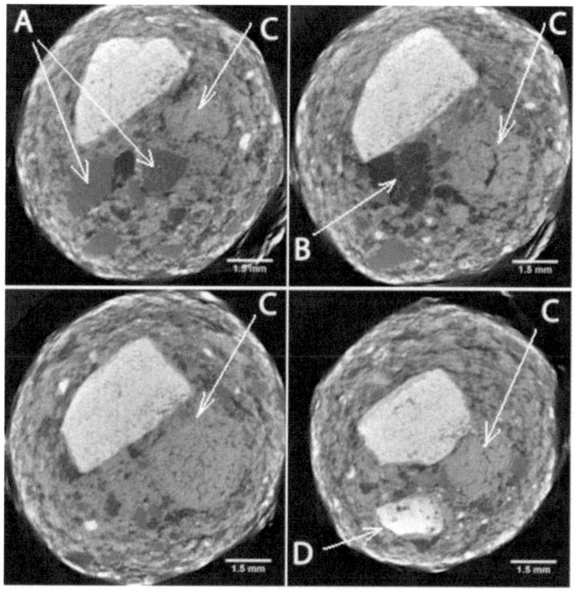

Fig. 7. Reconstructed slices of a granule (sample #6 in Table 1, four parallel sections): A - limestone, B - coke breeze, C - clod of iron ore concentrate, D -ore.

Application of X-ray tomography method enabled authors to study 3D structure of the granules and to perform their systematic typological analysis. In particular, it was found that the granules could be classified by the contribution of the share of adhering fines layer, formed at the surface of the nucleus, in the total radius of the granule. Three types of granules, different by adhering fines layer thickness, are represented in Fig.8 - both schematically and with the real samples.

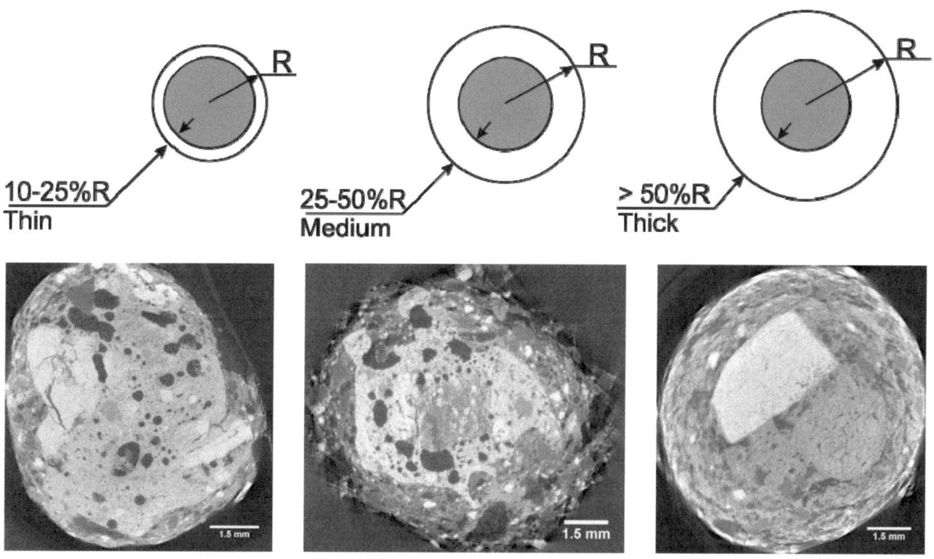

Fig. 8. Classification of the granules by the thickness of the adhering fines layer and real examples of each type.

Statistics of the shares of three types of granules in the granulated sinter mixtures versus the concentrate/ore ratio is shown in Fig.9. The increase of the concentrate/ore ratio in a sinter mixture leads to the decrease of the share of the granules with a thin and a medium thickness of the adhering fines layer. Correspondingly, the share of the granules with a thick adhering fines layer increases. Obviously, this effect is caused by the decrease of the number of coarse particles in a sinter mixture, able to serve as nuclei for the future granules with a simultaneous growth of the share of fines able to adhere to the nuclei or to the granules surface.

Introduction of the increased amounts of concentrate to the sinter mixture makes granulation more complex: superfine particles not only adhere to coarser ones but may also form durable nuclei while binding particles of intermediate size and the more superfines are used in the sinter mixture the less predictable is the size distribution of granules (Korotich, 1978).

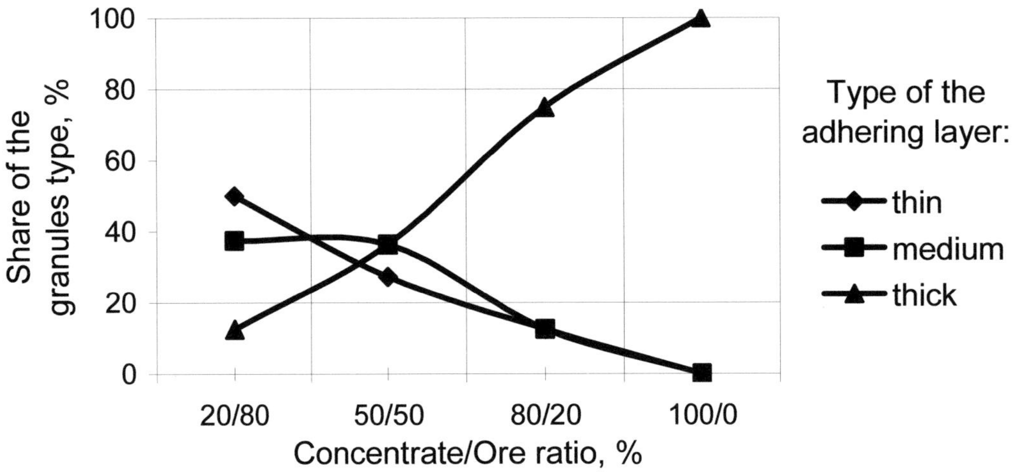

Fig. 9. Influence of the Concentrate/Ore ratio in a sinter mixture on the distribution of granules by the adhering fines layer thickness.

X-ray scanning confirms that sometimes the mechanism of granules formation is more complex than it is generally considered. For example, some of the granules, particularly those with a thick adhering layer, contain almost undistorted spherical granules of smaller (< 2 mm) diameter embedded in this layer. This means that some of the initially formed microgranules may not develop into independent granules. However, they may act as pseudo-particles, adhering to other granules and finally becoming a part of bigger granules. In some cases, granules formed as a result of the coalescence of two or three granules of similar size were observed. Examples of the granules with abnormal morphology are represented in Fig. 10. Effect of their presence on the sintering behaviour and quality of the final product requires further study.

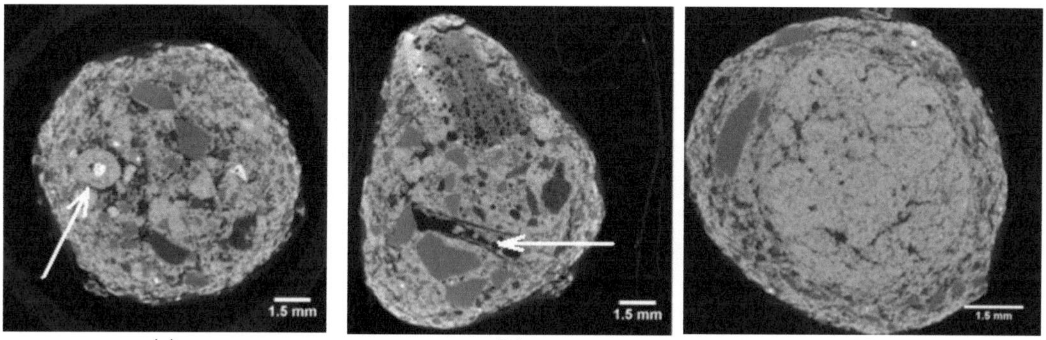

(a) (b) (c)

Fig. 10. Examples of granules with abnormal structure: (a) small granule embedded inside the adhering layer; (b) remains of the boundary of two granules coalesced to form a bigger granule; (c) pellet-like granule with no nucleus.

3.2 Porosity of sinter

Development of pores in iron ore sinter is very complex process involving generation of gaseous phase due to combustion of fuel, decomposition of limestone, vaporisation of water etc under conditions when part of material smelts, forming liquid slag prone to foaming (e.g. due to reaction of FeO with carbon).

In particular, it was demonstrated by Yang & Standish (1991) that pores may form at the sites of limestone particles due to complex process, involving decomposition, reaction of lime with iron ore particles, and formation of calcium ferrites, which then get melted and the melt reacts with other particles. The porosity of the studied sinters as a function of the basicity is shown in Fig.9 separately for open and closed pores.

In our study a significant drop of open porosity for all sinters was observed with an increase of basicity from 0.8 to 1.4 which is in agreement with the opinion (Kokubu et al, 2004; Yang & Standish, 1991) that porosity decreases with increasing basicity due to the increased amount of the melt. However, further increase of basicity to 2.0 leads to a less significant change of open porosity. Moreover, for the sinter series produced with maximum concentrate share in the mixture porosity reduces, while for the others there is a tendency to increase porosity.

For the slag systems representing the melt produced in the sintering processes a basicity of 1.4 lies near the eutectic area and as basicity increases slag liquidus temperature also augments (Slag atlas, 1995). Hence, further basicity increase under the same heat conditions might be followed by formation of a more heterogeneous and consequently more viscous and less mobile melt. This may explain the observation that the sinter porosity does not alter as basicity increases from 1.4 to 2.0.

The tendency of porosity to reduced drop only for the sinter produced with the maximum share of concentrate as basicity increases to 2.0 might be explained by additional heat produced from oxidation of concentrate's magnetite accompanied with generation of larger quantity of the melt.

For the closed porosity no clear relationship with the basicity was observed. Probably formation of this type of the pores is less affected by the melt properties or governed by some other phenomena. E.g. there is the hypothesis (Korotich, 1978) that the closed pores are formed due to the evolution of the carbon monoxide from direct reduction of iron oxides as a result of the interaction with the fine solid fuel particles rolled-in to microgranules. Later study of Yang & Standish (1991) confirmed that pores are formed at sites of coke particles after their burn out.

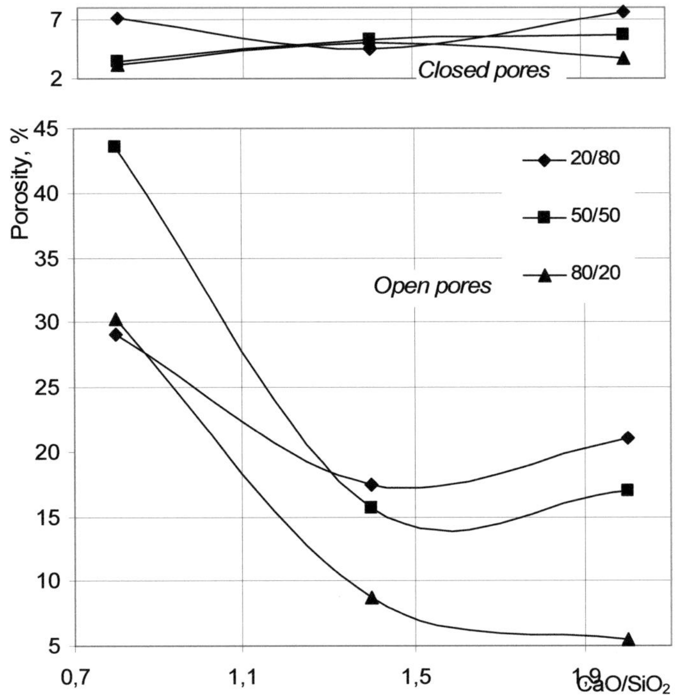

Fig. 11. Effect of sinter basicity at different concentrate/ore ratios on open and closed porosity.

The comparison of the data on open porosity obtained using 3D tomography and with the mercury test (sinter obtained from burden with 50% of concentrate at the ore/concentrate mix) is shown in Fig.12. The higher values obtained using 3D tomography can be explained by the narrow range of the measured pore intrusion diameter - from 329 to 0.01 μm - provided by the porosimeter used in this study. This means that a very significant portion of the sinter pores was not measured in the mercury intrusion test.

Fig.13 and Fig.14 compare the porosity measurement by the share of the volume related to the pores of given size, obtained by both methods for the similar range of pore sizes (only pores under 300 μm were taken into account for the tomography method). They demonstrate inherent incomparability of the results: the tomography method enables characterisation of the pores by their equivalent diameter, while mercury porosimetry - by their intrusion diameter -and large pores may have relatively small intrusion diameter (i.e. access window).

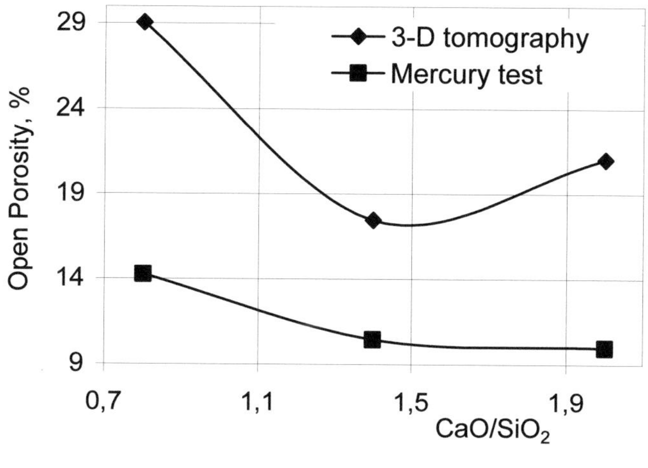

Fig. 12. Comparison of the data on open porosity obtained using 3-D tomography and mercury test (sinter produced from the mixture with ore/concentrate ratio of 20/80).

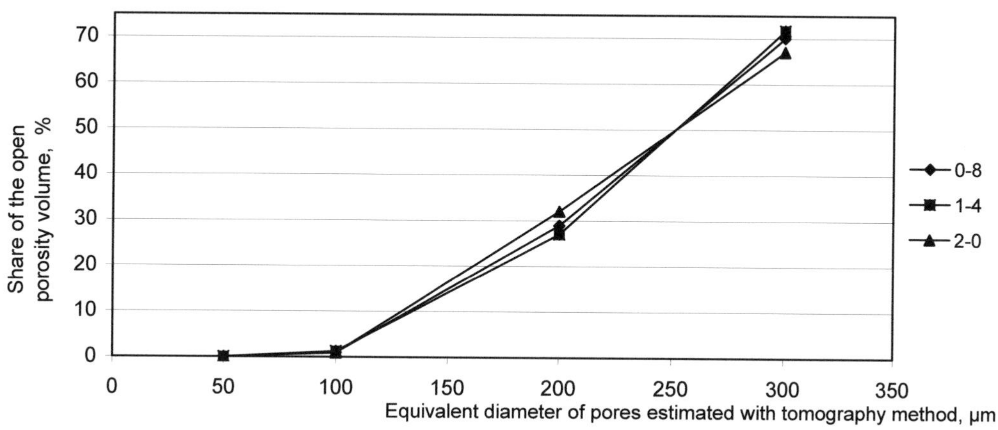

Fig. 13. Open porosity volume, related to pores with equivalent diameter under 300 µm measured by tomography.

The frequency distributions of the equivalent diameter of the open pores are shown in Fig. 15. Demonstration of the pore size distribution by the volume occupied by pores of certain size appears less informative: over 85-90% of the porosity volume belongs to very few pores larger than 500 µm. The frequency distributions are right-skewed with single peaks for a pore diameter of about 200 µm for almost all types of sinter, which is quite close to the average size of the pores in the range 148-159 µm (Hosotani et al, 1996), though data for open and closed pores were not distinguished in the given reference.

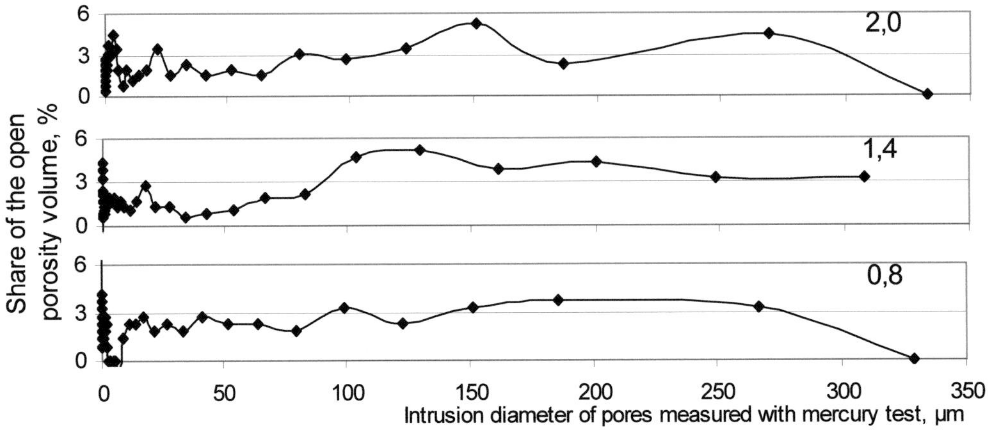

Fig. 14. Open porosity volume, related to the open pores with given intrusion diameter measured by the mercury porosimetry method.

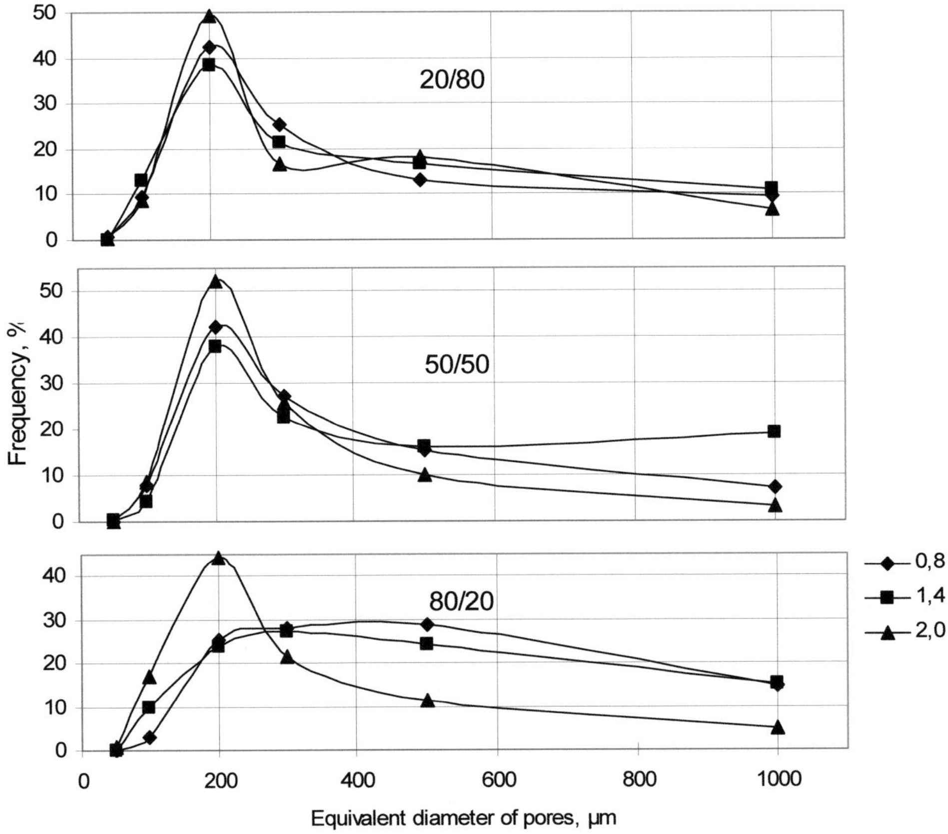

Fig. 15. Frequency distribution for equivalent diameter of open pores at basicities of 0.8, 1.4 and 2.0 for sinters produced from mixtures with concentrate/ore ratios of 20/80, 50/50 and 80/20.

Used resolution was sufficient to obtain information about the pore or piece of solid material with a lateral side of approximately 20 μm. However, during the watershed procedure, part of the open pores with equivalent diameter equal to 2-3 pixels (i.e. 40-60 μm) and smaller could be lost. Therefore it is possible to assume that open porosity under 60 μm is underestimated in the present study. According to Higuchi et al (2006) micro-porosity below 15 μm accounts for about 13% and Bhagat et al (2006) report that the micro-porosity below 10 μm accounts for 5.23-10.38% of the total porosity.

It must be mentioned that pores with a diameter larger than the mean free path of the reducing gases molecules (approximately 0.0004 μm in average at 1000 C° and under a pressure of 10^5 Pa (Lide, 2004) are considered as important in terms of their influence on the reducibility (Manchinskiy & Shkodin, 1963). Therefore further research using 3D tomography facilities with higher resolution would be helpful to measure the morphological parameters of the micro-porosity.

The sharpest peaks are always observed for the sinters with a basicity of 2.0. The curves corresponding to sinters of 0.8 and 1.4 basicities are very similar. The open pore size distribution curves for the sinters, obtained from mixtures with concentrate/ore ratios of 20/80 and of 50/50, are also similar. An increase of the concentrate share to 80% is accompanied by a softening of the peak for the sinter with basicity 2.0 and a shift to larger diameters.

3.3 Effect of porosity on the reducibility of sinter

Three sinter samples with different basicites from the series with concentrate/ore ratio of 50/50 were used for the reducibility test. It was considered as being very important for this test to use the samples previously analysed by tomography. Hence, it was not possible to determine the exact chemical composition of particular samples. Therefore, instead of the reduction degree, data on the weight loss (current in per cent of total) were used (Fig.16) as the reducibility characteristics.

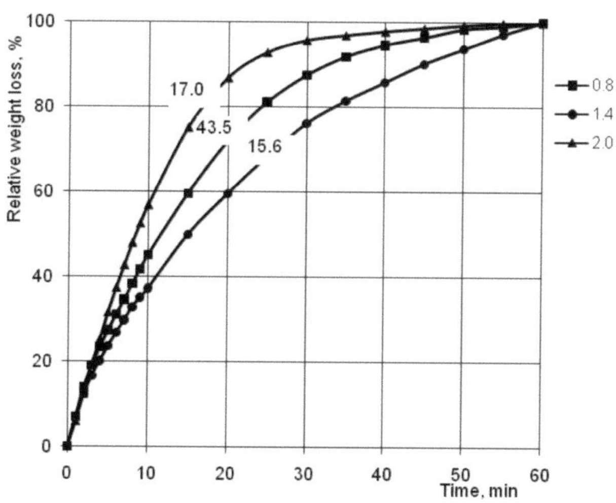

Fig. 16. Relative weight loss during reduction test for sinters with different basicities: (figures at the curves - open porosity in %).

The most intensive weight loss, observed for the basicity of 2.0, is in agreement with the existing knowledge and can be explained by the formation of more reducible calcium ferrites. However, more intensive reduction of sinter at the basicity of 0.8 to compare with those at 1.4 was less expected because in this range, an increase of reducibility with the increased basicity is normally observed for industrial sinters produced from the same raw materials (Katsman et al, 1986). This result might be explained by much higher porosity level observed for the sinter with basicity of 0.8. Positive effect of sinter porosity on its reducibility has been stated by many authors (Higuchi et al, 2004; Hosotani et al, 1996; Bhagat et al, 2006).

4. Conclusions

The X-ray tomography method was applied for the investigation of the internal structures of sinter mixture granules and sinter. The method has important advantages over the methods used so far for this kind of materials, allowing non-destructive characterisation of the microstructure in the bulk. Following main results were obtained:

1. Various components of the sinter mixture like iron ore, concentrate, coke breeze and limestone can be identified in the granules and it is possible to estimate their size and spatial distribution.
2. A novel approach for the classification of granules by the thickness of the adhering fines layer in a proportion to the whole granule radius has been suggested. It was shown that the growth of concentrate's share in a sinter mixture leads to a statistically increased presence of granules with a thicker adhering layer (more than a half of the granule's radius).
3. Deviation of the granules structure from the generally accepted granulation model was observed when the share of concentrate in the mixture was increased. Abnormal structures of some granules have been observed: microgranules embedded in the adhering layer of bigger granules; granules formed as a result of the coalescence of two or more granules of similar size; granules formed without coarse nucleus particle, with a structure similar to this of pellets.
4. For the sinter product tomography method allows obtaining of the data on the volume and equivalent diameter of each pore, thus providing the possibility to analyse open and closed pores separately.
5. Influence of the sinter basicity on the open porosity is associated with the quantity and properties of the melt formed during sintering. For the closed porosity clear relationship neither with the sinter basicity nor with the share of concentrate was observed.
6. Porosity characterisation using mercury porosimetry and 3D tomography methods gives inherently incomparable results: the former characterises the pores by their intrusion diameter and the latter – by their equivalent spherical diameter.
7. After the non destructive tomography analysis the samples could be used for further research. This opens new possibilities to analyse the effect of the porosity in relation to the mechanical and physicochemical properties of the sinter. Study of the reducibility of the sinter samples, previously analysed using tomography method, confirms predominant influence of the open porosity on iron ore sinter reducibility.

5. References

Bhagat, R. P.; Chattoraj, U. S. & Sil, S. K. (2006): Porosity of Sinter and its relation with the sintering indices. *Iron and Steel Institute of Japan International*, Vol.46, 2006, pp. 1728–1730

Debrincat, D.; Loo, C. E. & Hutchens, M. F. (2004). Effect of Iron Ore Particle Assimilation on Sinter Structure. *Iron and Steel Institute of Japan International*, Vol. 44, 2004, pp. 1308–1317

Ellis, B. G.; Loo, C. E. & Witchard, D. (2007). Effect of ore properties on sinter bed permeability and strength. *Ironmaking & Steelmaking*, Vol. 34, 2006, pp.99-108.

Haga, T.; Nakano, M.; Kasama, S. & Bergstrand, R. (2005). Effects of Metallic Iron Bearing Resources on Iron Ore Sintering. *Proceedings of the 5th European Coke and Ironmaking Congress*. Stockholm, Sweden, June 2005, Vol.2, pp. 2:2-1-2:2-9.

Higuchi, K.; Naito, M.; Nakano, M. & Takamoto, Y. (2004): Optimization of Chemical Composition and Microstructure of Iron Ore Sinter for Low-temperature Drip of Molten Iron with High Permeability. *Iron and Steel Institute of Japan International*, Vol. 44, (2004), pp. 2057–2066

Hosotani, Y.; Konno, N.; Yamaguchi, K.; Orimoro, T. & Inazumi, T. (1996): Reduction properties of sinter with fine dispersed pores at high temperatures of 1273 K and above. *Iron and Steel Institute of Japan International*, Vol. 36, 1996, pp. 1439-1447

Hunt, G. BHP Billiton's iron ore business - current status and future outlook. *Proceedings of the 7th AJM Annual Global Iron Ore And Steel Forecast Conference*, 23- February 2004, Brisbane, Australia

Katsman, V.; Shatokha, V. & Emelyanov, V. (1986). Studies on partition of slag and metal during melting of sinter and pellets. *Izvestiya VUZov. Chernaya Metallurgiya*. № 9, 1986, pp. 11-13

Khosa, J. & Manuel, J. (2007). Predicting Granulating Behaviour of Iron Ores Based on Size Distribution and Composition. *Iron and Steel Institute of Japan International*, Vol 47, 2007, pp. 965-972.

Kokubu, H.; Kodama, T.; Itaya, H. & Oguchi, Y.(1986). Formation of pores in iron ore sinter *Transactions of the Iron and Steel Institute of Japan International*, Vol. 26, 1986, pp. 182-185

Korotich, V. I. (1978). *Fundamentals of the theory and technology of ironmaking burden preparation*, Metallurgiya, Moscow, USSR.

Lide, D.R. (2004). *CRC Handbook of Chemistry and Physics: A Ready-reference Book of Chemical and Physical Data*, CRC Press, 2712 p.

Maire, E.; Buffière, J.-Y.; Salvo, L.; Blandin, J. J.; Ludwig, W. & Létang, J. M.(2001): On the application of x-ray microtomography in the field of materials science. *Advanced Engineering Materials*, Vol. 3, 2001, pp. 539 - 546.

Makkonen, H.T.; Heino, J.; Laitila, L.; Hiltunen, A.; Pöyliö, E. & Härkki, J. (2002). Optimisation of steel plant recycling in Finland: dusts, scales and sludge. *Resources, Conservation and Recycling*. Vol. 35, 2002, pp.77-84

Manchinskiy, V. G. & Shkodin, K. K. (1963). Overview on the development and modern state of the reduction processes kinetics. In: *Modern Studies of the Blast Furnace Process*, pp.153-166, Metallurgiya, Moscow, USSR

Mou, J.-L.; Sun, Y-M.; Chen, Y-C.; Liao, C.W. & Tarng, Y-S. (2007). Optical Line Scan Inspection System for Pseudo-particle Analysis. *Iron and Steel Institute of Japan International*, Vol.47, 2007, pp. 1780–1283

Naakano, M.; Yamakawa, K.; Hayakawa, N. & Nagabuchi, M. (1998). Effects of Metallic lron Bearing Resources on lron Ore Sintering. *Iron and Steel Institute of Japan International*, Vol. 38, 1998, pp. 16-22

Shatokha, V.; Gogenko, O. & Kripak, S. (2011). Utilising of the oiled rolling mills scale in iron ore sintering process. *Resources, Conservation and Recycling*, Vol.55, 2011, pp.435–440

Shatokha, V.; Korobeynikov, I.; Maire, E. & Adrien, J. (2009). Application of 3D X-ray tomography to investigation of structure of sinter mixture granules. *Ironmaking & Steelmaking*, Vol. 36, 2009, pp.416-420.

Shatokha, V.; Korobeynikov, I.; Maire, E.; Gremillard, L. & Adrien, J. (2010). Iron ore sinter porosity characterisation with application of 3D X-ray tomography. *Ironmaking & Steelmaking*, Vol. 37, 2010, pp.313-319

Slag atlas (1995). Ed. VDEh, 2nd Edition, Verlag Stahleisen GmbH, Dusseldorf, Germany

Yang, Y.-H. & Standish, N., (1991). Mechanisms of Pore Formation in Iron Ore Sinter and Pellets. *Iron and Steel Institute of Japan International*, Vol. 31, 1991, pp. 468-477
http:/rsb.info.nih.gov/ij

11

Effect of the Additives of Nanosized Nb and Ta Carbides on Microstructure and Properties of Sintered Stainless Steel

Uílame Umbelino Gomes, José Ferreira da Silva Jr. and Gislâine Bezerra Pinto Ferreira
Federal University of Rio Grande do Norte
Brazil

1. Introduction

Many applications on modern technologies try to combine uncommon properties that can't be attended by conventional metallic alloys, ceramics and polymeric materials. On this way, engineers of aerospace, for example, are ever looking for structural materials that have low densities, be stronger, with wear resistance and that have high resistance to corrosion (Gordo et al., 2000). These characteristics combine amazing properties for just one material. Normally, very dense materials are harder with high resistance to wear. Thus, composite materials are designed to provide this mixing of controlled properties.

The principle of combined action of different properties from two or more distinct materials characterizes the composite materials. Then, it's the central principle of composite materials that is enhancing one or more properties my mixing materials in various ways. Thus a very elementary example of a composite would be mud mixed with straw; still a very widely used material in the construction of houses. These innovative materials open up unlimited possibilities for modern material science and development.

Basically there are three kinds of composite materials, they are: polymer matrix composite (PMC), for example fiberglasses embedded in a thermoplastic resin of polyethylene; ceramic matrix composite (CMC) with carbon fibers reinforcing SiC matrix; and metallic matrix composite (MMC) to reinforce a matrix of steel with refractory ceramic hard particles.

This last one kind of composite material will be discussed in this chapter with focus on effect of the dispersion of nanosized carbides (NbC and TaC) in the sintered microstructure of the stainless steel 316L. Powder metallurgy technique will be applied as well as mechanical alloying and the effects promoted by hard particles in a metal matrix. A practical case study will be shown.

2. Metal Matrix Composites (MMC)

Metal matrix composites are an attractive end goal for many materials engineers as an alternative to the traditional pure ceramics or pure metal material. The goal is to create a material that has nearly the same high strength as the ceramics, but also high fracture toughness due to the contribution from the metal, this has lead to the development of a new family of metallo-ceramic composites. Classical examples are the precipitation and nucleation of carbides in steel for obtaining good strength while still maintaining most of the toughness from the metal, or the fabrication of hard metals such as a matrix cobalt containing a large percentage of grains of hard tungsten or titanium carbides or nitrides, which during the powder processing grow together and form a continuous skeleton (Breval, 1995).

In principle, wrought irons and most conventional steels could be treated as MMCs since they have a metallic matrix which is reinforced with a dispersed phase that could be oxides, sulphides, nitrides, carbides, etc. Even if we take a definition through "microstructural design", many conventional engineering alloys, such as some steels involving dispersions created at a moving alpha-gamma interface, really ought to be included. Although, composite material is a definition very embracing, it is quite tendentious to think that a directional solidification of a eutectic microstructure is within the framework of an MMC definition (Ralph, 1997).

The reinforcement of metals can have many different objectives. The reinforcement of light metals opens up the possibility of application of these materials in areas where weight reduction has first priority. The precondition here is the improvement of the component properties. The development objectives for light metal composite materials are:

- Increase in yield strength and tensile strength at room temperature and above while maintaining the minimum ductility or rather toughness,
- Increase in creep resistance at higher temperatures compared to that of conventional alloys,
- Increase in fatigue strength, especially at higher temperatures,
- Improvement of thermal shock resistance,
- Improvement of corrosion resistance,
- Increase in Young's modulus,
- Reduction of thermal elongation.

To summarize, an improvement in the weight specific properties can result, offering the possibilities of extending the application area, substitution of common materials and optimization of component properties. With functional materials there is another objective, the precondition of maintaining the appropriate function of the material. Objectives are for example:

- Increase in strength of conducting materials while maintaining the high conductivity,
- Improvement in low temperature creep resistance (reactionless materials),
- Improvement of burnout behavior (switching contact),
- Improvement of wear behavior (sliding contact),
- Increase in operating time of spot welding electrodes by reduction of burn outs,
- Production of layer composite materials for electronic components,
- Production of ductile composite superconductors,
- Production of magnetic materials with special properties.

For other applications different development objectives are given, which differ from those mentioned before. For example, in medical technology, mechanical properties, like extreme corrosion resistance and low degradation as well as biocompatibility are expected. Although increasing development activities have led to system solutions using metal composite materials, the use of especially innovative systems, particularly in the area of light metals, has not been realized. The reason for this is insufficient process stability and reliability, combined with production and processing problems and inadequate economic efficiency. Application areas, like traffic engineering, are very cost orientated and conservative and the industry is not willing to pay additional costs for the use of such materials (Kainer, 2006).

Reinforcements for metal matrix composites have a manifold demand profile, which is determined by production and processing and by the matrix system of the composite material. The following demands are generally applicable:

- Low density,
- Mechanical compatibility (a thermal expansion coefficient which is low but adapted to the matrix),
- Chemical compatibility,
- Thermal stability,
- High Young's modulus,
- High compression and tensile strength,
- Good processability,
- Economic efficiency.

On the other hand, the matrix has the role of distribute and transfer the tensile and to protect the surface of reinforcement. Reinforcements can be used as particles, whiskers or fibers. Unfortunately, high strength fibers and their processing methods are extremely expensive, and this limits their wide industrial application. Moreover, continuous fiber reinforced composites do not usually allow secondary forming process, which is used in the original shape in which they were manufactured. As a result of these limitations, over the last few years, new efforts on the research of non-continuous reinforced composites have been applied. Composites with non-continuous reinforcement do not have the same level of properties as continuously reinforced composites, but their cost is lower, their processing methods are more adaptable to conventional ones and their performance is acceptable (Torralba et al., 2003).

The selection of suitable matrix alloys is mainly determined by the intended application of the composite material. With the development of light metal composite materials that are mostly easy to process, conventional light metal alloys are applied as matrix materials. In the area of powder metallurgy special alloys can be applied due to the advantage of fast solidification during the powder production. Those systems are free from segregation problems that arise in conventional solidification. Also the application of systems with oversaturated or metastable structures is possible (Kainer, 2006).

Normally, the process to produce MMC involves at least two steps: consolidation and synthesis (i.e. introduction of the reinforcement into the matrix), following by a modeling operation. There are a lot of techniques to consolidate; MMC reinforced with discontinuous fibers are susceptible to modeling through mechanical forming (i.e. forging, extrusion and lamination).

3. Powder Metallurgy (P/M) used to obtain composite materials

Between lots of ways to produce composite materials, powder metallurgy (P/M) is an easy way to manufacture composite materials with metallic matrix and offer some advantages compared with ingot metallurgy or diffusion welding. The main advantage is the low manufacture temperature that avoids strong interfacial reaction, minimizing the undesired reactions between the matrix and the reinforcement. In other cases, P/M allows production of some materials which cannot be obtained by any other alternative route (for example, SiC reinforcing Ti alloys). Composites that use particles or whiskers as reinforcement can be obtained easier by P/M than by other routes. Another advantage of P/M is the homogeneous distribution of reinforcements. This uniformity not only improves the structural properties but also the reproducibility level in properties (Torralba et al., 2003).

In short, powder metallurgy is the process of blending fine powdered materials, pressing them into a desired shape (compacting), and then heating the compressed material in a controlled atmosphere to bond the material (sintering). Sintering is a term applied to bond the material and it describes the process in which aggregated or compacted powders acquire a solid structure, coherent through internal displacements of mass by diffusion process of atoms activated by temperature (Gomes, 1995). P/M process generally consists of four basic steps: (1) powder manufacture, (2) powder blending, (3) compacting and (4) sintering. Compacting is generally performed at room temperature, and the elevated-temperature process of sintering is usually conducted at atmospheric pressure. Hot Isostatic Pressure (HIP) is another way to press and to sinter at the same time. This technique avoids some troubles in the pressing step. Optional secondary processing often follows to obtain special properties or enhanced precision such as thermal treatments. Each step has specific parameters that must be studied and controlled to reach the performance desired.

Among some techniques to manufacture powders, such as liquid metal atomization, chemical and thermal decomposition reactions, electrolyte deposition, sol-gel process and milling, this last one will be shortly discussed in the next section.

4. Mechanical alloying and milling

Two kinds of millingcould be distinguished - conventional low energy milling (e.g. in a ball mill) and high energy milling. In the last case, the most used devices are planetary ball mill, SPEX ball mill and attritor ball mill. High energy milling (HEM) and mechanical alloying (MA) are terms utilized to name the synthesis of alloys and compounds by means of intensive milling that causes mixing at an intimate level and deformation of the materials microstructure (Costa et al. as cited by Benjamin, 1976 and Suryanarayana, 2001). MA is a solid-state powder processing technique involving repeated welding, fracturing, and re-welding of powder particles in a high-energy ball mill. Originally developed to produce oxide-dispersion strengthened (ODS) nickel- and iron-base superalloys for applications in the aerospace industry, MA has now been shown to be capable of synthesizing a variety of equilibrium and non-equilibrium alloy phases starting from blended elemental or prealloyed powders. The non-equilibrium phases synthesized include supersaturated solid solutions, metastable crystalline and quasicrystalline phases, nanostructures, and amorphous alloys. Recent advances in these areas and also on disordering of ordered intermetallics and mechanochemical synthesis of materials have been critically reviewed after discussing the process and process variables involved in MA (Suryanarayana, 2001). The process of MA consists of loading the powder mix and the grinding medium (generally hardened steel or tungsten carbide balls) in a container sealed usually under a protective atmosphere (to avoid/minimize oxidation and nitridation during milling) and milling for the desired length of time. There are some parameters that must be controlled in the milling process, such as ball to powder ratio, length of milling time, atmosphere and ball sizes (Suryanarayana, 2001).

Nowadays, mechanical alloying has been successfully applied to prepare Ni-based, Fe-based, and Al-based oxide dispersion strengthened (ODS) alloys and carbide dispersion strengthened (CDS) alloys. During preparation of these materials, the strengthened phase is directly put into the matrix powders as a raw material. After a long time of ball milling, mechanical alloying can be achieved to some extent, but it is unlikely to produce the optimum combination between the strengthened phase, and the matrix. Element powders which can form oxide or carbide are mixed with the matrix powders and then subjected to ball milling, the supersaturated solid solution will form first, from which the dispersion strengthened alloy can be fabricated by a precipitation of second phase during heat treatment, or by mechanical-chemical reaction. This is a new method to produce CDS and ODS alloys by mechanical alloying, which can be used to optimize the strengthened action of the second phase to the maximum (Wang et al., 2001). During mechanical alloying, the mechanical energy transfers from the balls to the alloy powders, which results in the severe plastic deformation and the introduction of dislocations, lattice distortion and vacancies. When hard particles are incorporated into the matrix they promote an effect of hardening.

5. Mechanism of reinforcement

The characteristics of metal matrix composite materials are determined by their microstructure and internal interfaces, which are affected by their production and thermal mechanical prehistory. The microstructure covers the structure of the matrix and the reinforced phase. The chemical composition, grain and/or sub-grain size, texture, precipitation behavior and lattice defects are of importance to the matrix. The second phase is characterized by its volume percentage, its kind, size, distribution and orientation. Locally varying internal tension - due to the different thermal expansion behavior of the two phases - is an additional influencing factor. With knowledge of the characteristics of the components, the volume percentages, the distribution and orientation, it might be possible to estimate the characteristics of metallic composite materials. The approximations usually proceed from ideal conditions, i.e. optimal boundary surface formation, ideal distribution (very small number of contacts of the reinforcements among themselves) and no influence of the component on the matrix (comparable structures and precipitation behavior). However, in reality a strong interaction arises between the components involved, so that these models can only indicate the potential of a material. The different micro-, macro- and meso-scaled models proceed from different conditions and are differently developed (Kainer, 2006 as cited by Clyner & Withers, 1993 and McDaniels et al., 1965).

6. Effect of ceramic particles in a metal matrix

Ceramic particles are some of the most widely used materials for reinforcing metal matrix. In this way, the aging process can accelerate in aging alloys and the hardness can be improved considerably with direct impact and wear behavior (Torralba, 2003). Plastic deformation of a metallic material is associated to sliding of crystalline planes of crystalline structure and, therefore, there is a hardening and increase of mechanical resistance of the material when obstacles are placed to this movement. The particles of the reinforcement distributed uniformly into the grain serve as the obstacles to displacement of the discordances. The mechanisms of hardening occur when this discordance goes through the reinforcements or surrounds these reinforcements to continue its movement. In this sense, these mechanisms promote the hardening of the material during the plastic deformation. Obviously, the closer the particles are, the better tensile strength is improved. This is due the difficulty in movement of the discordances through or around the particles.

Compared with their unreinforced alloys, steel matrix composites offer higher hardness and wear resistance and their elastic modulus is the highest among all machinable materials and materials which can be hardened. On the other hand, as in all metal matrix composites, reductions in the fracture properties (ductility, toughness) are to be expected. Finally, the incorporation of dispersed ceramics particles inside the matrix can improve sintering, corrosion and oxidation properties of composites.

7. Case study – Effect of the dispersion of nanosized carbides (NbC-TaC) in the sintered microstructure of the stainless steel 316L

Composites containing different hard phases have already been studied (Tjong and Lau, 1999). Hard phases added to the steel matrix were mainly refractory carbides such as NbC, TaC and VC (Gordo et al., 2000). WC-Co (Ban and Shaw, 2002), Cr_2Al, $TiCr_2$, SiC and VC (Abenojar et al., 2002 and Abenojar et al., 2003) were also investigated as a potential reinforcement of the steel matrixes. This case study proposes a practical analysis to improve density and hardness of hard particles, used as additive. Recently a possibility to produce very homogeneous compacts with relative density among 95% without exaggerated grain growth was reported. These results were attributed to the incorporation of nanosized carbides' particles into stainless steel matrixes (Gomes et al., 2007).

7.1 Experimental procedure

This case study was carried out with starting powders of water-atomized stainless steel 316L manufactured by Höganäs. Pure sample and samples with addition of up to 3% of NbC and with 3% of TaC produced in the UFRN (Federal University of Rio Grande do Norte, Brazil) laboratory, previously characterized were mechanically milled in a conventional ball mill for 24 hours and axially cold pressed in a cylindrical steel die at 700 MPa. Sintering was carried out in vacuum. Samples were heated up to 1290°C with heating rate 20°C/min and isothermally held for 30 and 60 minutes. Sintered samples were characterized by x-ray diffraction, scanning electron microscopy and density and hardness were measured.

7.2 Results

Figures 1, 2, 3, 4, 5 and 6 show the x-ray diffraction patterns and micrographs of NbC and TaC used in this work. Only carbides characteristics peaks were identified. The carbides produced by UFRN show wider peaks than supplied by others due to the crystal refinement in accordance with previous work (Souza et al., 1999 and Medeiros, 2002). The carbides synthesized by UFRN are finer than those supplied by other manufacture. The small crystallite size and lattice strain values, provided by Rietveld's Method presented in the table 1, confirm it. Both carbides synthesized in UFRN, NbC (MCS = 16.67nm) and TaC (MCS = 13,78nm) present similar medium crystallite size (MCS).

Powder	Size (nm)								Lattice Strain (%)
	NbC (2θ)								
	34.90	40.49	58.40	69.90	73.40	87.48	97.70	MCS (nm)	
NbC-U	17.65	17.64	17.10	16.43	16.22	15.27	16.38	16.67	0.0009
NbC-C	75.77	75.65	77.22	79.44	80.61	84.73	80.35	79.11	0.000145
	TaC (2θ)								**Lattice Strain (%)**
	34.88	**40.48**	**58.60**	**70.06**	**73.66**	**87.61**	**98.03**	**MCS (nm)**	
TaC-U	15.91	15.47	14.14	13.39	13.18	12.43	11.97	13.78	0.00156
TaC-C	42.68	41.62	40.19	39.79	39.73	40.04	40.61	40.66	0.000102
MCS = Crystallite medium size; NbC-U and TaC-U = carbides synthesized by UFRN; NbC-C and TaC-C = carbides supplied by Johnson Matthey Company and Sigma Aldrich.									

Table 1. Crystallite size and lattice strain obtained for the carbides.

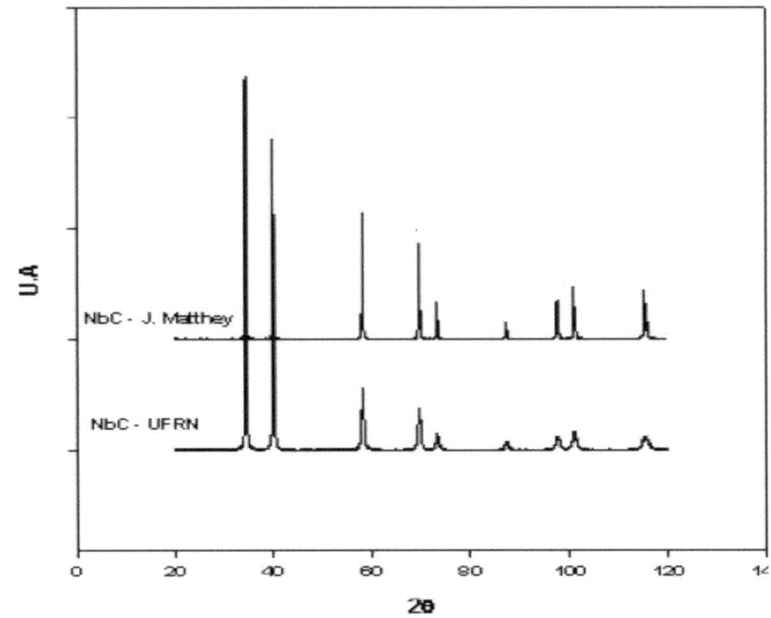

Fig. 1. X-ray diffraction patterns of Niobium Carbide produced at UFRN and from Johnson Matthey Company.

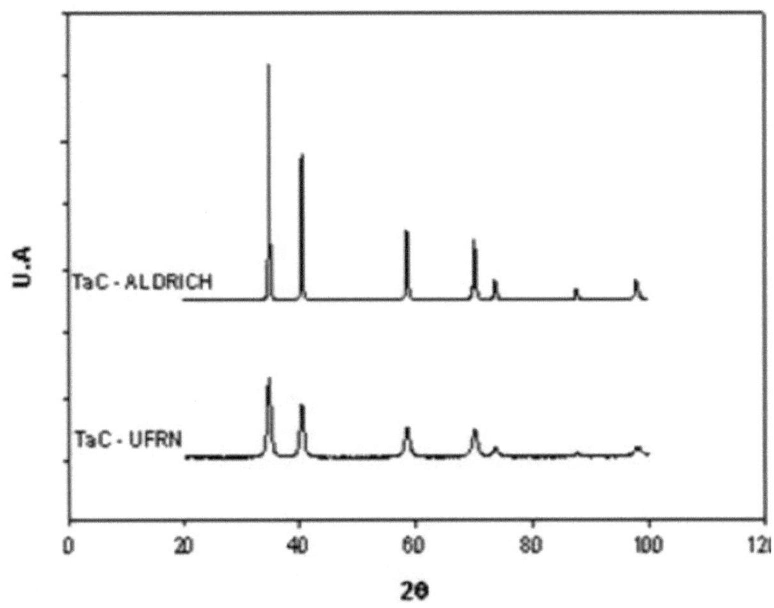

Fig. 2. X-ray diffraction patterns of Tantalum Carbide produced at UFRN and from Sigma Aldrich.

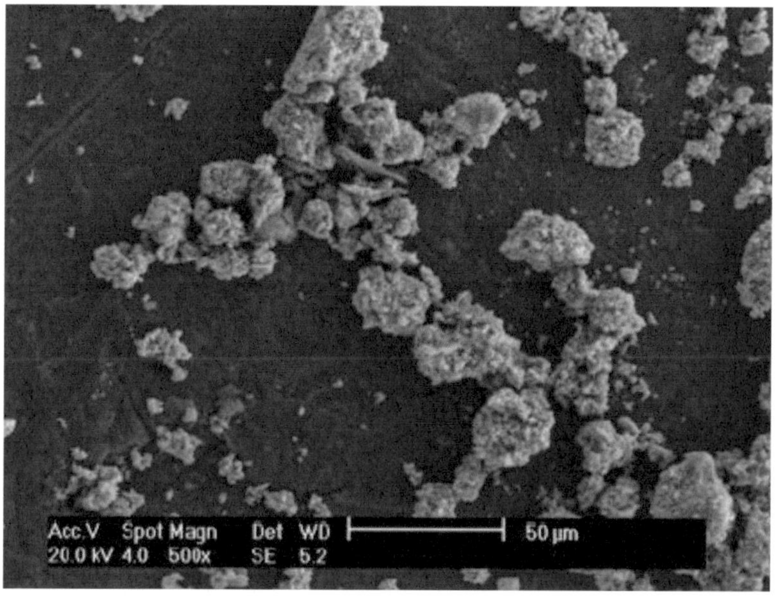

Fig. 3. Scanning electron micrographs of NbC as supplied from UFRN - 500x.

Fig. 4. Scanning electron micrograph of NbC as supplied from UFRN - 5000x.

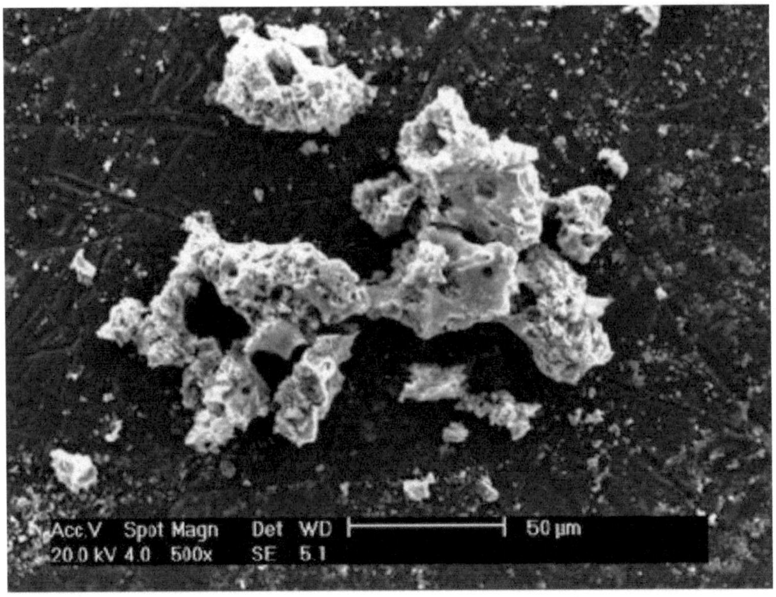

Fig. 5. Scanning electron micrographs of TaC as supplied from UFRN – 500x.

Fig. 6. Scanning electron micrographs of TaC as supplied from UFRN – 5000x.

Figures 7 and 8 show the micrographs of composite powders milled mechanically 24 hours, with addition (3%) of NbC (Fig. 7) and TaC (Fig. 8). The arrows indicate the carbide particles.

Fig. 7. Scanning electron micrographs of stainless steel powders reinforced with 3% of NbC from UFRN - 1000x.

Fig. 8. Scanning electron micrographs of stainless steel powders reinforced with 3% of TaC from UFRN - 1000x.

Figures 9 and 10 present the scanning electron micrographs of samples with NbC-U and sintered for 30 minutes and 60 minutes, respectively.

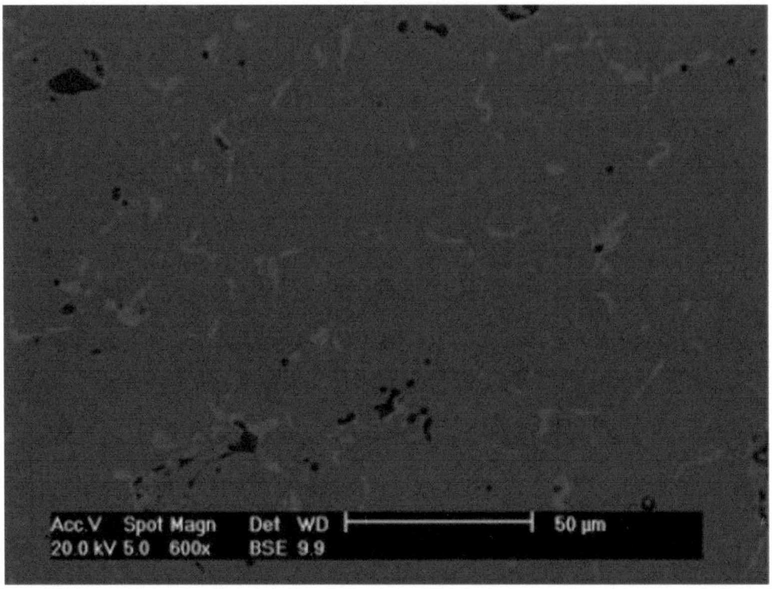

Fig. 9. Scanning electron micrographs of 316L stainless steel reinforced with 3% of NbC from UFRN sintered at 1290°C with isotherm of 30 minutes.

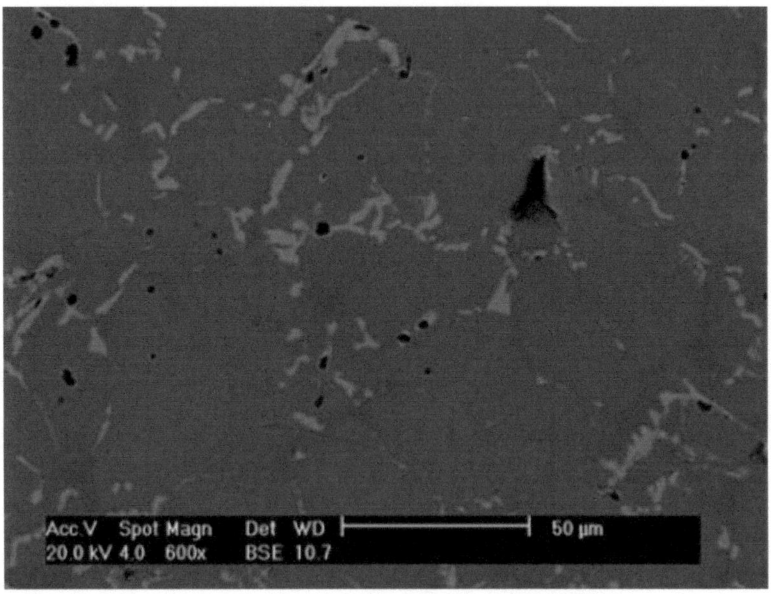

Fig. 10. Scanning electron micrographs of 316L stainless steel reinforced with 3% of NbC from UFRN sintered at 1290°C with isotherm of 60 minutes.

Figures 11 and 12 present the scanning electron micrographs of samples with TaC-U and sintered for 30 minutes and 60 minutes, respectively.

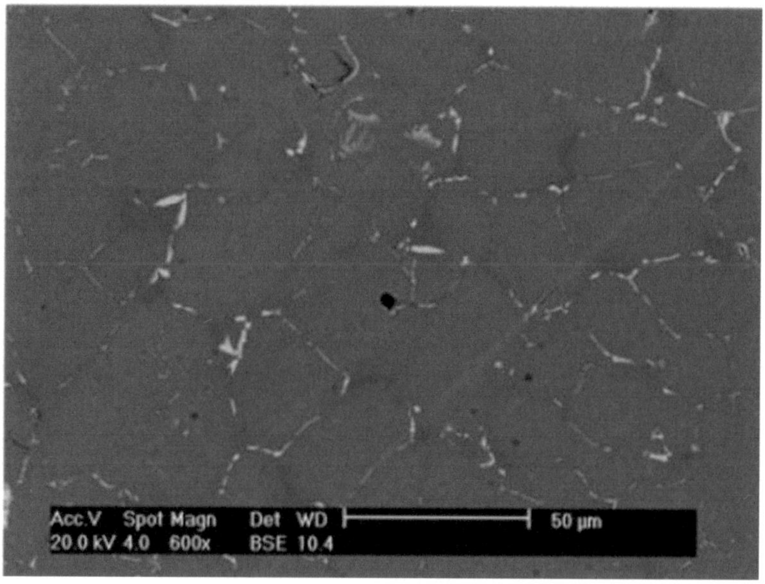

Fig. 11. Scanning electron micrographs of 316L stainless steel reinforced with 3% of TaC from UFRN sintered at 1290°C with isotherm of 30 minutes.

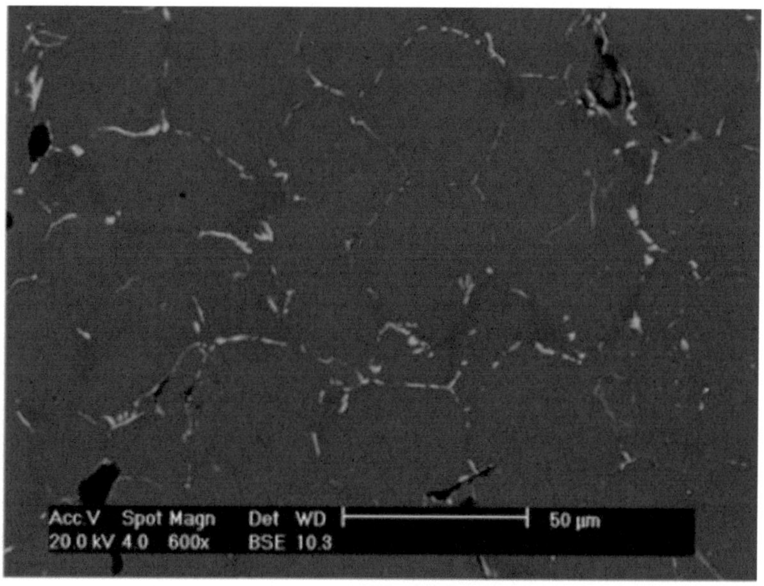

Fig. 12. Scanning electron micrographs of 316L stainless steel reinforced with 3% of TaC from UFRN sintered at 1290°C with isotherm of 60 minutes.

The table 2 presents values of densities and hardness. The carbides have influence on densification. Samples reinforced with NbC show higher densification. However, the great influence on the hardness values which increase from 76HV to 140 HV with addition of nanosized particles of NbC was observed.

Sample	Sintering time (min)	Density		Hardness (1000gf)
		(g/cm³)	(%)	
316L pure	30	7.10±0.07	89.0	76.0±5.2
316L pure	60	7.23±0.02	91.0	71.0±5.0
316L+NbC-U	30	7.53±0.05	94.6	133.0±8.3
316L+NbC-U	60	7.57±0.05	95.0	140±8.2
316L+TaC-U	30	7.57±0.07	93.0	94.0±9.8
316L+TaC-U	60	7.58±0.06	93.0	115.0±9.9

Table 2. Hardness and densities values from sintered samples.

7.3 Conclusions

The effects of nanosized particles of refractory carbides (NbC and TaC) on the sintering mechanism, denseness and hardness of stainless steel were investigated. The action of hardening mechanisms owed differences in fine particles size (10-20nm) results in great increase in the hardness values of the samples reinforced with carbides, especially for NbC.

The reinforced samples present very homogeneous composite raising the relative density to 93% for TaC and 95% for NbC. Great influence on the hardness that increase from 76.0 HV up to 115HV and even to 140HV, for the samples with nanosized reinforcement, was observed, owing to dispersion and precipitation of fine carbides in metallic matrix. The addition of carbides increases significantly the hardness of the stainless steel due to grain size reduction of the metallic matrix during sintering. Thus, the hardness does not increase only as a function of the density but mostly due to nanosized particles segregation on grain boundaries in the sintered microstructure.

8. Acknowledgment

The authors wish to thank CAPES, CNPq, LMCME-UFRN and NEPGN-UFRN.

9. References

Abenojar, J. ; Velasco, F. ; Torralba, J.A. ; Calero, J.A. & Marcè, R. (2002). Reinforcing 316L stainless steel with intermetallic and carbide particles. Materials Science and Engineering A. Vol. 335. pp. 1-5.

Abenojar, J. ; Velasco, F. ; Bautista, A. ; Campos, M. ; Bas, J.A. & Torralba, J.M. (2003). Atmosphere influence in sintering process of stainless steel matrix composites reinforced with hard particles. Composites Science and Technology. Vol. 63. pp. 69-79.

Ban, Z.G. & Shaw, L.L. (2002). Synthesis and processing of nanostructured WC-Co materials. Journal of Materials Science. Vol. 37. pp 3397-3403.

Benjamin, J.S. (1976). Mechanical alloying. Materials Science. Vol. 13. pp. 279-300.

Breval, E. (1995). Synthesis route to metal matrix composites with specific properties : a review. Composites Engineering . Vol. 5. pp. 1127-1133.

Clyne, T.W. & Withers, P.J. (1993). An introduction to metal matrix composites. Cambridge University Press. Cambridge.

Gomes, U.U. (1995). Tecnologia dos pós - Fundamentos e aplicações. Vol. 1. Federal University of Rio Grande do Norte.

Gomes, U.U. ; Furukava, M. ; Soares, S.R.S. ; Oliveira, L.A. & Souza, C.P. (2007). Carbide distribution effect on sintering and mechanical properties of stainless steel 316L. Materials - IV International Symposium. Portugal.

Gordo, E. ; Velasco, F. ; Atón, N. & Torralba, J.M. (2000). Wear mechanisms in high speed steel reinforced with (NbC)p and (TaC)p MMCs. Wear. Vol. 239. pp. 251-259.

Costa, F.A. ; Silva, A.G.P. ; Silva Jr, J.F. & Gomes, U.U. (2008). Composite Ta-Cu powders prepared by high energy milling. International Journal of Refractory Metals and Hard Materials.

Kainer, K.U. (2006). Metal matrix composites - custom-made materials for automotive and aerospace engineering. Wiley-VCH Velargh GmbH & Co, Weiheim. ISBN : 3-527-31360-5.

McDanels, D.L. ; Jech, R.W. & Weeton, J.W. (1965). Analysis of stress-strain behavior of tungsten-fiber-reinforced copper composites. Trans. Metall. Soc. AIME. Vol. 223. pp. 636-642

Medeiros, F.F.P. (2002). Synthesis of tungstein of tungsten and niobium carbides at low temperature, through gas-solid reaction in fixed bed reactor. Doctoral Theses. Federal University of Rio Grande do Norte, Brazil.

Pagounis, E. & Kindros, V.K. (1998). Processing and properties of particles reinforced steel matrix composites. Materials Science and Engineering A. Vol. 246. pp. 221-234.

Ralph, B. ; Yuen, H.C. & Lee, W.B. (1997). The processing of metal matrix composites - an overview. Journal of Materials Processing Technology. Vol. 64. pp. 339-353.

Souza, C.P. ; Favotto, C. ; Satre, P. ; Honoré, A. & Roubin, M. (1999). Preparation of tantalum carbide from an organometallic precursor. Brazilian Journal of Chemical Engineering. V. 16. pp. 3397-3403.

Suryanarayana, C. (2001). Mechanical alloying and milling. Progress in Materials Science. Vol. 46. pp. 1-184.

Suryanarayana, C. ; Ivanov, E. & Boldyred, V.V. (2001). The science and technology of mechanical alloyng. Materials Science & Engineering A. Vol. 304-306. pp. 151-158.

Tjong, S.C. & Lau, K.C.J. (1999). Sliding wear of stainless steel matrix composite reinforced with TiB_2 particles. Materials Letters. Vol. 41. pp. 153-158.

Torralba, J.M. ; Costa, C.E. & Velasco, F. (2003). P/M Aluminum matrix composites : an overview. Journal of Materials Processing Technology. Vol. 133. pp. 203-206.

Wang, C.G.; Qi, B.S.; Bai, Y.J.; Wu, J. & Yang, J.F. (2001). Dispersion strengthened alloy due to the precipitation of carbide during mechanical alloying.

12

Properties and Structure of Sintered Boron Containing Carbon Steels

G. Bagliuk
Institute for Problems of Materials Science
Ukraine

1. Introduction

The iron-carbon alloys are the most wide-spread powder structural materials due to relative simplicity of technologies and accessibility of input materials for their fabrication (Dorofeev, 1986; Metals Handbook , 1978).

However, the need for increasing of sintered steels mechanical and operational characteristics level causes the need of their supplementary alloying. The basic alloying elements, which have found a wide application in a practice of powder metallurgy, are copper, nickel, manganese, chromium, vanadium and molybdenum (Zhang, 2004; Dorofeev, 1986).

Herewith, the significant capacities for manufacturing of low-alloy steels, which operational characteristic often are not inferior of the level of the same for steels, produced with application of traditional alloying modes, makes it possible the use of boron as basic alloying element (Dudrova, 1993; Gülsoy, 2004; Napara-Volgina, 2011; Suzuki, 2002). In our previous articles it has been demonstrated that boron addition to the base material is very efficient way for improving basic mechanical properties of sintered steels as well positive effect on activation of sintering was observed too.

It is well known that boron activates sintering of iron and steels because of formation of persistent liquid phase. In addition, boron has high affinity to oxygen. Thereby while using boron as one of the alloying elements for production of sintered steels improved mechanical properties with excellent combination of hardness and ductility could be obtained.

Furthermore it is necessary to take into account that solubility of boron in the iron is exceedingly small, so its addition in powder steels will cause the number of difficulties, that make it necessary to optimize just but one boron composition in the sintered alloy, but the method of its addition in the initial powder mixture as well. The most often used method of boron addition is based on application of boron carbide powder as boron containing element. But sintering of Fe-B_4C powder mixtures is attended with formation of derivative porosity in consequence of dissolution of boron carbide particles in a contact with iron matrix. To eliminate the noted effect as boron containing additive Fe-B-C system master alloy can be used (Napara-Volgina, 2008; Xiu, 1999)..

So, in order to evaluate the ability of boron to activate sintering of the base powder preforms with aim of increasing of the mechanical properties and wear resistance, study was carried on influence of boron composition and modes of its addition in the initial powder mixture on basic mechanical and tribotechnical characteristics of sintered steel.

2. Effect of powder mixture content and modes of alloy addition on the structure and properties of sintered steels

In (Kazior, 2002) the authors had studied the influence of boron addition in different forms (elemental boron powder, boron carbide B_4C powder, and mixture of boron and carbide elemental powders) and different weight percentages on density and mechanical properties of prealloyed Astaloy CrM powders alloys.

The specimens were prepared by mixing of commercial water atomized Astaloy CrM powder with above-mentioned boron containing powders and were compacted at a pressure of 600 MPa to obtain an average density of 6.9 g/cm^3. The green parts were sintered at 1250 ^{0}C for 30 min. in laboratory furnace in pure dry hydrogen.

The obtained results had shown, that the additive of boron caused densification, but the effect depends on the amount and form of boron additives. For 0,2 (wt.) % B only for boron carbide small increase in density can be observed, whereas for the other modes of boron addition density is of the same level as for boron-free sintered steel (fig. 1). A significant increase in sintered density was obtained with higher content of boron (0.4 wt. %), reaching 7.44 g/cm^3 for iron–boron carbide powders mixture.

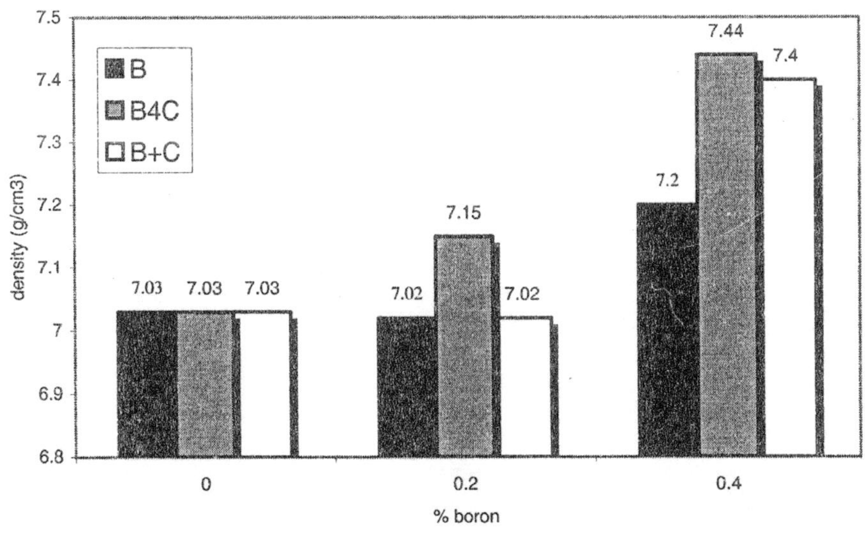

Fig. 1. Sintered density as a function of boron content and mode of its addition (Kazior, 2002).

Figure 2 shows the pore morphology of the boron-free and with 0.4 (wt.) % B_4C sintered alloys. It is evident, that boron addition considerably changed the pore morphology: while the boron-free sintered alloy has in its structure the irregular pores, typical for solid phase sintering (fig. 2, a), otherwise the boron containing sample has the large well rounded pores, which are typically formed during liquid phase sintering (fig. 2, b).

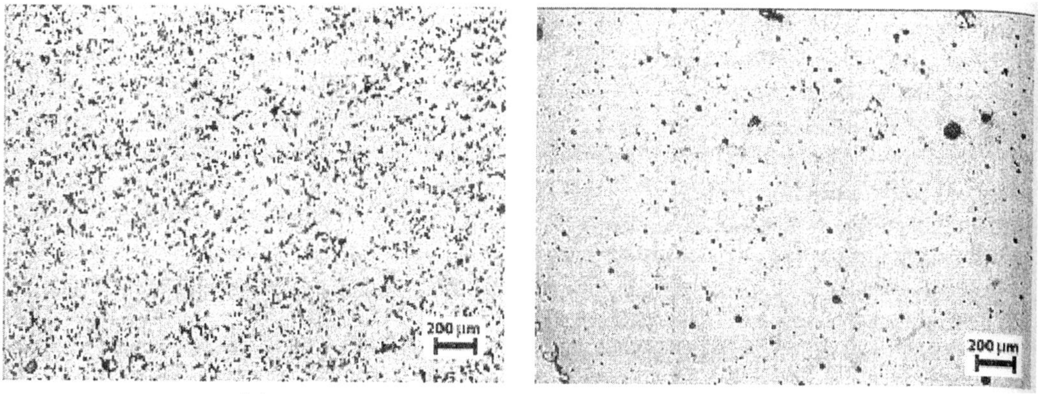

(a) (b)

Fig. 2. Optical micrograph of unetched boron-free (a) and with 0.4 (wt.) % boron (B_4C) alloyed sintered Astaloy CrM.

The unalloyed sintered Astaloy CrM has a homogeneous ferritic microstructure (fig. 3,a). The microhardness of the ferritic grains is 150-170 $HV_{0,02}$. The representative microstructure of the boron alloyed sintered specimens is shown in figure 3,b. Boron causes hardening of the matrix, and its microhardness is very sensitive of the form of boron. For elemental boron with increasing boron content the microhardness of matrix ranges between 270-300 and 330-370 $HV_{0,02}$ respectively. For B_4C with increasing boron content the microhardness of matrix ranges 330-370 and 400-470 $HV_{0.02}$ respectively. Beside, as a consequence of the eutectic reaction between austenite matrix and borides, the microstructure contains second constituent, which forms for higher boron content a continuous network on the grain boundaries. Figure 3,b shows the typical representative morphology of the second constituent, observed in 0.4 wt % boron (B_4C) alloyed sintered specimen. Two different borides phases were found in the second constituent. The gray component is the Fe-Cr rich boride, while the white part of the component is the Fe-Mo rich boride.

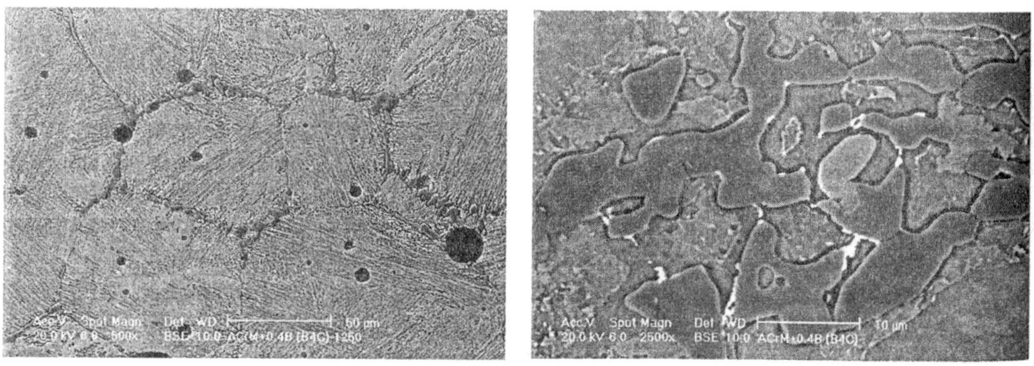

(a) (b)

Fig. 3. SEM micrograph of 0.4 (wt.) % boron (B4C) alloyed sintered Astaloy CrM. ×500 (a); ×2500 (b).

Table 1 summarizes the effect of boron on mechanical properties of the studied materials. The general trend is that, the higher boron content, the higher are the yield, strength, hardness and lower elongation of course. However, the results indicate that form of boron has significant effect on the tensile properties. For alloyed sintered specimens, both by boron carbide or elemental boron and graphite mixture, the microhardness of the matrix and in consequences tensile properties are higher as compared with elemental boron additive. Furthermore, it is clear that the tensile properties are not simply related to density. Boron causes both hardening of the matrix and densification, but the prevailing effect on properties is hardening. For example for 0.2 (wt.) % boron contents, despite of the increase in tensile strength, density does not increase significantly.

The above considerations are confirmed by fracture surface analyses. Figure 4 shows the representative fracture surface of tensile specimens of unalloyed and 0.4 (wt.) % boron (B4C) alloyed specimens, respectively. The fracture surface of the unalloyed material (fig. 3,a) presents the typical fracture morphology of a sintered material with interconnected porosity, while 0.4 (wt.) % boron (B_4C) alloyed material fracture surface has a brittle character and crack propagation occurs along the eutectic constituent (fig. 3,b).

Material	0.2 % offset yield stress (MPa)	MTS (MPa)	Elongation (%)	Hardness HV 10
Astaloy CrM	157	287	4.36	182
Astaloy CrM + 0.2 % B	463	613	1.99	187
Astaloy CrM + 0.4 % B	515	688	1.43	247
Astaloy CrM + 0.2 % B (B_4C)	509	666	1.74	272
Astaloy CrM + 0.4 % B (B_4C)	592	741	1.27	302
Astaloy CrM + 0.2 % B (B_4+C)	515	725	1.76	268
Astaloy CrM + 0.4 % B (B_4+C)	662	806	1.57	320

Table 1. Mechanical properties of boron-free and boron alloyed sintered Astaloy CrM (Kazior, 2002).

The research of structure formation at sintering of Fe - B_4C powder mixtures had shown, that active interaction of the components takes place already at temperatures of 850÷900 ^{0}C with generation of Fe_2B and FeB (at ~1100 ^{0}C) phases (Turov, 1991). During heating of powder mixture yet before the beginning of diffusion of the boron to iron matrix contact surface of B_4C particle with iron diminishes owing to considerable difference of their coefficients of thermal expansion (fig. 5,a). Despite of that, at 900÷950 ^{0}C it is beginning the formation of the new phases in neighborhood of boron carbide particles (fig. 5,b). This phase has a typical for boride layers needle-shaped structure. The X-ray analysis of insoluble residue, which was isolated by means of electrolytic etching, has shown the presence in a structure of the material of FeB and Fe_2B phases with microhardness of 14÷18 GPa.

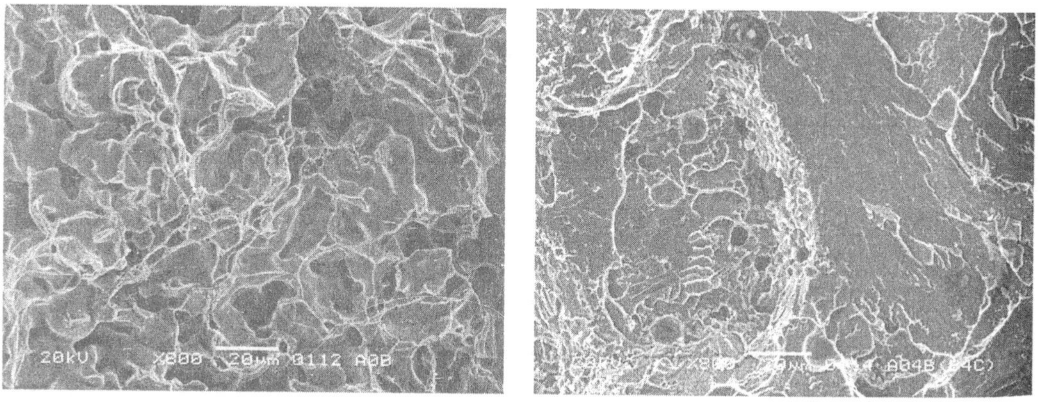

(a) (b)

Fig. 4. Fracture surface of unalloyed (a) and 0.4 % boron alloyed (b) sintered Astaloy CrM.

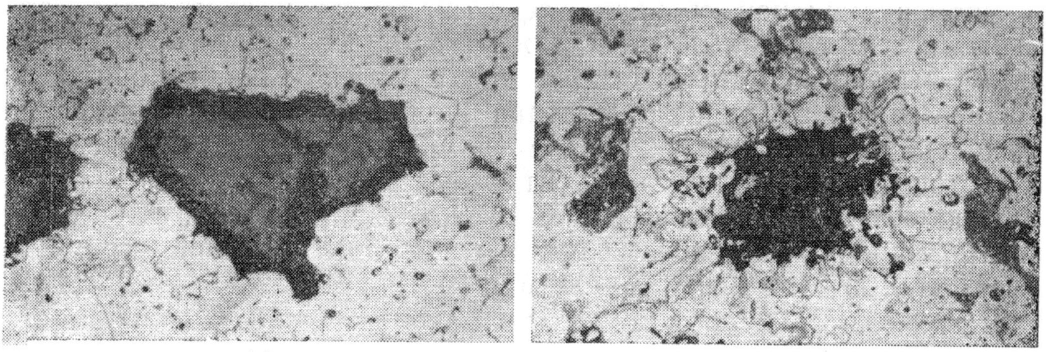

(a) (b)

Fig. 5. Optical micrograph of sintered at 900 ^{0}C for 10 (a) and 120 (b) min.
Fe+3 % (wt.) B_4C powder mixture.

With increasing of sintering temperature to 1050 ^{0}C around B_4C particle the new phase appears, which composition was identified by microX-ray spectrum analysis (fig. 6) as boron cementite with ~75 % (atomic weight) Fe, ~15 % C and ~10 % B.

The above presented results testify, that boron carbide is the compound, that is easily dissociates in a contact with iron at relatively low temperatures. Herewith B_4C is a source of atomic boron and carbon, which are reacting with iron with establishing of new strengthening phases. Boron carbide is a sintering activator as well. However, thereupon that B_4C is the compound relatively poor with carbon, the opportunity of supplementary strengthening of Fe-B_4C system by additive of carbon powder probably exists.

To evaluate both boron and carbon content in the initial powder mixture on structure and properties of sintered steel mixtures were prepared by mixing of water atomized pure iron powder with particles size smaller than 160 μm, 1÷3 % (weight) commercial boron carbide and 0,5÷2,0 % graphite powders.

Powders were compacted at a pressure of 800 MPa to obtain cylindrical specimens with Ø10 and h≈10÷12 mm. The green parts were sintered at 1100÷1200 ^{0}C for 60 min. in the container with fusible glass bath gate (Fedorchenko, 1972). To estimate the influence of carbon on properties of the material the same specimens were prepared with boron carbide but without graphite additive. After sintering some specimens were quenched from 1000 ^{0}C in a water with subsequent tempering at 250 ^{0}C.

Density, hardness and compression strength of all specimens were examined. For the purpose of determining changes of structure and identifying phases, structural investigations by optical microscopy and phases microhardness estimations were carried out.

Figure 7 shows the influence of graphite content in the powder Fe-B_4C mixtures on porosity and hardness of sintered at 1150 ^{0}C materials. As it can be seen from fig. 7,a porosity of compositions rises with increase of graphite content, and very considerably for Fe-3 % B_4C in consequence of worse powder compressibility at compaction of B_4C rich mixtures.

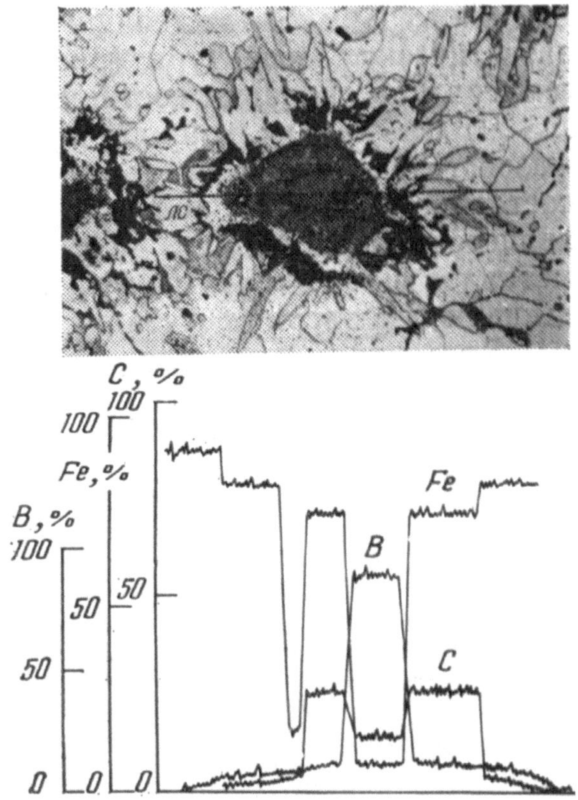

Fig. 6. Microstructure of sintered at 1050 ^{0}C Fe+3 % (wt.) B_4C steel and elements distribution in the area of B_4C particle.

Hardness of the sintered materials with 1 and 3 % of B_4C with addition of graphite depends on two factors: values of their porosity and microstructure nature. So, with rise of B_4C and graphite content hardness of sintered steels increase for all mixture compositions except of

Fe-3 % B_4C, which hardness for mixtures with 1.5 and 2.0 % additive graphite considerably decreases (fig. 7,b, plot 3) owing to respective increase of specimens porosity (fig. 7,a).

The investigation of structure of sintered at relatively low temperatures (1100 ºC) materials had shown, that for all steel compositions on a basis of ferrite-pearlite matrix structures contain multiple isolated inclusions, located in pores, that correspond to the configuration of boron carbide particles (fig. 8,a). Herewith, quantity of such inclusions increases with rise of boron carbide content in the mixture.

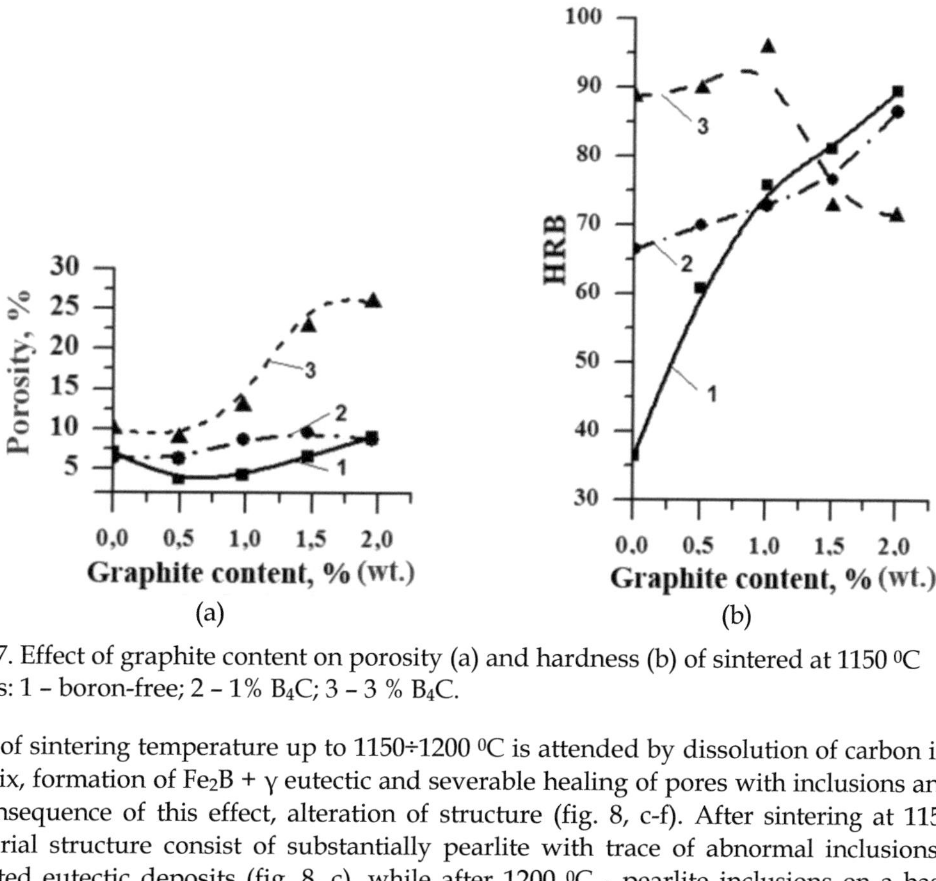

Fig. 7. Effect of graphite content on porosity (a) and hardness (b) of sintered at 1150 ºC steels: 1 - boron-free; 2 - 1% B_4C; 3 - 3 % B_4C.

Rise of sintering temperature up to 1150÷1200 ºC is attended by dissolution of carbon in the matrix, formation of $Fe_2B + \gamma$ eutectic and severable healing of pores with inclusions and, as a consequence of this effect, alteration of structure (fig. 8, c-f). After sintering at 1150 ºC material structure consist of substantially pearlite with trace of abnormal inclusions and isolated eutectic deposits (fig. 8, c), while after 1200 ºC - pearlite inclusions on a basis of eutectic and separate pearlite areas with cementite net (fig. 8,e).

With rise of carbon content the microhardness and amount of eutectic in sintered material increase. So, at enhancing of graphite content in a mixture from 0.5 to 2.0 % the eutectic microhardness increased from 7.4 to 9.2 GPa, while its amount - from 50 to 80 %.

After sintering of all examined materials at 1150 ºC some free carbon remain in their structure; its amount rises from 0.07 to 0.39 % in proportion to graphite content in the initial mixture.

Thermal treatment of the sintered materials results in essential advance of hardness, which values for the steels with 1 % B_4C acquire 35÷40 HRC and with 3 % B_4C - 50÷56 HRC, while for the boron-free specimens - from 25÷27 to 40÷43 HRC (subjected to carbon content) (fig. 9).

Investigation of sintered and heat treated materials compression strength had shown, that ultimate values of durability have the steels with 1 % B_4C and 1÷1.5 % additive carbon (fig. 10, plot 1). The least strength is for materials with 3 % B_4C and carbon content more than 1.0 % (fig. 10, plots 4 and 4a). That kind of compositions are characterized by fragile nature of destruction, that can be explained by relatively high porosity (see fig. 7,a) and presence of significant amount of fragile eutectic.

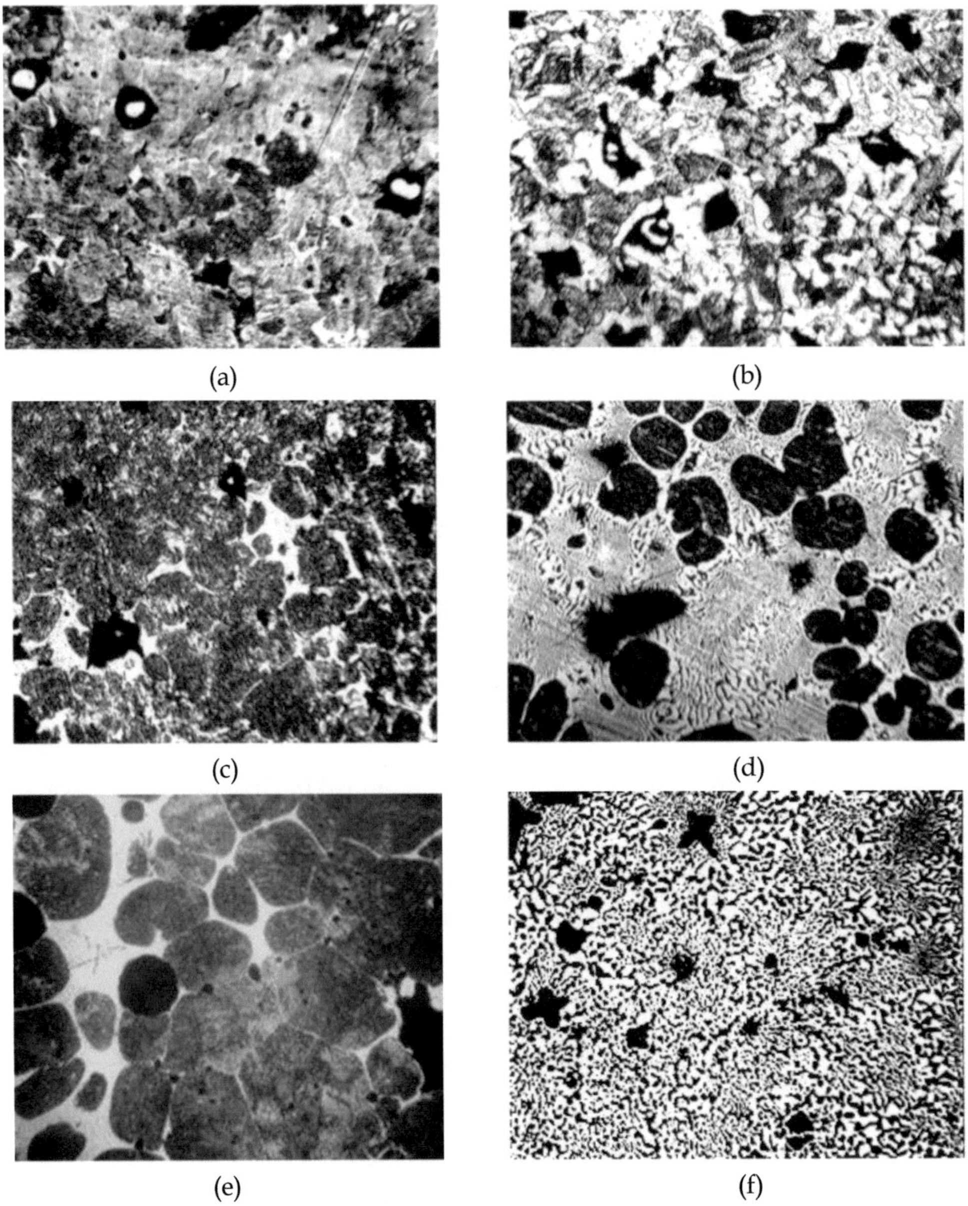

Fig. 8. Structure of the materials sintered at 1100 (a, b); 1150 (c, d, f) and 1200 ^{0}C (e); composition of mixtures: Fe + 1 % B_4C + 2 % C (a, c, e); Fe + 3% B_4C + 2 % C (b, d, f).

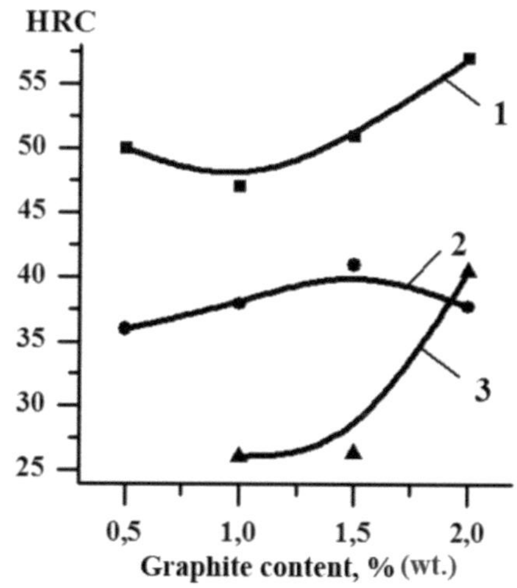

Fig. 9. Effect of graphite content on HRC hardness of sintered and thermal treated steels: 1 - Fe + 3 % B_4C; 2 - Fe + 1,5 % B_4C; 3 - boron-free Fe.

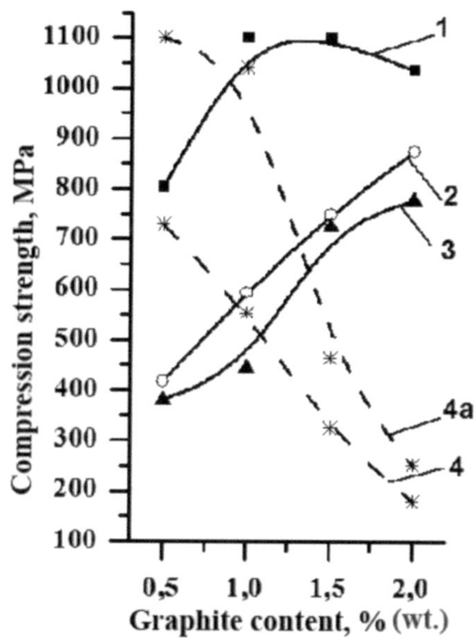

Fig. 10. Effect of graphite content on compression strength of sintered (1-4) and thermal treated (4a) steels with 1 % B_4C (1), 3 % B_4C (4, 4a) and boron-free Fe (2,3). Sintering temperature: 1, 2 - 1200 ^0C; 3, 4 - 1150 ^0C.

3. Application of boron containing master alloys for manufacture of sintered steels

As it can be seen from the above presented results, sintering of Fe-B_4C powder mixtures is attended with formation of derivative porosity in consequence of dissolution of boron carbide particles in a contact with iron matrix. To eliminate the noted effect as boron containing additive Fe-B-C system master alloy can be used. The last one can increase the boron uniformity in the volume of sintered steels too, that probably result in improvement of their mechanical properties.

For estimation of the proposed approach several routes of boron addition to a powder mixture on the amount from 0,3 to 2,4 % and some mechanical properties of the sintered steels had been investigated. As boron-containing components boron carbide, amorphous boron of high purity and Fe-B-C system master alloy, produced by thermal synthesis from Fe and B4C powder mixtures, were used. All of the noted boron-containing components were mixed with water atomized iron powder with maximal particles size of 160 μm. Samples of ∅10×(10÷12) mm size were compacted by uniaxial die pressing at 800 MPa and sintered at 1100, 1150 and 1200 ^{0}C for 1,5 h in the container with fusible glass bath gate.

Three types of powder master alloys chemical compositions (table 2) were used for estimation of sintered steels mechanical properties. Boron content in the powder mixtures for the samples manufacturing was 0,4; 0,8; 1,2 and 1,6 % (weight) for every mode of boron addition (table 2).

Sintered iron samples with 0,5; 1,0; 1,5 and 2,0% boron carbide additives, that provided the same total boron in powder mixtures, were used as the reference ones.

Some sintered specimens were heat treated, i.e. water quenched from 1000 ^{0}C with subsequent tempering at 250 ^{0}C over a period of 3 hours.

No.	Modes of boron addition	Boron content, % (weight)	Carbon content, % (weight)
1	Master alloy	3.6	1.7
2	Master alloy	5.9	2.3
3	Master alloy	8.7	3.5
4	B_4C	80.0	20.0

Table 2. Chemical composition of master alloys and boron carbide, used for manufacturing of sintered steels.

Values of die pressed green compacts and sintered at 1200 ^{0}C preforms porosity for sintered steels, produced from different raw materials (powder mixtures with master alloy, B_4C and amorphous boron powders) with different boron contents were investigated (fig. 11). It can be seen, that with increasing of boron concentration in the powder mixture preform green density decreased for all kinds of boron-containing components.

Maximal porosity values were observed for the green compacts, produced from master alloy containing powder mixtures (fig. 11, curve 3). It can be explained by significant its amount in the source mixture in comparison with other boron containing components and, owing to that, deterioration of such mixtures compressibility.

No.	Mixture composition, %			Carbon content in a mixture, %	Boron content in a mixture, %
	Master alloy		Content of B_4C in a mixture, %		
	No. of master alloy (from table.1)	Content of master alloy in a mixture, %			
I	1	11,0	-	0,19	0,4
	2	6,8	-	0,17	
	3	4,6	-	0,16	
	4	-	0,5	0,10	
II	1	22,2	-	0,37	0,8
	2	13,6	-	0,35	
	3	9,2	-	0,32	
	4	-	1,0	0,20	
III	1	33.3	-	0,56	1,2
	2	20,3	-	0,52	
	3	13,7	-	0,48	
	4	-	1,5	0,30	
IV	1	44,4	-	0,76	1,6
	2	27,1	-	0,69	
	3	18,4	-	0,64	
	4	-	2,0	0,40	

Table 3. Composition of powder mixtures, used for manufacturing of boron containing sintered steels.

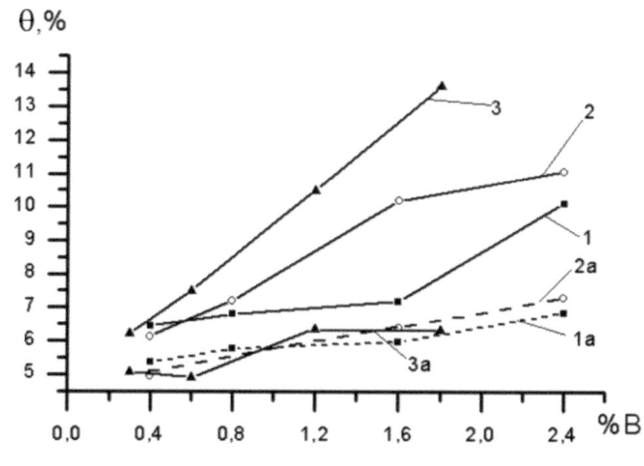

Fig. 11. Effect of boron content on porosity θ of green compacts (1, 2, 3) and sintered at 1200 ^0C preforms (1a, 2a, 3a) for different modes of his addition: B_4C (1, 1a); amorphous boron (2, 2a); master alloy with 6% of B (3, 3a).

However, after sintering at 1200 ^{0}C porosity of the samples, which contains master alloy admixtures, don't differ virtually from that, produced with use of the other modes of boron addition, since sintering of master alloy containing powders is accompanied by noticeable increased value of shrinkage. The denoted data are confirmed as well by phenomenon of shrinkage of the samples with master alloy for all sintering temperatures, unlike the materials with other boron containing components in the initial powder mixture, which are characterized by dilatation after sintering (fig. 12). The latter phenomenon is explained by the effect of boron carbide high temperature dissociation in contact with iron and high reaction activity of amorphous boron at process of sintering accompanied by formation of nonhealing "derivative" porosity in the areas of their location.

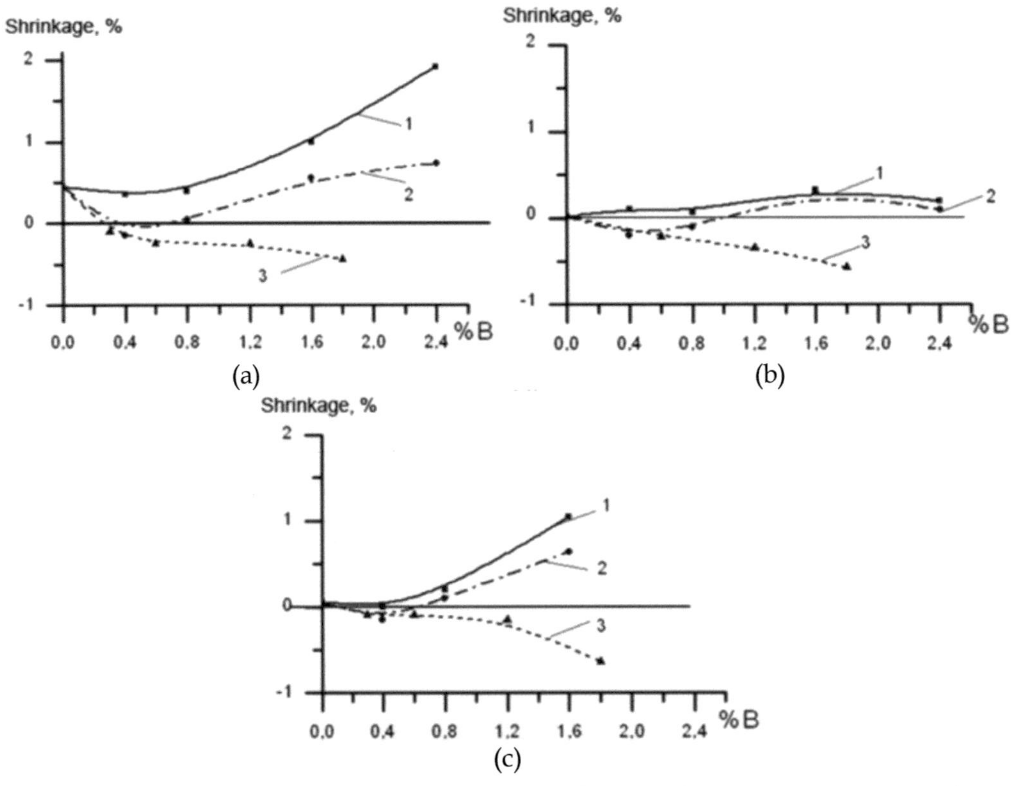

Fig. 12. Effect of boron content on linear shrinkage of billets after sintering at 1100 ^{0}C (a), 1150 ^{0}C (b) and 1200 ^{0}C (c) for different modes of boron insertion: B_4C (1); amorphous boron (2); master alloy with 6 % of B (3).

After sintering of powder preforms with boron carbide, the last easily dissociates in contact with iron at relatively low temperatures and is a source of monatomic boron and carbon, which generate at interaction with Fe hard compounds, that reinforce the sintered metal. After sintering at 1050 ^{0}C it can be seen in the microstructure light-coloured and relatively hard (about 14 GPa) phase, located in the vicinity of incompletely dissociated boron carbide particles (fig. 13,a). Similar phase can be seen also after sintering at 1150-1200 ^{0}C, but its morphology is different: it is situated along the grain boundaries, shaping the developed framework (fig. 13, b), which can appear only as a result of liquid phase appearance in consequence of eutectic contact melting.

According to X-ray micrography data, the composition of interface region, which appears around the boron carbide particle, is similar to boron cementite composition (about 75 % of Fe, 15 % C and 10 % B).

The feature of structure of sintered steels, manufactured from powder mixtures with boron carbide, unlike those of steels, made with use of master alloy, is the availability of diffusion porosity, generated owing to dissociation of boron carbide particles (fig. 14).

Mechanical properties testing results of sintered steels, produced from powder mixtures with different types of master alloys and boron carbide (table 2), had shown, that materials with 0.8-1.2 % B, produced from powder mixtures with low-alloy additives, provided higher values of hardness, than that, made with use of boron carbide as boron containing additives after both sintering and thermal treatment (fig. 15, plots 1 and 2).

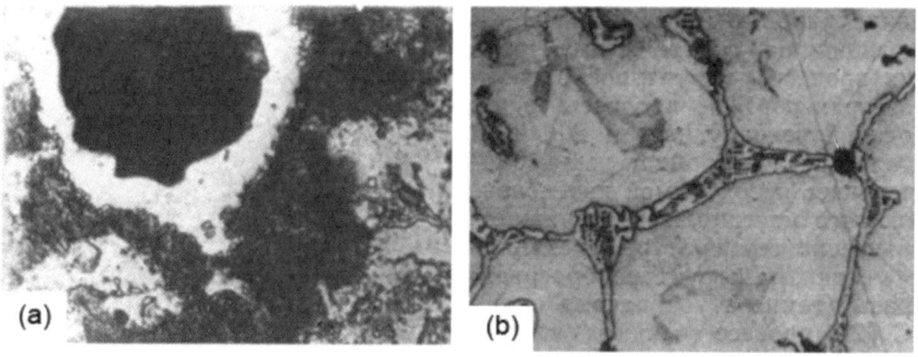

Fig. 13. Microstructure of steels with 1.0 % B_4C sintered at 1050 (a) and 1200 ^0C (b).

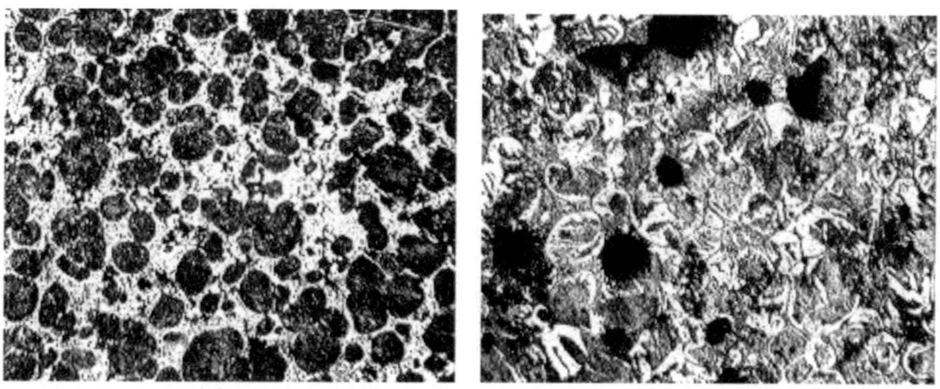

(a) (b)

Fig. 14. Microstructure of sintered steels with 1,2 % B after sintering at 1150 ^0C, manufactured with application of master alloy (Fe - 5,3 % B - 2,3 % C) (a) and boron carbide (b); ×150.

Similar behavior for relation of compression strength with boron content for different modes of boron addition in the powder mixtures (fig. 16) had been registered.

Indeed, hardness values of the same materials, produced with application of master alloy with 8,7 % B and boron carbide (fig. 15, plots 3 and 4) are considerably lower than that, produced with low-boron master alloy.

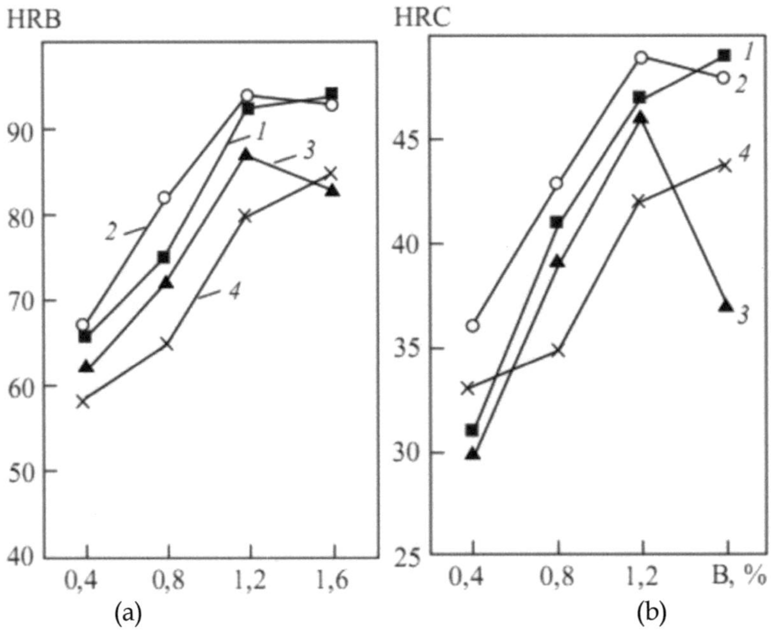

Fig. 15. Effect of boron content on hardness of sintered at 1150 ⁰C (a) and heat-treated (b) steels, manufactured with use of master alloys of three compositions (No. 1–3, table 2) and boron carbide (4).

This difference in the mechanical properties of sintered steels with different kinds of boron source can be explained by more uniform distribution of low-boron containing elements in a balk of material, particularly - of eutectic constituent, in case of employment of master alloys in the source powder mixtures and higher content of carbon in the master alloy as compared with boron carbide.

The drawn conclusions were confirmed by the results of sintered steels metallographic study, which had shown, that microhardness of structural components in the materials, manufactured with use of master alloy with 8.7 % B4C (no. 3, table 2) is somewhat lower than that, produced with low-boron containing master alloys (No. 1 and 2) with 3.6 and 5.8 % of B.

So, for example, in a steel, sintered at 1150 ^{0}C with use of master alloy with 8.7 % of B, microhardness of eutectic is in the range of 6.3÷7.7 GPa, whereas the same for a steel, produced with master alloy with 5.3 % B – 8.58÷9.56 GPa.

Similar relationship can be seen too for microhardness of base material for steels, produced with use of master alloys.

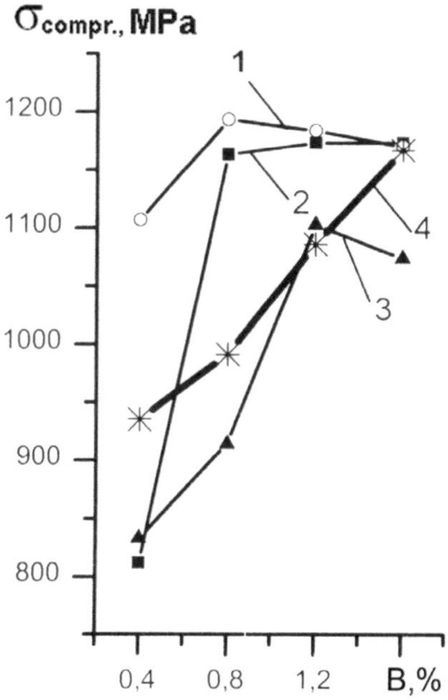

Fig. 16. Effect of boron content on compression strength for steels sintered at 1080 ^{0}C, manufactured with use of master alloys of different compositions
(No. 1–3, table 2) and boron carbide (4).

The wear tests of the sintered steels with 1.2 % B (table 4) were performed in a pin on disc test bed in conditions of unlubricated friction with utilization of step loading method at loads 10-40 N/cm^2 and sliding speed 5 m/s in air. As opposite element 65Mn steel with hardness 45 HRC had been used, while as a pattern material - 100Cr6 bearing steel (HRC 52). Evaluation for magnitude of wear was carried out by assessment of weight loss after every stage of testing and translation the data into µm/km.

Wear testing results for three compositions of sintered steels are presented on fig. 17. It can be seen, that wear resistance for all kinds of sintered boron steels considerably exceeds that for 100Cr6 cast and rolled bearing steel. However, in spite of sufficiently near compositions of boron and carbon in the sintered steels under test, wear resistance of materials, manufactured with use of master alloys with increased composition of boron carbide (No. 2

and 3) at moderate loads (50 and 75 N/cm²) considerably exceeds the same for the steels, synthesized with relatively "poor" master alloy (No. 1).

No.	Composition of master alloy	Composition of B and C in steels, %			
		B	C	B+C	B/C
1	3,6 % B + 1,7 % C	1,2	0,86	2,06	1,4
2	5,9 % B + 2,3 % C	1,2	0,82	2,02	1,5
3	8,7 % B + 3,5 % C	1,2	0,78	1,98	1,5

Table 4. Compositions of sintered and heat-treated boron steels for wear testing.

With increasing of loads to 100 N/cm² wear characteristics for sintered steels of all compositions are flattening.

High wear resistance of boron sintered steels as compared with 100Cr6 bearing steel can be explained by their heterophase structure and, particularly, by presence of hard eutectic component, which amount growth with increase of boron and carbon composition in the source powder mixture.

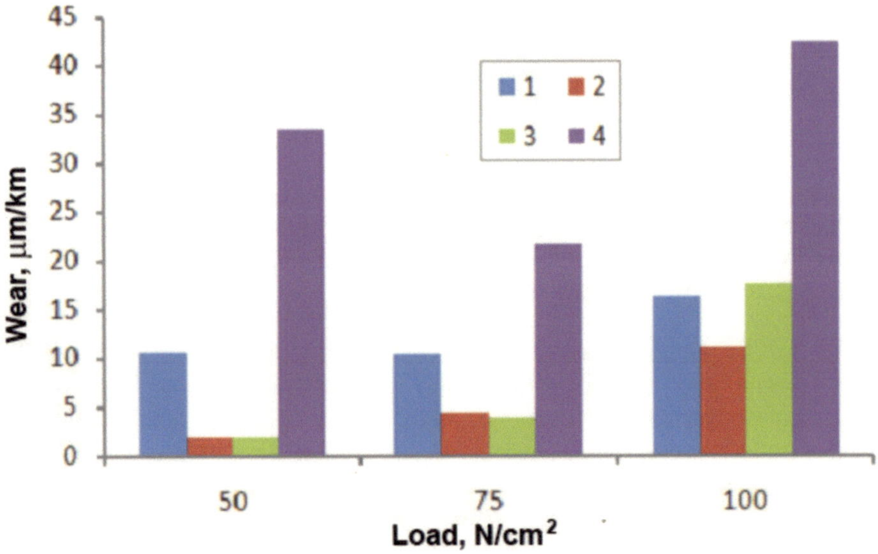

Fig. 17. Effect of load on wear at unlubricated friction of sintered boron steels (No. 1-3, table 3) and 100Cr6 bearing rolled steel (4).

4. Conclusions

The presented results had shown that application of boron containing master alloys provides a superior level of sintered and heat-treated boron steels mechanical properties as compared with boron carbide.

Boron content in the sintered steels that provides an optimum combination of their properties with application of master alloys of different composition, is 1.2 % (weight). This content of B corresponds to ultimate values of hardness and strength for both sintered and heat-treated materials and high level of wear resistance as well.

Comparison of the mechanical and tribotechnical characteristics of the tested materials testify that there is no direct relation among values of their strength, hardness and wear resistance properties. However, every of this steels, in spite of presence of residual porosity, while possessing by commensurable mechanical characteristics with 100Cr6 bearing rolled steel, is distinguished by considerably raised wear resistance.

5. References

Causton, R. J. (2001). Hardenability of sintered boron-carbon steels. *Metal Powder Report*, Vol. 56, Issue 6, June 2001, p.40.

Dorofeev, Yu. G. Marinenko, L. G. (1986). *Structural powder materials and production.* Metallurgia, Moscow, 1986. (In Russian).

Dudrova, E. et al. (1993). Properties of liquid phase sintered low alloy steel containing boron. *Pokroky Praskove Metalurgie.* No 3, 1993, pp. 63–74. (In Czech).

Fedorchenko, I. M. et al. (1972). Technology of sintering metal-ceramic materials without use of protective flowing atmosphere. Powder metallurgy. 1972, no. 5, pp.26-32. (In Russian).

Gülsoy, H.Ö. et al. (2004). Effect of FeB additions on sintering characteristics of injection moulded 17-4PH stainless steel powder. *Journal of Materials Science.* 2004, August, vol. 39, no. 15, pp. 4835-4840.

Kazior, J. et al. (2002). The influence of boron on the mechanical properties of prealloyed CrM powders. *Deformation and Fracture in Structural PM Materials. DF PM 2002.* Vol. 1, pp.125-131.

Metals Handbook. Amer Society for metals. (1978). *Ferrous Powder Metallurgy Materials in Properties and Selection: Irons and steels.* Vol. I, 9th ed., 1978.

Napara-Volgina, S. G. et al. (2008). Investigation of features of various modes of alloying by boron of powder structural steels. *Metal science and metal treatment.* 2008, no. 4, pp. 34-37. (In Ukrainian).

Napara-Volgina, S. G. et al. (2011). Structure and Properties of Sintered Iron-Boron-Carbon Alloys with Different Carbon Contents. *Powder Metallurgy and Metal Ceramics*, 2011, Vol. 50, No. 1-2, pp.67-72.

Suzuki, H. Y. et al (2002). Effect of boron on alloying and pore morphology in PM carbon steel. *J. Jpn. Soc. Powder Metal.*, 2002, vol. 49, no. 7, pp. 600-606. (In Japanese).

Turov, Yu. V. et al. (1991). Structure formation at sintering of powder composition iron-boron carbide. *Powder metallurgy.* 1991, no. 6, pp.25-31. (In Russian).

Zhang, Z. et al (2004). Mechanical properties of Fe–Mo–Mn–Si–C sintered steels. *Powder Metallurgy*, 2004, Vol. 47, No. 3, pp. 239-246.

Xiu, Z.M., et al. (1999). Fe-Mo-B-C - sintered steels produced by addition of master alloy powders. *Acta Met. Sin.* 1999. vol. 12, no. 5. - pp. 1198-1201.

13

Influence of Sintering Temperature on Magnetotransport Behavior of Some Nanocrystalline Manganites

G. Venkataiah[2], Y. Kalyana Lakshmi[1] and P. Venugopal Reddy[1,*]
[1]*Department of Physics, Osmania University, Hyderabad,*
[2]*Materials and Structures Laboratory, Tokyo Institute of Technology, Nagatsuta, Midori-ku, Yokohama,*
[1]*India*
[2]*Japan*

1. Introduction

The colossal magnetoresistance (CMR) in hole doped manganese oxides, widely known as manganites with formula $Ln_{1-x}Ae_xMnO_3$ (where Ln is a trivalent rare earth and Ae is a divalent alkaline earth ion; x=0-1), has been intensively studied over the last two decades. In general, these systems comprise a strong competition between charge, orbital, lattice and spin degrees of freedom all of which make them an intriguing subject of research (Dogotto, 2003; Goodenough, 2003; Ramirez, 1997; Tokura, 2000). The end-members of this system are antiferromagnetic (AFI) insulators. Partial substitution of the Ln with divalent alkaline-earth ion of the ABO_3 structure introduces mixed valence Mn^{4+}/Mn^{3+} on the B site. With decreasing temperature the system undergoes a transition from the paramagnetic to the ferromagnetic state accompanied by a metal-insulator transition. A large number of studies on CMR materials of different forms such as single crystals (Okuda et al., 1998; Zhou et al., 1997), thin film (Kwon et al., 1997; Rao et al., 1998; Suzuki et al., 1997) and ceramics (Hwang et al., 1995; Mahendiran et al., 1996a, 1996b) for the basic research point of view and also the possible future device applications view point. The Zener (1951) has proposed a simplest model, known as double-exchange (DE) model to explain the electrical behavior in ferromagnetic metallic region below Curie Temperature (T_C). However, due to various interactions among charge, spin and lattice makes these materials more complex and DE alone cannot explain the entire electrical transport behavior. Later on various theoretical models have been proposed by considering, electron-lattice and spin lattice interaction and even today there is no comprehensive model to explain transport phenomena in manganites (Millis et al., 1995; Tokura, 2000).

These doped perovskite manganites show large magnetoresistance (MR) around metal to insulator transition temperature (T_P) at high magnetic fields and below room temperatures which restricts their applicability to hands on devices. Improved MR could be achieved for

* Corresponding Author

nanosized perovskite manganite samples prepared through the sol–gel process. The sol–gel process also has potential advantages over other traditional processing techniques such as better homogeneities, low processing temperature and improved material properties. Although there are several reports on the synthesis of nanoscale manganites, none of them were carried out systematically in the context of the influence of sintering temperature on the electrical, magnetic and magneto transport behavior of manganites. It is well know that when the size of the particles is reduced to a few nanometers some of the basic properties such as magnetoresistance, superparamagnetism, coercivity, Curie temperature and Saturation magnetization are effected when compared with their bulk counterparts. Several works on different nanoscale systems indicate that the surface is responsible for their apparent anomalous behavioral changes with reduced dimensions. In fact, the preparation procedure and sintering conditions determines the nature of the surface region of nanosize grains, which plays a very crucial role in electrical transport, magnetic, and magnetotransport behavior of nano dimensional systems. Therefore, it is felt that a detailed review on the influence of sintering on magnetotransport behavior of manganites is essential and important.

The senior author of the paper (P. V. Reddy) along with his group of students investigated the structural, magnetic, electrical and magnetotransport properties of a number doped manganites over a period of more than a decade and published a series of papers in a number of International Journals (Kalyana Lakshmi and Reddy 2008, 2009a, 2009b, 2010, 2011; Kalyana Lakshmi et al., 2009; Venkataiah et al., 2005, 2007a, 2007b, 2007c, 2008, 2010; Venkataiah and Reddy, 2005, 2009a, 2009b). A summary of the work based on the influence of sintering on various properties of the manganites substituted with both divalent and monovalent ions is presented here.

2. Experimental procedure

The samples with different particle sizes were prepared using sol-gel route followed by heat treatment at four different sintering temperatures (T_S) (Venkataiah et al., 2005, 2007, 2008, 2010; Venkataiah and Reddy 2005, 2009a, 2009b; Kalyana Lakshmi and Reddy 2009a). Five series of samples i.e., $La_{0.67}Ca_{0.33}MnO_3$, $La_{0.67}Sr_{0.33}MnO_3$, $La_{0.67}Ba_{0.33}MnO_3$, $Nd_{0.67}Sr_{0.33}MnO_3$ and $La_{0.67}Na_{0.33}MnO_3$ were prepared to study the effect of sintering temperature on various properties. The samples are hereafter designated as LCMO, LSMO, LBMO, NSMO and LNMO respectively and the samples sintered at 800, 900, 1000 and 1100 C are designated with 8, 9, 10 and 11 numerical representations followed by compositional abbreviations. The synthesis procedure of sol-gel method is outlined in Fig. 1.

The stoichiometric amounts of oxides / carbonates were taken and all of them were converted into metal nitrates and the resultant nitrate mixture solution was converted into citrate by adding 1:1 ratio of the citric acid to the metal atoms present in the composition. To carry out the reaction at greater speed, a proper reaction environment is to be created and hence pH of the solution was adjusted between 6.5 and 7 by adding ammonia solution. Finally, the powder was pressed into pellets and sintered in air at four different temperatures i.e., 800, 900, 1000 and 1100^{0}C for 4 hr. For a direct comparison of the properties of the samples as a function of sintering temperature (grain size) it is important to ensure that the Mn^{4+} concentration remains similar in different samples, since it is crucial factor in controlling the transport and magnetic properties. Mn^{4+} concentration has been determined by redox titrations (Vogel, 1978). The phase purity of the samples was checked by X-ray diffraction (XRD) and grain distribution was estimated by scanning electron microscopy (SEM). The electrical resistivity and magnetoresistance studies were undertaken by a Janis "supervaritemp" cryostat in applied fields of 0–7 T, over a temperature range of 77–300 K using a four-point probe method. a.c susceptibility measurements were carried out using a dual channel lock-in amplifier, while thermoelectric power studies were carried out by a two-probe differential method over a temperature range 77–300 K.

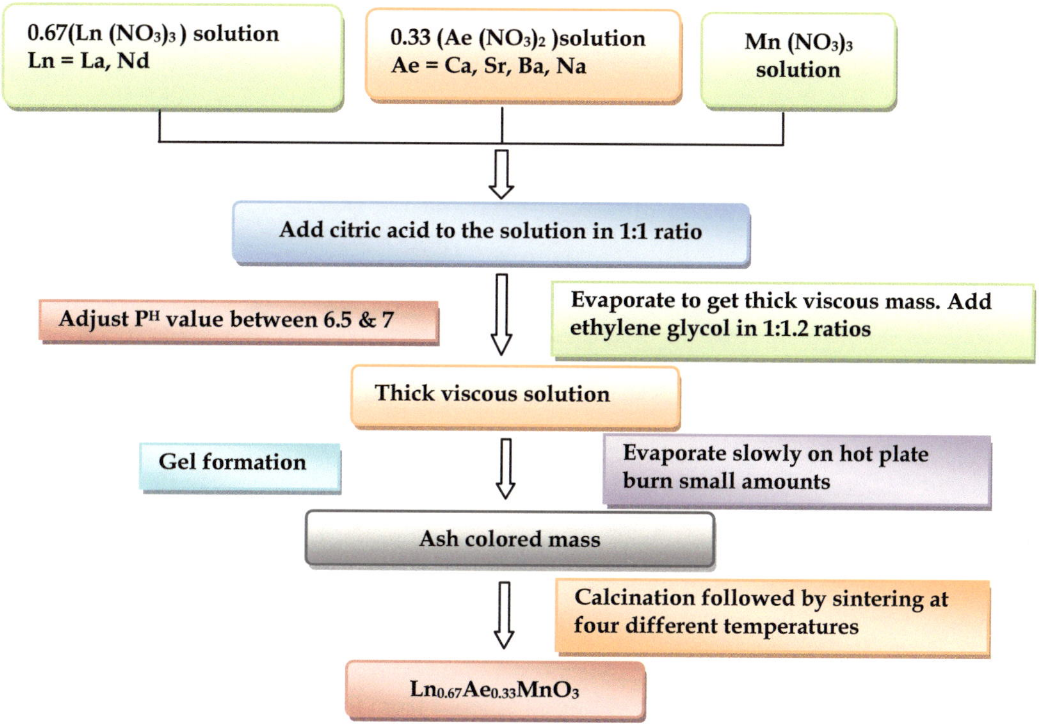

Fig. 1. A flow chart of Sol-gel method for the preparation of CMR manganites.

3. Results & discussions

3.1 Structural properties

Most of the manganites crystallize in the various derivatives of the so-called perovskites structure named after the mineral perovskite, $CaTiO_3$.Figure 2 illustrates the ideal cubic perovskite structure, with the general formula ABO_3. The structure may be conceived as a close-packed array formed of O^{2-} anions and A cations with B cations located at the octahedral interstitial sites. The BO_6 octahedra make contact to each other by their vertices and form a three-dimensional network. In manganites the A-sites of the perovskite structure are occupied by trivalent rare earth ion while B-sites with Mn ions. The stability of the perovskite structure depends strongly on the size of the A-site and B-site ions. If there is a size mismatch between the A-site and B-site ions and the space in the lattice where they reside, the perovskite structure will become distorted. Such a lattice distortion of perovskite structure is governed by Goldschmidt's tolerance factor (τ) (Goldschmidt, 1953).

$$\tau = (r_A + r_O)/\sqrt{2}(r_B + r_O) \tag{1}$$

where r_A and r_B are the mean radii of the ions occupying the A-sites and B-sites, respectively, and r_O is the ionic radius of oxygen. For an ideal cubic perovskite $\tau=1$.

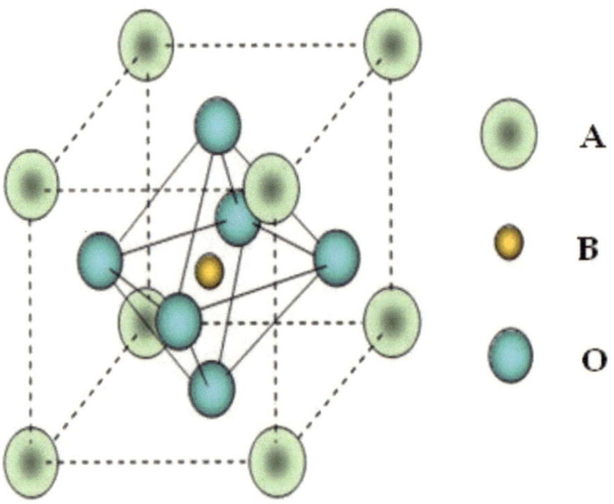

Fig. 2. Unit cell structure of prototype cubic perovskite, ABO_3, with A ions situated at the corners, B ions in the center and oxygen ions at the centers of the faces.

If τ differs slightly from unity the atoms are displaced from their ideal positions to minimize the free energy and a distorted perovskite structure is formed. For $0.96<\tau<1$ the lattice structure transforms to the rhombohedral structure and then to orthorhombic structure for $\tau<0.96$, in which B-O-B bond angle gets distorted from 180°. Structure like tetragonal, hexagonal and monoclinic etc. are reported for manganites for different x values. Among the studied samples the LSMO, LBMO and LNMO systems crystallize in the rhombohedral with $R\bar{3}c$ space group (Venkataiah et al., 2007a, 2010; Kalyana Lakshmi and Reddy 2009a) while LCMO and NSMO group of samples crystallize in to orthorhombic structure with *Pbnm* space group (Venkataiah et al., 2005; Venkataiah and Reddy 2009b). A typical XRD patterns of LSMO & NSMO samples are shown in Figure 3. The X-ray linewidths are used to estimate average crystallite size <s> values through the Scherrer's formula modified by Williamson and Hall (1953).

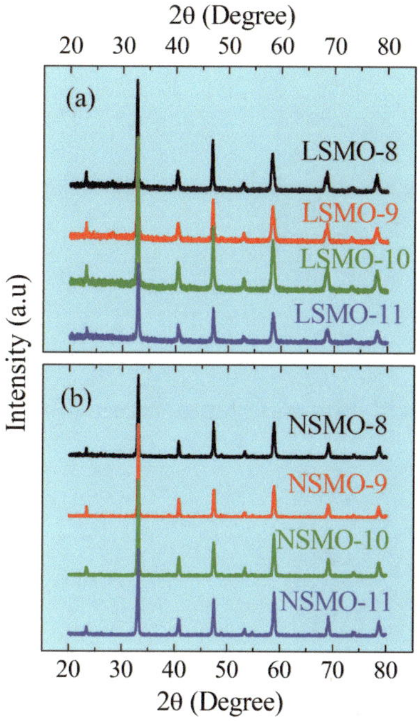

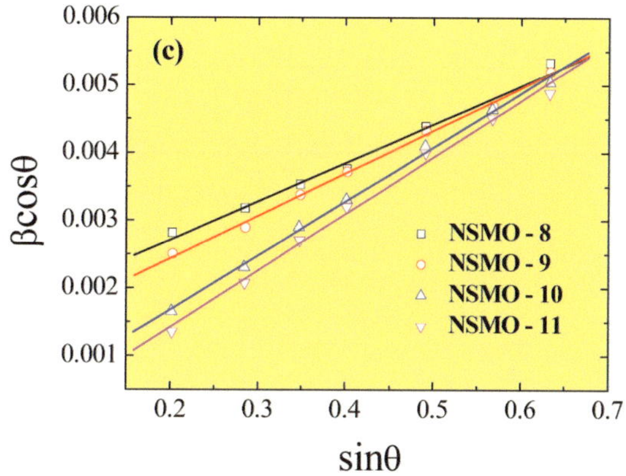

Fig. 3. (a, b) XRD patterns of LSMO & NSMO group of manganites at various sintering temperatures (T_S) (c) Williamson-Hall plot of $Nd_{0.67}Sr_{0.33}MnO_3$ manganites.

In this method, the experimental x-ray peak broadening is given by $\beta=\varepsilon\tan\theta+K\lambda/$ st $\cos\theta$, where β is the full width at half maximum (FWHM) of the XRD peaks after subtracting FWHM of the standard sample (SiO_2), θ is the Bragg angle, K is the grain shape factor (0.89), λ is the wavelength of Cu Kα radiation, and ε is the strain in the sample. A plot between $\beta\cos\theta$ versus $\sin\theta$ is shown in Figure 3(c). The average crystallite sizes were calculated from the intercept of the straight line with the vertical axis, while the slope of the line gives the lattice strain (ε). The average crystallite sizes are found to increase with increasing sintering temperature and are in nanometer range. The variation of crystallite size with sintering temperature is shown in Figure 4. In fact, the computed crystallite sizes are comparable with those obtained from SEM. Figure 5 shows the representative SEM images of LSMO samples sintered at 800 and 1100°C. The SEM observation reveals that there is a uniform distribution of grain sizes for the samples and as the sintering temperature increases, the grain size increases and porosity decreases.

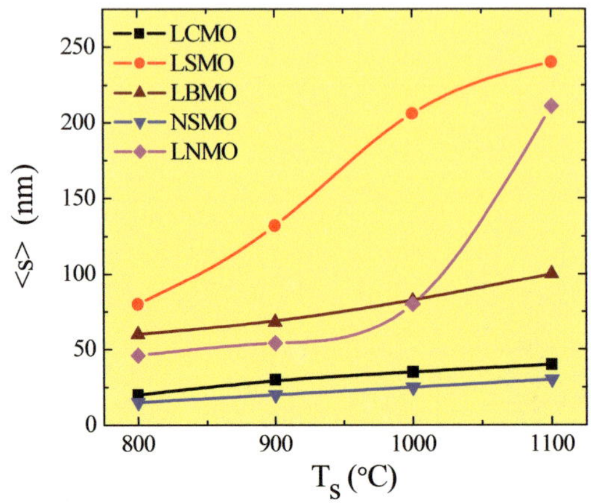

Fig. 4. Variation of Crystallite size (<s>) in nanometers with Sintering temperature (T_S °C).

Fig. 5. SEM images of LSMO samples sintered at 800 & 1100°C.

3.2 Magnetic properties

Mixed-valence manganites are complex materials showing a rich variety of structural, magnetic and electronic phases. Phase transitions may be induced by changing the composition, temperature or sometimes by applying an external magnetic field. Usually structural, magnetic and electronic phase transitions occur simultaneously. The magnetic and electronic properties of lanthanum manganites, $La_{1-x}Ae_xMnO_3$ (Ae = Ca, Sr, etc) are strongly related to the simultaneous presence of manganese in different valence states. The end-members, $x = 0$ and $x = 1$, containing Mn in only one valence state (for divalent ion at A-site substitution leads to Mn in 3+ and 4+ valence states), are usually antiferromagnetic insulators (AFI), but intermediate compositions, which have mixed Mn valence, may be ferromagnetic and have good conducting properties. Figure 6 illustrates the magnetic and electrical properties of $La_{1-x}Ca_xMnO_3$ (x=0-1) system with change in composition and temperature (Schiffer et al., 1995, Urushibara et al., 1995). Ferromagnetic behavior starts to manifest itself when x is 0.1 and the composition upto x~0.3 have both antiferromagnetic (AFM) and ferromagnetic (FM) characteristics. The x=0.3 composition is clearly FM, while the x>0.5 compositions are AFM. The coexistence of metallic conductivity and ferromagnetic coupling in these materials at low temperature has been explained in terms of a double exchange mechanism, proposed by Zener in 1951. The double exchange mechanism involves simultaneous transfer of an electron from Mn^{3+} to O^{2-} and from O^{2-} to Mn^{4+}. Moreover, the properties of manganites are not only sensitive to the manganese valency, but also are strongly affected by other factors such as average cationic radius $\langle r_A \rangle$ of A-site and A-site cationic mismatch quantified by σ^2 which effects the magnetic interaction. Due to the long history of work on these compounds, most of the studies have been performed on bulk ceramics and thin films. Apart from this it was demonstrated that the properties of manganites are strongly affected by particle size. It was found that the sintering temperature is one of the key factors which influences the crystallization and microstructure of the perovskite samples. The basic magnetic properties such as spontaneous magnetization, the Curie temperature and coercivity are strongly influenced with sintering temperature. Mahesh et al., (1996) were the first to explore the size effects in polycrystalline $La_{0.67}Ca_{0.33}MnO_3$prepared by citrate-gel route and observed a close relationship between magnetotransport properties and microstructure.

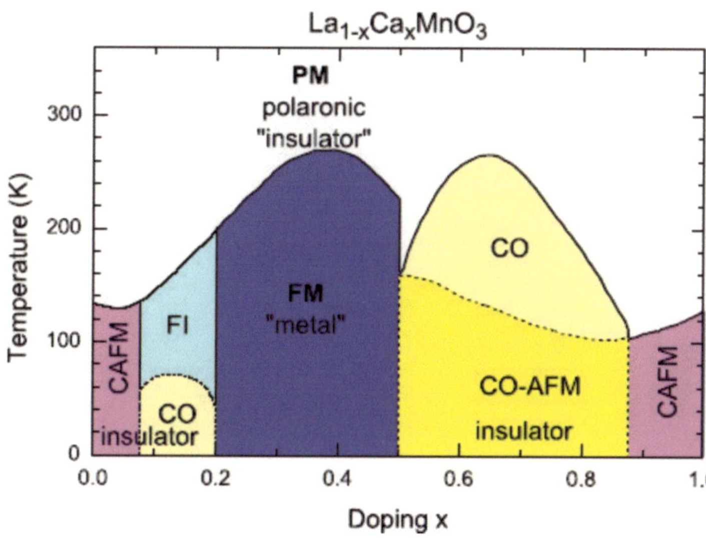

Fig. 6. Magnetic and electronic phase diagram versus doping x for $La_{1-x}Ca_xMnO_3$ manganites (Schiffer et al., 1995, Urushibara et al., 1995).

The results obtained by the authors of the present investigation on five series of samples sintered at various temperatures are presented in this section. Out of these samples, the a.c. susceptibility (χ) versus temperature plots of two representative samples viz., LSMO and LBMO sintered at different temperatures are displayed in Figure 7. All the five groups of samples exhibit para to ferromagnetic transition with decreasing temperature and the width of transition broadens on lowering sintering temperature indicating magnetic inhomogeneity in the samples. The variation of T_C with sintering temperature (T_S) for the samples is shown in Figure 8. T_C values do not vary much in the case of LSMO and LBMO systems (Venkataiah et al., 2007a, 2010) while after an initial increase they remain almost constant with further increase in sintering temperature for $La_{0.67}Na_{0.33}MnO_3$manganites (Kalyana Lakshmi & Reddy 2009a). The observed behavior may be explained qualitatively as follows. In general variations in T_C values take place whenever there is a change in Mn-O-Mn bond angle as well as Mn-O bond length, thereby indicating the presence of magnetic inhomogeneity in the sample as it approaches the transition temperature. In contrast to these arguments, perhaps due to magnetic homogeneity especially as sample approaches the phase transition, T_C might be remaining constant, thereby giving an impression that T_C might be independent of grain size. In fact a similar conclusion was arrived at earlier in the case of $La_{0.7}Sr_{0.3}MnO_3$ (Gaur & Verma 2006) and $La_{0.8}Ca_{0.2}MnO_3$ manganites (Balcells et al., 1998; Fu 2000; Rivas et al., 2000).

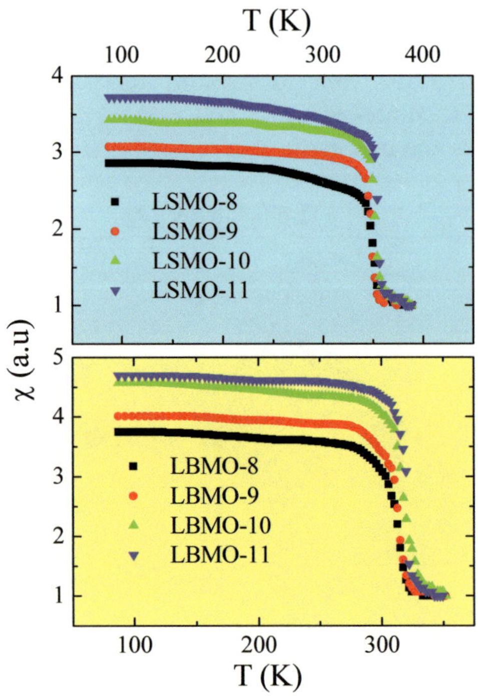

Fig. 7. Variation of susceptibility (χ) with temperature for (a) LSMO & (b) LBMO group of manganites.

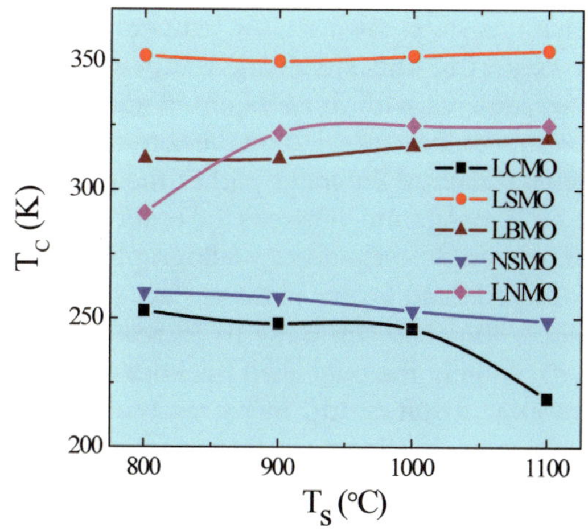

Fig. 8. Variation of Curie temperatures (T_C) as a function of sintering temperature (T_S).

In contrast to the above results, T_C values decrease with increasing sintering temperature in the case of LCMO and NSMO manganites (Venkataiah et al., 2005; Venkataiah & Reddy 2005). The observed reduction of ferromagnetic transition temperature (T_C) with increasing particle size is related to increase in unit cell volume. The increased unit cell volume enhances Mn-O bond length thereby deviating Mn-O-Mn bond angle from 180^0, which in turn decreases the overlap integral decreasing the bandwidth. Thus the lattice effects arising from increasing particle size are expected to decrease in bandwidth, which in turn decreases T_C. In addition, the dependence of T_C on $\langle s \rangle$ ($T_C \propto 1/\langle s \rangle$) can also be rationalized if the size reduction can be thought as producing an internal pressure (P_I) such as the hydrostatic pressure (P_E): Size reduction due to interface stress can lead to an internal pressure $\Delta P_I \propto A_g / V_g$, where A_g and V_g are the surface area and the volume of a nanocrystal respectively. For a nearly spherical particle, the relationship between A_g and V_g is given by a relation, $A_g / V_g \propto 1/\langle s \rangle$. This gives $\Delta P_I \propto 1/\langle s \rangle$. As per the external, hydrostatic pressure, $dT_C = dP_I \approx$ constant, one can write as $\Delta T_C \propto 1/\langle s \rangle$ (Shankar et al., 2004; Venkataiah and Reddy, 2009b).

3.3 Electrical studies

A fundamental characteristic of mixed-valence manganites is the close relationship between electronic transport and magnetism. Most of the $La_{1-x}Ae_xMnO_3$ compositions are paramagnetic insulators at room temperature and exhibit an increase in electrical resistivity with a decrease in temperature. Compositions that are ferromagnetic show insulating ($d\rho/dT < 0$) behavior above T_C, but the resistivity decreases with decreasing temperature, as in metals ($d\rho/dT > 0$), when they are cooled below T_C. This insulator-metal transition is therefore associated with a peak in resistivity at a temperature, T_P.

The variations of electrical resistivity with temperature for all the five groups of samples show similar behavior. The behavior of two representative samples viz., LCMO and LSMO is shown in Figure 9. All the samples show metal to insulator transition at T_P and their variation with sintering temperature for the investigated samples is shown in Figure 10 (a). The metal to insulator transitions shifts towards low temperature side and the magnitude of resistivity increases on reducing the sintering temperature or particle size.These observations are in good agreements with those reported earlier (Mahesh et al., 1996; Gaur & Verma 2006). This behaviour can be explained by the core-shell model proposed by Zhang et al., (1997). In the core-shell structure the inner part of the grain, i.e. the core, would have the same properties as the bulk manganite, whereas the outer layer, i.e. the shell (thickness t), would contain most of the oxygen defects and crystallographic imperfections, which would lead to a magnetically disordered dead layer. As the surface to volume ratio becomes larger, i.e., the grain size is reduced, the shell thickness (t) increases. Basically, the net intercore barrier thickness ($s = 2t + d$), namely the total shell thickness (2t) of two neighboring grains together with the intergranular distance (d), increases with the reduction of grain size. Further, with the decrease in grain size, core separation ($s \sim 2t$) increases significantly with the thickness of the shell (t), even if we consider the grains to be in intimate contact ($d = 0$) for all grain size samples. Another important fact is that in the absence of magnetic field the contributory portion of each individual grain to the magnetization is the core and not

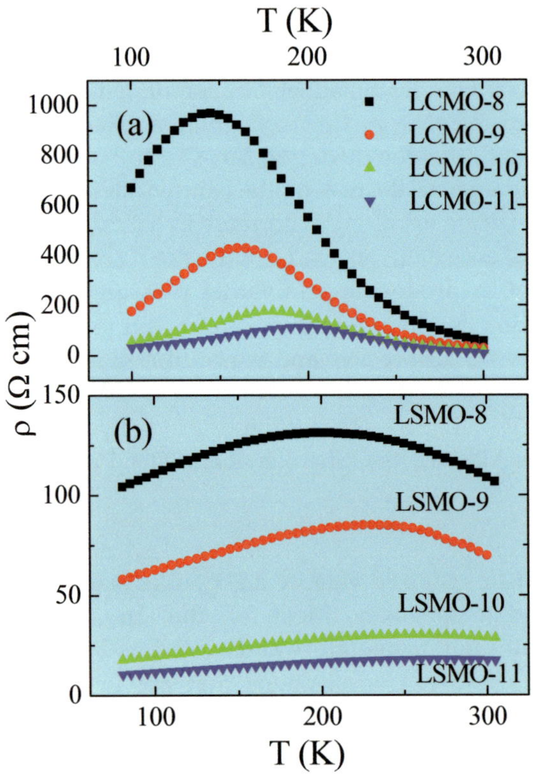

Fig. 9. Temperature dependence of resistivity for (a) LCMO & (b) LSMO samples sintered at various temperatures.

the shell, in the absence of applied magnetic field the net magnetization of the shell is considered to be zero. Since the surface would contain most of the oxygen defects and faults in the crystallographic structure that will lead to a magnetically disordered state, which may lead to formation of spin canting or antiferromagnetic state at the manganite grain surface, thereby decreasing T_P values and increasing the magnitude of resistivity (Figure 11).

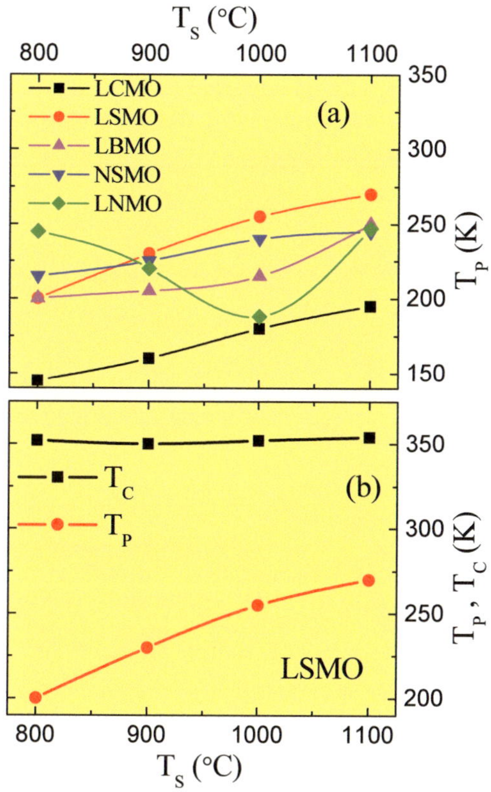

Fig. 10. (a) Variation of T_P (K) with sintering temperature (T_S) (b) T_C, T_P (K) variation of LSMO with T_S.

It can also be noticed from the Figure 10(b) (in the case of LSMO system) that there exists a large difference between T_C and T_P and the difference increases with decreasing sintering temperature. The large difference in the magnetic and electrical transitions for all the samples is thought to be due to the existence of the disorder and is in fact a common feature of the polycrystalline manganites (Gaur & Verma 2006). The T_C being an intrinsic characteristic does not show significant change as a function of sintering temperature. On the other hand T_P is an extrinsic property that strongly depends on the synthesis conditions and microstructure (e.g. grain boundary density). The suppression of the T_P as compared to T_C is caused by the induced disorders and also by the increase in the non-magnetic phase fraction, which is due to enhanced grain boundary densities as a consequence of lower sintering temperature. This also causes the increase in the carrier scattering leading to a corresponding enhancement in the resistivity. Thus lowering of sintering temperature reduces the metallic transition temperature and hence the concomitant increase in resistivity.

In contrast to the above results, it has been observed that T_P values of sodium doped samples, are found to decrease with increasing grain size except in the case of the sample sintered at 1100°C (Kalyana Lakshmi & Reddy, 2009a). The observed variation may be interpreted on the basis of oxygen deficiency. According to Malavasi et al., 2002, 2003a, 2003b) the presence of oxygen vacancies deeply affect the transport properties resulting in both Mn^{4+} reduction and point defect creation within the structure thereby shifting T_P towards low temperature side. This behavior is in close agreement with the results reported in the case of LCMO samples (Hueso et al., 1998) where T_C remained constant while T_P decreased with oxygen deficiency. This suggests that the combined effect of crystallite size and oxygen deficiency might be responsible for the observed resistivity behavior (Hueso et al., 1998; Malavasi et al., 2002, 2003a, 2003b).

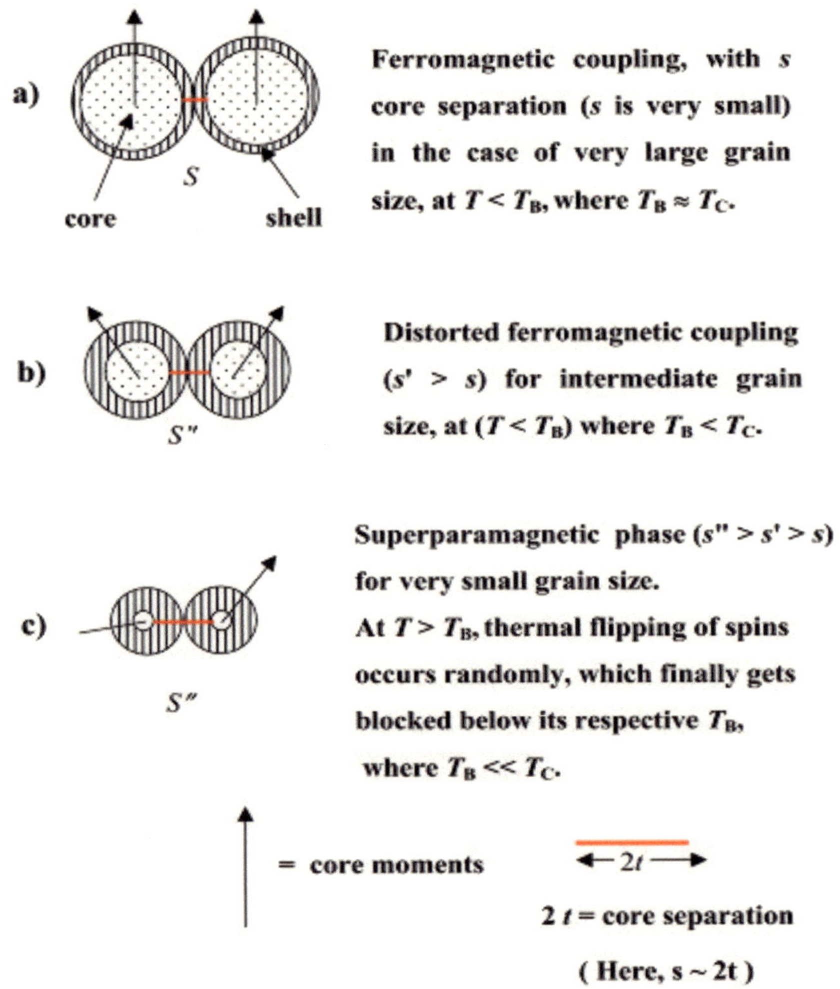

Fig. 11. Phenomenological demonstrative representation of the possible ordering of core moments in the core-shell structure of nanometric manganite grains with the grain size as a variable parameters in different temperature ranges [Dey and Nath 2006].

3.4 Thermopower studies

Measurement of emf (electromotive force) induced by a temperature gradient across a sample provides complementary information to the resisitivty. The Seebeck coefficient (S) is defined as $\Delta V/\Delta T$, the thermoelectric voltage per degree of temperature difference. Among various transport properties, thermopower is a simple and sensitive one for detecting the scattering mechanism that dominates the electronic conduction and also provides insight into changes of band structure near the metal-insulator transition. The thermoelectric power (TEP) of manganites shows strong temperature dependence and it can have positive or negative sign depending on the temperature and degree of substitution of divalent/monovalent cation that determines the carrier density in these systems. This behaviour suggests that either both types of carriers (electrons and hole) may be involved in the charge transport mechanism or additional scattering phenomenon may contribute to thermopower. Because no current flows, the thermopower does not depend on the connectivity of conducting regions, and the thermopower of individual grains are additive. There have been many measurements of the thermopower of samples with $x \approx 0.3$ which exhibit a metal-insulator transition. These data, on ceramics, thin films and single crystals, are not very consistent and results are sensitive to details of sample preparation and composition (Jaime etal., 1996; Mahendiran et al., 1996b; Mandal 2000).

3.4.1 Low temperature behavior ($T<T_P$)

Most of the studies on thermopower were reported on the manganites with varying doping concentration [Battacharya et al., 2003; Jirak et al., 1985; Mahendiran et al., 1996b; Volger 1954). In order to get further insight into the effect of particle size on thermopower, the variations of thermopower as a function of sintering temperature on various manganites systems were carried out. This section describes thermopower results obtained by the authors of the present investigation on various manganite systems doped with both divalent and monovalent ions sintered at different temperatures. Figure 12 shows the variation of seebeck coefficient (S) with temperature in LCMO and NSMO manganites. It has been observed that the magnitude of S is found to increase with decreasing particle size and positive throughout the temperature range of investigation, thereby representing that the charge carriers are holes for LSMO, LBMO, NSMO, LNMO, LCMO-8, LCMO-9 and LCMO-10. In contrast to this, the sign of S for LCMO-11 changes from positive to negative value indicating the coexistence of two types of carriers. Banerjee et al., (2003) has also found a similar variation of S with varying particle size in lead doped manganites. One can notice from the figure that as the temperature of the samples is decreased from room temperature to liquid nitrogen temperature, S values increase continuously attaining a maximum value with a sharp peak in the vicinity of T_C. The peak observed at T_C may be attributed to enhancement in spin polarization caused by the magnetic transition (Das et al., 2004). Apart from this, except LCMO-8 and LCMO-9 all are found to exhibit a broad peak around 120 K and similar behavior was reported earlier (Das et al., 2004; Urushibara et al., 1995). It was reported earlier that (Banerjee et al., 2003) phonon drag (S_P) and magnon drag (S_M) contributions are present in the low temperature region and that these are proportional to their respective specific heat contribution so that $S_P \propto T^3$ and $S_M \propto T^{3/2}$. This suggests that in the low temperature ferromagnetic region, a magnon drag effect is produced due to the presence of electron - magnon interaction along with the phonon drag due to electron – phonon interaction.

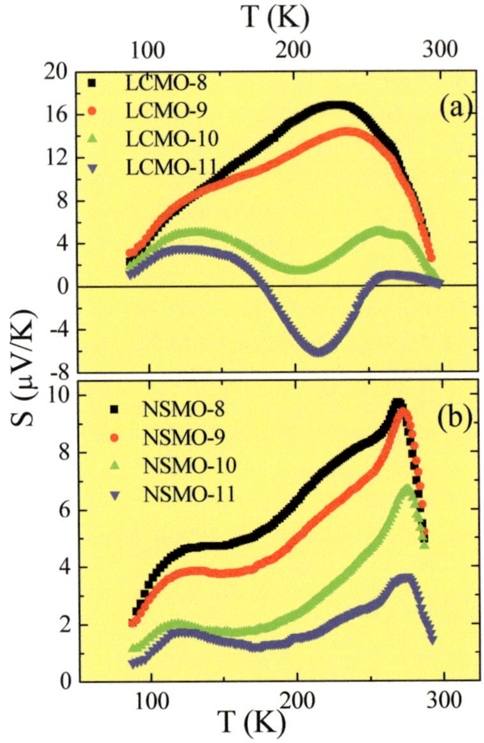

Fig. 12. Variation of seebeck coefficient (S) with temperature of LCMO & NSMO manganites.

In general, to understand the influence of various contributions to TEP in ferromagnetic region, the following equation was used earlier (Mandal 2000)

$$S = S_0 + S^{3/2}T^{3/2} + S_4T^4 \tag{2}$$

where S_0 term has no physical origin and is inserted to account for the problem of truncating the low temperature thermopower data. The second term of the equation represents the single-magnon scattering process. Although the origin of S_4T^4 term is still not clear, it is generally believed that it may be due to spin wave fluctuation. It has been observed that the eq. (2) doesn't fit the TEP data in the entire ferromagnetic region. Therefore, the above equation has been modified by taking into account the diffusion and phonon drag contributions and the equation is as follows, (Kim et al., 2008),

$$S = S_0 + S_1T + S^{3/2}T_{3/2} + S_3T^3 + S_4T^4 \tag{3}$$

Here, the second and fourth terms represents the diffusion and phonon drag contributions to TEP respectively. The experimental data in the low temperature region were fitted to the above equation and the solid curve in Figure 13 (a) (in the case of NSMO) represents the best fit to the experimental data in FMM region and the best fit parameters for NSMO sample is given in Table 1.

Sample code	S_0 (μV/K)	S_1 (μV/K^2)	$S_{3/2}$ (μV/K$^{5/2}$)	S_3 (μV/K^4)	S_4 (μV/K^5)	E_P (meV)	E_S (meV)	α′
NSMO-8	-62.936	2.3350	-0.1849	2.0x10^{-5}	-7.46x10^{-11}	140.99	30.56	-0.117
NSMO-9	-50.745	1.8982	-0.1505	2.0x10^{-5}	-5.84x10^{-11}	130.92	21.35	-0.107
NSMO-10	-38.148	1.3672	-0.1071	1.0x10^{-5}	-3.59x10^{-11}	125.95	17.60	-0.065
NSMO-11	-29.817	1.1270	-0.0897	0.8x10^{-5}	-3.13x10^{-11}	118.83	12.12	-0.047

Table 1. The best fit parameters obtained from thermoelectric power data for NSMO system (EP & ES represent the activation energies obtained from resistivity and thermopower data)

All the fitting parameters decrease with increasing particle size. This might be due to the magnetic domain and grain boundary scattering mechanism discussed previously in analyzing resistivity data of the samples.

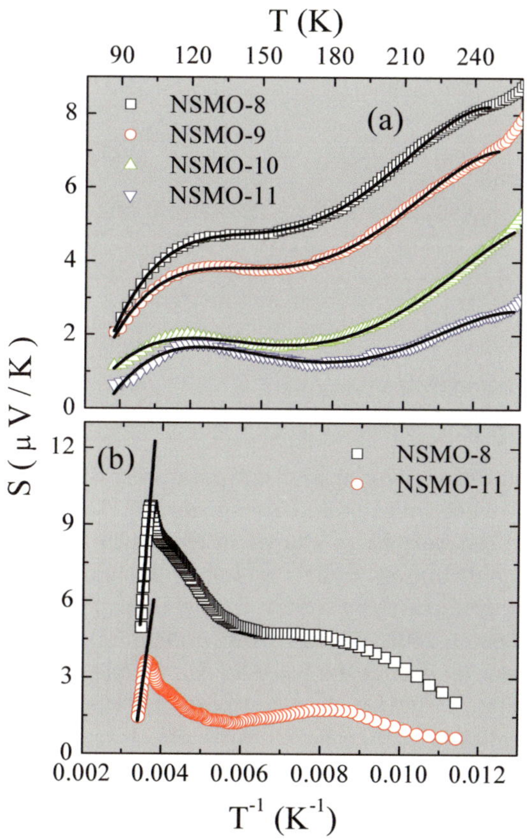

Fig. 13. (a) The temperature dependence of TEP of NSMO group of samples. The solid line represents the fitting with eq (3). (b) Variation of S vs $T^{-1}(K^{-1})$ of NSMO manganites. The solid line represents the best fit to the small polaron hopping model.

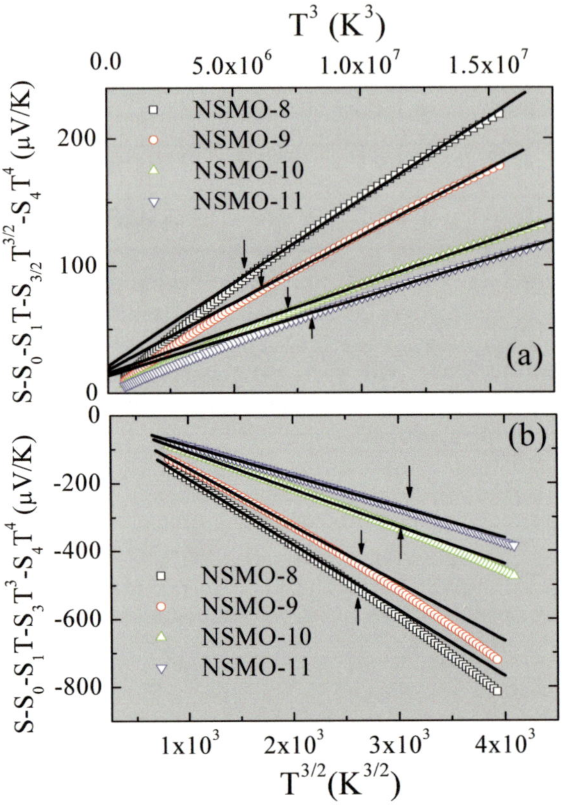

Fig. 14. (a) Variation in phonon drag component with T^3 and (b) variation in magnon drag component with $T^{3/2}$. The arrow marks represents the deviation of linear fit to the data.

In order to explain the origin of peak at low temperatures, the phonon and magnon drag contributions to the TEP were investigated using eq. (3). The variation of phonon drag component with T^3 for all the samples is shown in Figure 14 (a). It is clear from the figure that the phonon drag contribution (S_3T^3) is found to vary linearly up to a certain temperature and deviates (indicated by arrow in the figure) below 250K. As the phonon drag contribution deviates at 250K, it has been concluded that the origin of the low temperature peak might not be due to the phonon drag effect. It is interesting to note from Fig. 14(a) that the extrapolated phonon drag component is approaching zero as the sample approaches to T = 0 indicating that the phonon drag effect might be disappearing probably due to depleting number of phonons at T = 0. It can also be seen from the figure that the deviation temperatures occurring from phonon and magnon drag are almost equal in the case of NSMO-8 and NSMO-9 samples indicating that the magnon and phonon drag effects are occurring in different temperature regions. The magnon drag component was calculated from eq (3) and is shown in Figure 14(b). It is interesting to note that the magnon drag component fits well below 200K, indicating that the low temperature peak might have arisen due to magnon drag effect (Venkataiah & Reddy, 2009b).

3.4.2 High temperature behavior (T>T_P)

The charge carriers in the insulating region are not itinerant and the transport properties are governed by thermally activated carriers because the effect of Jahn-Teller distortions in manganites results in strong electron-phonon coupling and hence the formation of polarons. Therefore, the thermoelectric power data of the present samples in the insulating regime are fitted to Mott's polaron hopping equation,

$$S(T) = \frac{k_B}{e}\left[\frac{E_S}{k_B T} + \alpha\right] \quad (4)$$

where e is the electronic charge; E_S is the activation energy obtained from TEP data and α is a constant. α<1 implies the applicability of small polaron hopping model whereas α>2 is for large polaron hopping (Das et al., 2004). It has been found that the above equation fits well with the experimental data of the present investigation. The best fit curves of S versus 1/T of NSMO samples are shown in Figure 13(b). The calculated values of α for the samples of the present investigation are less than one, therefore it may be concluded that the small polaron hopping mechanism is more appropriate to explain the thermopower data in the high-temperature regime.

3.5 Magnetoresistance

One of the important properties of the doped perovskite manganites is their mangetoresistace (MR).The percentage of MR is defined as the relative change in the electrical resistivity of a material on the application of an external magnetic field and is given by a relation,

$$MR\% = \frac{\rho_H - \rho_0}{\rho_0} X100 \quad (5)$$

where ρ_H and ρ_0 are the resistivities at a given temperature in the applied and zero magnetic fields, respectively. MR can be positive or negative depending on increase or decrease in the resistivity values. The magneto resistance studies were undertaken on all the five groups of samples of present investigation. The variation of electrical resisitivity with temperature at different fields for LBMO sample sintered at two different sintering temperatures (T_S = 800 & 1100°C) is shown in Figure 15. It is also clear from the figures that the resistivity value decrease with increasing magnetic field and T_P shifts towards high temperature side. This may be due to the fact that the applied magnetic field induces delocalization of charge carriers, which in turn might suppress the resistivity and also causes the local ordering of the magnetic spins. Due to this ordering, the ferromagnetic metallic (FMM) state may suppress the paramagnetic insulating (PMI) regime. As a result value of T_P shifts to high temperature with application of magnetic field. The variation of MR% with temperature in the case of LBMO system is shown in Figure 16.The measurements of MR reveal that the maximum MR is observed for the samples sintered at the lowest temperature. All the samples are exhibiting high MR% at low magnetic fields and also remain high in low temperature regime and then decrease with increasing temperature. This is a clear signature of low field magnetoresistance (LFMR) behavior.

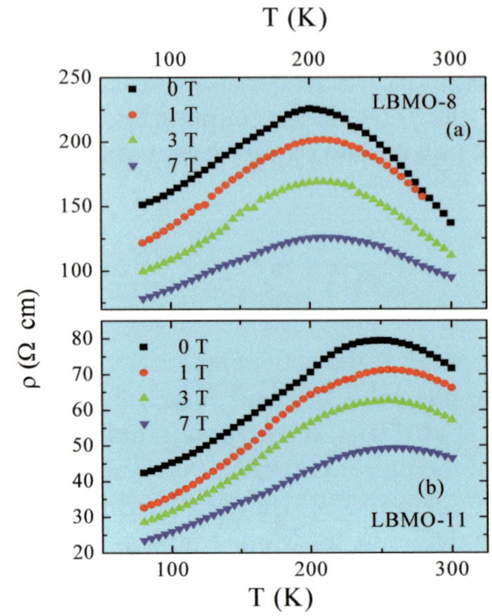

Fig. 15. ρ-T plots of (a) LBMO-8 &(b) LBMO-11 manganites at different magnetic fields.

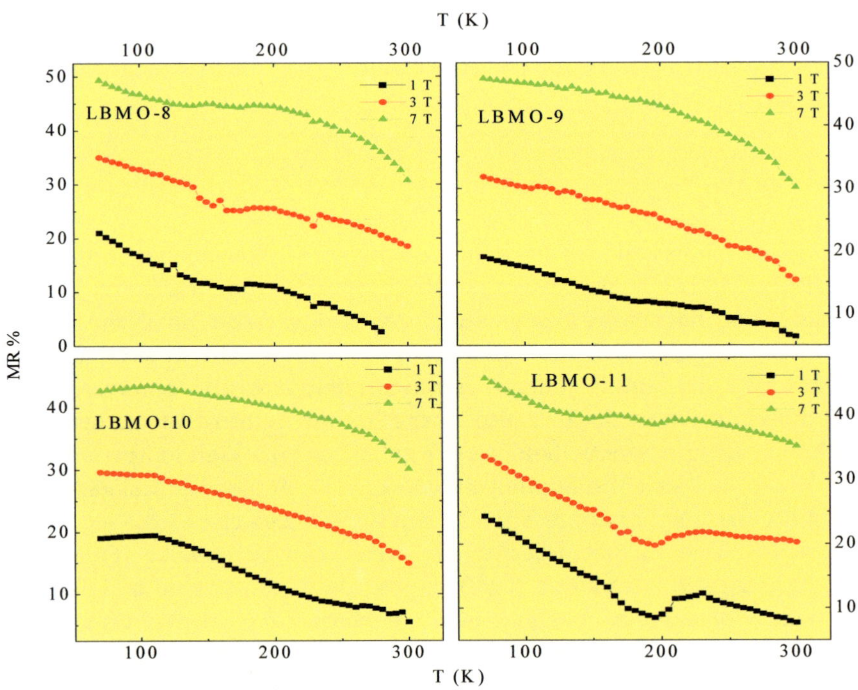

Fig. 16. Variation of MR% with temperature at different magnetic fields for LBMO group of manganites.

The observed behavior may be explained on the basis of a qualitative model. According to this model, magnetoresistance may be classified into two categories (Dutta et al., 2003; Hwang et al., 1996; Mandal et al., 1998). The first one is the intrinsic MR, which arises due to the suppression of spin fluctuations by aligning the spins on the application of magnetic field. This MR has highest value near the ferromagnetic transition temperature and is generally observed among single crystalline bulk as well as thin films. In the polycrystalline samples, there is an additional MR, which is extrinsic in nature; arises due to inter-grain spin-polarized tunneling (ISPT), across the grain boundaries (GBs). This MR contribution increases as the temperature decreases and is usually found in nanocrystalline materials. Rivas et al. (2000) pointed out that the sol-gel prepared samples sintered at low temperatures (900°C) with particle size less than 150nm show more extrinsic nature of magnetoresistance than the intrinsic one. Until recently, it was believed that the former mechanism is responsible for the MR at high fields and the later one at low fields. But recent experiments have shown that high field response is also due to the existence of GB. The nature of GB is a key ingredient in the mechanism of electrical transport, as it constitutes the barrier through which carriers cross or tunnel (Hsu et al., 2006). In view of this, one may conclude that the samples of the present investigation, might be exhibiting extrinsic nature of magnetoresistance arising due to inter-grain spin-polarized tunneling.

3.6 Conduction mechanism

Inspite of extensive experimental and theoretical work to understand the conduction mechanism of CMR materials in general and rare-earth manganites in particular, the present scenario is more confusing than earlier. Therefore, an effort has been made to understand the conduction mechanism of these materials by analyzing the experimental data of both the ferro-as well as the paramagnetic regions using various theoretical models.

3.6.1 Low temperature resistivity ($T<T_P$)

In order to understand the relative strengths of different scattering mechanisms originating from different contributions both as a function of particle size and magnetic field, the FMM ($T<T_P$) part of electrical resistivity data of the samples of the present investigation has been fitted to various empirical equations (De Teresa et al., 1996; Kubo & Ohata, 1972; Schiffer et al., 1995; Snyder et al., 1996; Urushibara et al., 1995). Except LCMO samples, the low temperature resistivity of LSMO, LBMO, NSMO and LNMO group of samples is found to fit well with the equation,

$$\rho = \rho_0 + \rho_2 T^2 + \rho_{4.5} T^{4.5} \tag{6}$$

where ρ_0 arises due to the grain or domain boundaries, $\rho_2 T^2$ indicates the electron–electron scattering, while $\rho_{4.5}T^{4.5}$ is attributed to two magnon scattering process. The two magnon process is more favorable in half-metallic band structure materials such as manganites. The best fit curve for a representative sample viz., LBMO sintered at 800 & 1100 °C various sintering temperatures is shown in Figure 17 (a, b) and fitting parameters are given in Table 2. On the other hand, in the case of LCMO system resistivity in metallic region is well fitted to an equation $\rho=\rho+\rho_{2.5}T^{2.5}$, the term $\rho_{2.5}T^{2.5}$ the resistivity due to single magnon scattering mechanism. It has been observed that the temperature independent residual resistivity term for all the samples is larger than that obtained for single crystals.

Sample code	ρ_0 (Ωcm)			ρ_2 (Ωcm K^{-2})			$\rho_{4.5}$ (Ωcm $K^{-4.5}$)		
	B=0T	B=3T	B=7T	B=0T	B=3T	B=7T	B=0T	B=3T	B=7T
LBMO-8	133.02	81.82	63.28	30.50x10-4	28.60x10-4	24.70x10-4	12.20x10-9	9.78x10-9	7.62x10-9
LBMO-9	88.95	59.47	46.47	28.10x10-4	23.10x10-4	21.10x10-4	3.10x10-9	1.98x10-9	1.44x10-9
LBMO-10	59.12	38.82	32.01	27.50x10-4	22.00x10-4	16.70x10-4	2.00x10-9	1.43x10-9	1.08x10-9
LBMO-11	36.28	21.86	18.95	10.20x10-4	9.30x10-4	7.30x10-4	1.30x10-9	3.01x10-10	2.16x10-10

Table 2. The best-fit parameters obtained from low temperature ($T<T_P$) resistivity data both in presence and in absence of magnetic field for LBMO group of mangnaites sintered at different temperatures.

Further, the fitting parameters are found to decrease continuously with increasing particle size as well as the magnetic field. One of the reasons may be due to decrease in the grain boundary size with increasing sintering temperature. The second reason may be due to the spins present in the domain wall align in the direction of applied magnetic field resulting in the enlargement of magnetic domains. Therefore, the parallel configuration of the spins present in the domain, suppresses various scattering contributions and as a result ρ_0, ρ_2, $\rho_{2.5}$ and $\rho_{4.5}$ are decreasing with the application of the magnetic field and sintering temperature. (Banerjee et al., 2001).

3.6.2 Insulating region ($T>T_P$)

The pair-density function (PDF) analysis (Toby & Egami, 1992; Mott & Davis 1971) of powder neutron scattering data on manganites shows that the high temperature semiconducting region ($T>T_P$) is mainly due to lattice polarons. The analysis clearly indicates that the doped holes in manganites are likely to be localized within one octahedron of Mn^{4+} ion, forming a single-site polaron (small polarons). These small polarons in the high temperature semi-conducting regime of electrical resistivity data are in good agreement with the theoretical prediction of entropic localizations. Therefore, an attempt has been made to fit the high temperature resistivity data of the samples of present investigation to small polaron hopping (SPH) model explained by the equation (Emin & Holstein 1969),

$$\rho = \rho_\alpha T \exp\left(\frac{E_\rho}{k_B T}\right) \tag{7}$$

where ρ_α is a pre-factor and E_ρ is the activation energy. $\rho_\alpha = 2k_B/3ne^2a^2\nu$, where k_B is Boltzmann's constant, e is the electronic charge, n is the number of charge carriers, a is site – to – site hopping distance and ν is the longitudinal optical phonon frequency.

The high temperature data of the samples of the present study is found to fit well with the SPH model. A typical plot of ln (ρ/T) versus T^{-1} in the case of LBMO-8 & 11 is shown in Figure 17 (c, d). From the best fit parameters the activation energy (E_ρ) values have been estimated and the values are found to decrease continuously with increasing particle size and magnetic field. The observed behavior may be due to the fact that the increasing particle size may increase the interconnectivity between grains, which in turn enhances the possibility

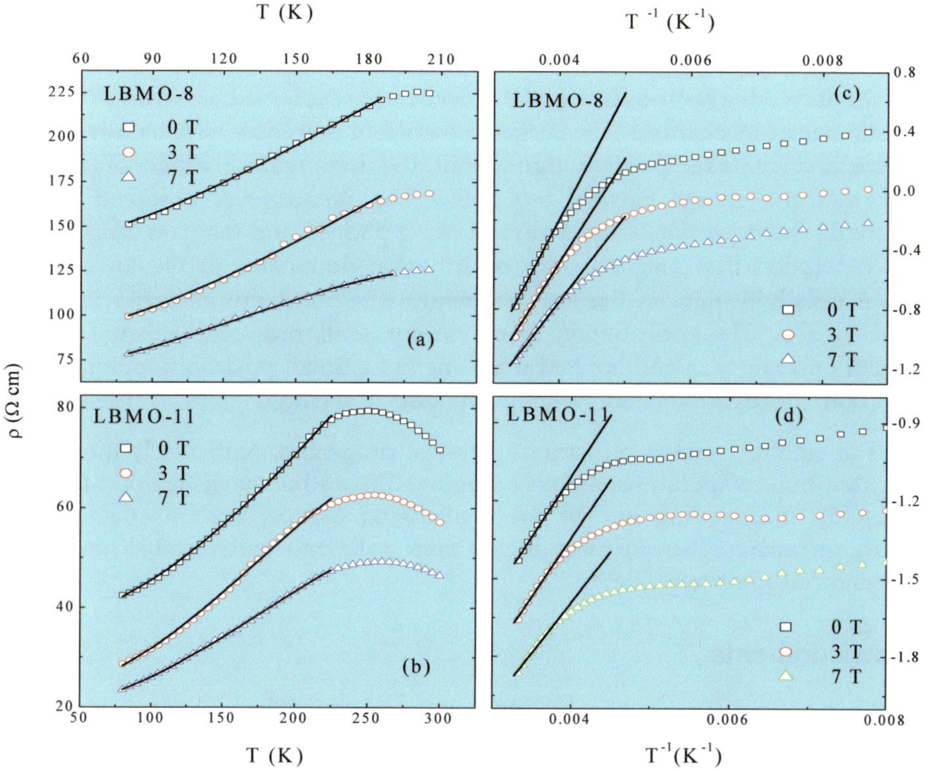

Fig. 17. (a,b) ρ-T plots of LBMO manganites sintered at 800 & 1100°C both in presence and absence of magnetic field. The solid line gives the best-fit to eq.(6) and (c,d) plots of ln(ρ/T) vs. T^{-1} (K^{-1}) in LBMO-8 & LBMO-11 samples. The solid line indicates the best fit to SPH model.

of conduction electron to hop to the neighboring sites, thereby conduction bandwidth increases and as a result the value of Eρ decreases. Therefore, the conduction bandwidth of the materials may be modified by proper tuning of their particle size. The increase in Eρ in the presence of magnetic field is mostly attributed to the increase of charge localization (Banerjee et al., 2002).

4. Summary and outlook

For over more than a decade, studies on manganites have been focused on both various details covering the underlying fundamental Physics and on applied aspects; it is the former which has received most of the attention. This chapter embodies the influence of sintering temperature on physical properties in doped manganites. Five series of doped manganties i.e., $La_{0.67}Ca_{0.33}MnO_3$, $La_{0.67}Sr_{0.33}MnO_3$, $La_{0.67}Ba_{0.33}MnO_3$, $Nd_{0.67}Sr_{0.33}MnO_3$ and $La_{0.67}Na_{0.33}MnO_3$, were synthesized using sol-gel method by sintering the samples at four different temperatures. The crystallite sizes of all the samples are in the nanometer range and are found to decrease on lowering the sintering temperature. The ferro to paramagnetic transition temperatures (T_C) remains unaltered, while the metal to insulator transitions (T_P) decrease with decreasing sintering temperature. The behavior was explained using core-shell model. The percentage of magentoresistance is high at low temperatures and varies linearly with decreasing temperature and the behavior is explained based on inter grain spin polarized tunneling mechanism. The Seebeck coefficient decreases with increasing sintering temperature and remains positive throughout the temperature range of investigation indicating that the charge carriers are holes. The broad peak observed in the low temperature thermopower data was analyzed using phonon and magnon drag terms and it has been concluded that magnon drag contribution dominates in the low temperature region. The resistivity data in the low temperature region were analyzed using various theoretical models. The contribution from various scattering mechanism are found to depend both on grain size and applied magnetic field. Small polaron hopping mechanism dominates both the resistivity and thermopower data in the high temperature region.

The trends of sintering effects on various physical properties, particularly the electric and magnetic transition temperatures are encouraging. The authors envisage that further work will be useful in achieving one of the challenging aspects such as the above room temperature ferromagnetism in these highly spin polarized half metallic manganites for future commercial spintronic devices.

5. Acknowledgments

The authors acknowledge Defence Research and Development Organisation (DRDO) for funding the part of the work. The authors also thank Dept. of Science & Technology, Govt. of India for assisting part of the work. One of the authors (YKL) thanks Council of Scientific and Industrial Research (CSIR), Govt. of India for sanctioning the fellowship.

6. References

Banerjee A, S. Pal, B.K. Chaudhuri, (2001). Nature of small-polaron hopping conduction and the effect of Cr doping on the transport properties of rare-earth manganite $La_{0.5}Pb_{0.5}Mn_{1-x}Cr_xO_3$. *J. Chem. Phys.* 115, 1550.

Banerjee A, S. Pal, S. Bhattacharya, B.K. Chaudhuri, (2002) Particle size and magnetic field dependent resistivity and thermoelectric power of $La_{0.5}Pb_{0.5}MnO_3$ above and below metal-insulator transition. *J. Appl. Phys.* 91, 5125.

Balcells L I, J Fontcuberta, B Martinez and X Obradors (1998) High-field magnetoresistance at interfaces in manganese perovskites. *Phys. Rev. B* 58, R14697

Battacharya S, S Pal, A. Banerjee, H. D. Yang, and B. K. Chaudhuri, (2003). Magnetotransport properties of alkali metal doped La-Ca-Mn-O system under pulsed magnetic field: Decrease of small polaron coupling constant and melting of polarons in the high temperature phase. *J.Chem. Phys.* 119, 3972

Das S, A. Poddar, B. Roy, and S. Giri, (2004). Studies of transport and magnetic properties of Ce-doped $LaMnO_3$. *J. Alloys Compd.* 365, 94

Dey P and Nath T.K (2006), Effect of grain size modulation on the magneto-and electronic-transport properties of $La_{0.7}Ca_{0.3}MnO_3$ nanoparticles: The role of spin-polarized tunneling at the enhanced grain surface. *Phys. Rev. B* 73, 214425

De Teresa J. M, M. R. Ibarra, J. Blasco, J. Garcia, C. Marquina and P A Algarabel (1996). Spontaneous behavior and magnetic field and pressure effects on $La_{2/3}Ca_{1/3}MnO_3$ perovskite. *Phys. Rev. B* 54, 1187

Dutta A, N. Gayathri, R. Ranganathan, (2003). Effect of particle size on the magnetic and transport properties of $La_{0.875}Sr_{0.125}MnO_3$. *Phys. Rev. B* 68, 054432

Dogotto E (2003). *Nanoscale Phase Separation and Colossal Magnetoresistance* Springer Series in Solid State Physics vol 136

Emin D, T. Holstein, (1969). Studies of small-polaron motion IV. Adiabatic theory of the hall effect. *Ann. Phys.* 53, 439

Fu Y (2000). Grain-boundary effects on the electrical resistivity and the ferromagnetic transition temperature of $La_{0.8}Ca_{0.2}MnO_3$. *Appl. Phys. Lett.* 77, 118

Goldschmidt V (1958). *Geochemistry* (Oxford: Oxford University Press)

Goodenough J B (2003). Rare earth manganese perovskites Handbook on the Physics and Chemistry of Rare Earth vol 33 ed KA Gschneidner Jr et al Amsterdam: Elsevier

Gaur A and Verma G. D. (2006). Sintering temperature effect on electrical transport and magnetoresistance of nanophasic $La_{0.67}Sr_{0.33}MnO_3$. *J. Phys.: Condens. Matter* 18, 8837

Hueso L. E, F. Rivadulla, R.D. Sanchez, D. Caeiro, C. Jardon, C. Vazquez-Vazquez, J. Rivas, M.A. Lopez-Quintela, (1998). Influence of the grain-size and oxygen stoichiometry on magnetic and transport properties of polycrystalline $La_{0.67}Ca_{0.33}MnO_{3\pm\delta}$ perovskites. *J. Magn. Magn. Mater.* 189, 321

Hsu C.Y, H. Chou, B.Y. Liao, J.C.A. Huang, (2006). Interfacial and quantum wall effects on ac magnetotranport of $La_{0.67}Sr_{0.33}MnO_3/La_{1.4}Sr_{1.6}Mn_2O_7$ composites. *Appl. Phys. Lett.* 89, 262510.

Hwang H. Y, S. W. Cheong, N. P. Ong, and B. Batlogg, (1996). Spin-polarized intergrain tunneling in $La_{2/3}Sr_{1/3}MnO_3$. *Phys. Rev. Lett.*77, 2041.

Hwang H. Y, S-W. Cheong, P. G. Radaelli, M. Marezio, and B. Batlogg, (1995). Lattice effects on the magneoresistance oin doped $LaMnO_3$. *Phys. Rev. Lett.* 75, 914.

Jirak, Z., Krupicka, S., Simsa, Z., Dlouha, M., and Vratislav, S., (1985). Neutron diffraction study of $Pr_{1-x}Ca_xMnO_3$ perovskites. *J. Magn. Magn. Mater.* , 53, 153

Jaime M, M.B. Salamon and K. Pettit et al., (1996). Magnetothermopower in $La_{0.67}Ca_{0.33}MnO_3$ thin films. *Appl. Phys. Lett.* 68, 1576

Kubo K & N. Ohata, (1972). A quantum theory of double exchange. *J. Phys. Soc. Japan,* 33, 21

Kwon C. W, M. C. Robson, K-C. Kim, J. Y. Gu, S. E. Lofland, S.M. Bhagat, Z. Trajanovic, M. Rajeswari, T. Venkatesan, A. R. Kratz, R. D. Gomez, and R. Ramesh, (1997). Stress-induced effects on epitaxial $(La_{0.7}Sr_{0.3})MnO_3$ films. *J. Magn. Magn. Mater.*172, 229.

Kim B. H, J. S. Kim, T. H. Park, D. S. Le, and Y. W. Park, (2008). Magnon drag effect as the dominant contribution to the thermopowe in $Bi_{0.5-x}La_xSr_{0.5}MnO_3$ $(0.1 \leq x \leq 0.4)$. *J. Appl. Phys.*103, 113717

Kalyana Lakshmi Y, G. Venkataiah, M. Vithal and P. Venugopal Reddy (2008). Magnetic and electrical behavior of $La_{1-x}A_xMnO3$ (A= Li, Na, K and Rb) manganites. *Physica B*. 430, 3059 a Kalyana Lakshmi Y and P. Venugopal Reddy, (2009). Influence of sintering temperature and oxygen stoichiometry on electrical transport properties of $La_{0.67}Na_{0.33}MnO_3$ manganites. *J. Alloys Compd.* 470, 67

b Kalyana Lakshmi Y and P. Venugopal Reddy, (2009). Electrical behavior of some silver-doped lanthanum-based CMR materials. *J. Magn. Magn. Mater.* 321, 1240

Kalyana Lakshmi Y, G. Venkataiah and P. Venugopal Reddy, (2009). Magnetoelectric behavior of sodium doped lanthanum manganites. *J. Appl. Phys.* 106, 023707

Kalyana Lakshmi Y and P. Venugopal Reddy (2010). Influence of silver doping on the electrical and magnetic behavior of $La_{0.67}Ca_{0.3}MnO_3$ manganites. *Solid State Sciences* 12, 1731

Kalyana Lakshmi Y and P. Venugopal Reddy, (2011). Investigation of ground state in sodium doped neodymium manganites. *Phys. Lett. A* 375, 1543

Mahesh R, Mahendiran R, Raychaudhuri A K and Rao C N R (1996). Effect of particle size on the gaint magnetoresistance of $La_{0.67}Ca_{0.33}MnO_3$. *Appl. Phys. Lett.* 68, 2291

Malavasi M, M.C.Mozzati, C.B. Azzoni, G. Chiodelli, G. Flor,(2002). Role of oxygen content on the transport and magnetic properties of $La_{1-x}Ca_xMnO_{3\pm\delta}$ manganites. *Solid State Commun.*123 321.

a Malavasi L, M. C. Mozzati, S. Polizzi, C. B. Azzoni, and G. Flor, (2003). Nanosized sodium-doped lanthanum manganites: Role of the synthetic route on their physical properties. *Chem.Mater.* 15, 5036

b Malavasi L, M. C. Mozzati, C. Tealdi, C. B. Azzoni and G. Flor. (2005). Influence of Ru doping on the structure, defect chemistry, magnetic interaction, and carrier motion of the $La_{1-x}Na_xMnO_{3+\delta}$. *J. Phys. Chem. B*, 109, 20707.

a Mahendiran, R., Tiwary, S., Raychaudhuri, A.,Mahesh, R., and Rao, C. N. R.,(1996). Thermopower and nature of the hole-doped states in $LaMnO_3$ and related systems showing gaint magnetoresistance. *Phys. Rev. B* 54, R9604.

b Mahendiran, R., R Mahesh, N Rangavittal, S Tewari, A K Raychauduri, T V Ramakrishnan and CNR Rao, (1996). Structure, electron-transport properties, and gaint magnetoresistance of hole-doped $LaMnO_3$ systems. *Phys. Rev. B* 53, 3348.

Mandal P (2000). Temperature and doping dependence of the thermopower in $LaMnO_3$. *Phys. Rev. B* 61, 14675

Mandal P, K. Barner, L. Haupt, A. Poddar, R. von Helmolt, A. G. M.Jansen, and P. Wyder, (1998). High-field magnetotranport properties of $La_{2/3}Sr_{1/3}MnO_3$ and $Nd_{2/3}Sr_{1/3}MnO_3$ systems. *Phys. Rev. B* 57, 10256

Millis A J, Littlewood P B and Shraiman B (1995). Double exchange alone does not explain the resistivity of $La_{1-x}Sr_xMnO_3$. *Phys. Rev. Lett.* 74, 5144

Mott N F, E.A. Davis, (1971). Electronic Process in Noncrystalline Materials,Clarendon, Oxford,.

Okuda T, A. Asamitsu, Y. Tomioka, T. Kimura, Y. Taguchi, and Y. Tokura, (1998). Critical behavior of the metal-insulator transition in $La_{1-x}Sr_xMnO_3$. *Phys. Rev. Lett.* 81, 3203

Ramirez A. P. (1997). Colossal magnetoresistance. *J. Phys.:Condens. Matter* 9, 8171

Rao R. A, D. Lavric, T. K. Nath, C. B. Eom, L. Wu, and F. Tsui,(1998) Three-dimensional strain states and crystallographic domain structures of epitaxial colossal magnetoresistive $La_{0.8}Ca_{0.2}MnO_3$ thin films. *Appl. Phys. Lett.* 73, 3294

Rivas J, L.E. Hueso, A. Fondado, F. Rivadulla, M.A. Lopez-Quintela, (2000). Low field magnetoresistance effects in fine particles of $La_{0.67}Ca_{0.33}MnO_3$ perovskites. *J. Magn.Magn. Mater.* 221, 57.

Schiffer P, A. Ramirez, W. Bao & S.-W. Cheong, (1995). Low temperature magnetoresistance and the magnetic phase diagram of $La_{1-x}Ca_xMnO_3$. *Phys. Rev. Lett.* 75, 3336

Shankar K. S, S. Kar, G.N. Subbanna, A.K. Raychaudhuri, (2004) Enhanced ferromagnetic transition temperature in nanocrystalline lanthanum calcium manganese oxide ($La_{0.67}Ca_{0.33}MnO_3$). *Solid State Commun.* 129, 479.

Suzuki Y, H. Y. Hwang, S. W. Cheong, and R. B. vanDover, (1997). The role of strain in magnetic anisotropy of manganite thin films. *Appl. Phys. Lett.* 71, 140

Snyder G.J, R.Hiskes, S.Dicarolis, M.Beasley & T.Geballe, (1996). Intrinsic electrical transport and magnetic properties of $La_{0.67}Ca_{0.33}MnO_3$ and $La_{0.67}Sr_{0.33}MnO_3$ MOCVD thin films. *Phys. Rev.B* 53, 14434

Toby B. H, T. Egami, (1992). Accuracy of pair distribution fuction analysis applied to crystalline and non-crystalline materials. *Acta Crystallogr. Sect. A* 48, 336.

Tokura Y (2000). *Colossal Magnetoresistive Oxides*, (Gordon and Breach,New York)

Urushibara A, Y. Moritomo, T. Arima, A. Asamitsu, G. Kido & Y. Tokura (1995). Insulator-metal transition and giant magnetoresistance in $La_{1-x}Sr_xMnO_3$. *Phys. Rev. B* 51, 14103

Volger, J., (1954). Physica, 20, 49

Venkataiah G, D. C. Krishna, M. Vithal, S. S. Rao, S. V. Bhat, V. Prasad, S. V. Subramanyam and P. Venugopal Reddy (2005). Effect of sintering temperature on electrical transport properties of $La_{0.67}Ca_{0.33}MnO_3$. *Physica B* 357, 370.

Venkataiah G and P. Venugopal Reddy, (2005). Electrical behavior of sol-gel prepared $Nd_{0.67}Sr_{0.33}MnO_3$.manganite system. *J. Magn. Magn. Mater.* 285, 343

a Venkataiah G, Y. Kalyana Lakshmi and P. Venugopal Reddy, (2007). Influence of particle size on electrical transport properties of $La_{0.67}Sr_{0.33}MnO_3$ manganite system. *J. Nanosci. Nanotech.* 7, 2000

b Venkataiah G, Y. Kalyana Lakshmi and P. Venugopal Reddy, (2007). Thermopower studies of $Pr_{0.67}Ca_{0.33}MnO_3$. manganite system. *J. Phys. D: Appl. Phys.* 40, 721

c Venkataiah G, V. Prasad and P. Venugopal Reddy, (2007). Anomalous variation of magnetoresistance in $Nd_{0.67-y}$ $Eu_ySr_{0.33}MnO_3$ manganites. *Solid State Commun.* 141, 73.

Venkataiah G, Y. K. Lakshmi and P. Venugopal Reddy, (2008). Influence of sintering temperature on resistivity, magnetoresistance and thermopower of $La_{0.67}Ca_{0.33}MnO_3$. *PMC Phys B* 1754, 1:7

a Venkataiah G and P. Venugopal Reddy, (2009). Variation of thermopower with crystallite size of $La_{0.67}Sr_{0.33}MnO_3$ manganites. *Phase Transitions* 82, 156

b Venkataiah G and P. Venugopal Reddy, (2009). Magnon drag contribution to thermopowe of $Nd_{0.67}Sr_{0.33}MnO_3$ nanocrystalline manganites. *J. Appl. Phys.* 106, 033706

Venkataiah G, J. C. A. Huang and P. Venugopal Reddy, (2010). Low temperature resistivity minimum and its correlation with magnetoresistance in $La_{0.67}Ba_{0.33}MnO_3$. nanomanganites. *J. Magn. Magn. Mater.* 322, 417

Vogel A. I, A text book of Quantitative Inorganic Analysis including Elementary Instrumental Analysis, (1978) 4th edn., Longman, London,.

Williamson G. K and W. H. Hall, (1953). X-ray line broadening from field aluminium and wolfram. *Acta Metall.* 1, 22.

Zhou J.-S, J. B. Goodenough, A. Asamitsu, and Y. Tokura, (1997). Pressure induced polaronic to itinerant electronic transition in $La_{1-x}Sr_xMnO_3$ crystals. *Phys. Rev. Lett.* 79, 3234

Zhang N, Ding W, Zhong W, Xing D and Du Y (1997). Tunnel-type gian magnetoresistance in the granular perovskite $La_{0.85}Sr_{0.15}MnO_{3.}$ *Phys.Rev. B* 56, 8138

Zenar C, (1951). Interaction between the d-shell in the transition metals. II. Ferromagnetic compounds of manganese with perovskite structure. *Phys. Rev.* 82, 403.

14

The Quantification of Crystalline Phases in Materials: Applications of Rietveld Method

Cláudia T. Kniess[1,2], João Cardoso de Lima[2] and Patrícia B. Prates[2]
[1]*Nove de Julho University – UNINOVE - São Paulo - SP*
[2]*Federal University of Santa Catarina – UFSC - Florianópolis - SC*
Brazil

1. Introduction

The materials have their properties defined by the chemical composition and the microstructure presented on them. The quantification of crystalline phases is a key step in determining the structure, properties and applications of a given material. Therefore, the study of the amount of crystalline phases present in a material represents an important parameter to control the microstructure and the correlation of the properties associated with the developed stage in the process. The distribution of phases or microstructure depends on the manufacturing techniques, sintering process, raw materials used, equilibrium reactions, kinetics and phase changes. The characterization of crystalline microstructure, regarding the density, atomic distribution and unit cell dimensions, contributes to the control of the manufacturing process. It is also the basis for phase identification, structure, porosity and evaluating the performance of materials. Within this context, the book's chapter entitled "Quantification of Crystalline Phases in Materials: Applications of the Rietveld Method" aims to present and discuss some topics related to methods for quantification of crystalline phases in materials using the technique of X-ray diffraction, especially in the case of ceramic materials. It will examine various methods of quantification of crystalline phases described in the literature, but the focus will be applications of the Rietveld Method.

2. Materials and microstructure

The modern construction techniques, the growing competitiveness in international markets and the technological renewal, demand products that meet the basic requirements of high quality and low cost. The knowledge of raw materials and the effect on the processing and properties stages of the final product are necessary to achieve these requirements (KNIESS, 2005).

The materials have their properties defined by the chemical composition and the microstructure they exhibit. These characteristics are influenced by the selection of raw materials and the manufacturing process, mainly the sintering process. The research advances in materials engineering aspects seek to correlate microstructual features, including crystalline and amorphous phases, porosity, etc., with properties of interest, like mechanical strength, coefficient of thermal expansion, density, among others. For most materials, the identification and quantification of amorphous and crystalline phases is crucial in determining the structure, properties, and applications of a material.

Solid materials can be classified according to the regularity by which their atoms or ions are arranged in relation to each other (PADILHA, 1997). The structure of solid material is a result of the nature of their chemical bonds, which defines the spatial distribution of atoms, ions or molecules. Solid materials can be crystalline or amorphous. The concept of crystal structure is related to organization of atoms geometry. The vast majority of materials, commonly used in engineering, particularly metals, shows a well defined geometric arrangement of their atoms, forming a crystalline structure.

A crystalline material is the one in which the atoms are located in a repeated arrangement or that is periodically over long atomic distances; that is, there is long-range order, so, when the solidification occurs the atoms will place themselves in a three dimensional repetitive pattern, in which each atom is connected to its nearest neighbor atoms. All metals, many ceramic materials and certain polymers were crystalline structures under normal solidification conditions (CALLISTER, 2002).

From the concept of crystal structure, where it is possible to describe a set of repetitive atomic, ionic or molecular positions, arises the concept of unit cell. A unit cell is defined as the smallest portion of the crystal that still retains its original properties. Through the adoption of specific values associated with the units of measure on the axes of references, defined as network parameters, and the angles between these axes, it can be obtained unit cells of various kinds.

The atomic structures of materials can not be regularly arranged as in crystalline networks. Such structures are called amorphous or vitreous (NAVARRO, 1991). This type of structure, however, is not completely disordered. It is formed by constitutive blocks arranged in a disorderly way. But the structure of these blocks is regular. Thus, it can be said that the glassy structures have long-range disorder and short-range order.

The ceramics and glasses are distinguished from other materials - metals and polymers - essentially by the type of chemical bonds that each one has. (KINGERY *et al.*, 1976). Regardless of the final products, the ceramic process begins with the selection of mineral or synthetic raw materials. We can classify the ceramic materials in two large groups: the traditional ceramics, using mostly natural raw materials, with a predominance of clays nature, and advanced ceramicas, in which raw materials are essentially synthetic (FONSECA, 2000).

The amount of crystalline phases present in a ceramic material is an important parameter to control the microstructure and the properties correlation associated to the phase developed in the process (BORBA, 2000). Another focus of the quantitative analysis of phases is the measure of crystallinity, in other words, the fraction of crystalline phases present in the sample. In the case of ceramic materials, where the thermal process generates different amounts of crystalline phases, the content of residual amorphous phase is an important parameter, not only as physical characterization of the product, but to correlate it with the mechanical properties and kinetic studies of crystallization.

3. Application of X-ray diffraction techniques to quantification of phases in materials

There are many techiniques for characterization of polycrystalline materials (X-ray diffraction, differential thermal analysis, thermogravimetric analysis, infrared spectroscopy, scanning electron microscopy, transmission electron microscopy, nuclear magnetic resonance spectroscopy, ultraviolet spectroscopy, etc), each of which is most suitable for a particular purpose and the others can be used to supplment the findings obtained by other technique. The characterization methods that make use of X-ray or neutron diffraction are particularly interesting for: (i) index of crystalline phases, (ii) refinements of unit cell, (iii) determination of crystallite size and network micro-deformations, (iv) quantitative analysis of phases, (v) determination of crystal structures, (vi) refinement of crystal structures, (vii) determination of preferred orientation (texture), etc. (PAIVA-SANTOS, 1990).

The technique of X-ray diffraction is still considered the most suitable for quantitative analysis of phases. However, the percentage of phases present in the material (amorphous and crystalline), quantification of the glassy phase, depends on the manufacturing techniques, sintering process, raw materials used, equilibrium reactions, kinetics and phase changes. The characterization of crystalline microstructure, with respect to the density, atomic distribution and dimensions of the unit cell, contributes to the control of the manufacturing process.

3.1 X-ray diffraction principles

The X-ray diffraction (XRD) represents the phenomenon of interactions between the incident X-ray beam and the electrons of the atoms of a material component, related to the coherent scattering.

The techinique consists in the incidence of radiation on a sample and in the detection of the diffracted photons, which constitute the diffracted beam.

The scattering and the resulting X-ray diffraction is a process that can be analyzed at different levels. In the most basic of them, there is the X-ray scattering by an electron. This scattering can be coherent or incoherent. In coherent scattering, the scattered wave has a definite direction, the same phase and energy in relation to the incident wave. It is an elastic collision. In incoherent scattering, the scattered wave has no definite direction (CULLITY, 1978). It does not keep the phase or energy (this is called Compton Effect). The collision is inelastic, and the energy for the difference between the incident wave and the wave scattered translates into temperature gain (vibration of the atom).

When an X-ray beam, with a certain frequency, focuses on an atom, it behaves as a scattering center, and vibrates at the same frequency as the incident beam, spreading in all directions. When the atoms are arranged in a lattice, this incident beam will undergo a constructive interference in certain directions, and destructive in others. The constructive interference of scattered radiation occurs when the path difference of successive planes scattered beam is equal to an integer number of wavelength (KLUG & ALEXANDER, 1954).

Bragg's law is a geometric interpretation of the diffraction phenomenon of atoms arranged in a lattice. In a material where the atoms are arranged periodically in space, characteristic of crystalline structures, the penomenos of X-ray diffraction occurs in the scattering directions that satisfy Bragg's Law, Equation (3.1). This Law is a consequence of the periodicity of the network and is not associated with each particular atom, or different atomic numbers (CULLITY, 1978).

Assuming that a monochromatic beam of a certain wavelength (λ) incident on a crystal at an angle θ, we have (KLUG & ALEXANDER, 1954):

$$n\,\lambda = 2\,d\,\text{sen}\,\theta \tag{3.1}$$

Where: n = integer number of wavelength, d = interplanar distance of the sucessive crystal planes; θ = angle measured between the incident beam and determined crystal planes.

Traditional instruments of measurement are the diffractometer and chambers of single crystals, application of the latter is now restricted to specific situations to determine the crystallographic parameters. In the study of polycrystalline materials in powder form, the radiation is monochromatic and the angle of incidence θ is variable. In traditional diffractometer, the diffracted uptake shaft is through a detector, according to a geometrical arrangement known as the Bragg-Brentano geometry (CULLITY, 1978). The X-ray beam falling on the sample, positioned at θ is diffracted at the periodicity of the network, and the signal is collected in a detector positioned at 2θ.

The diffracted beam is usually expressed through peaks that stand out from the background (or baseline), recorded in a spectrum of intensity per second (c.p.s.) versus the angle 2θ (or d), constituting the diffraction pattern of diffractogram.

The intensities obtained at 2θ angles, represented by the XRD patterns, correspond to the diffraction of the incident beam for a givem set of crystal planes, which have the same interplanar distance, each one with Müller hkl indices (hkl reflections). The relative intensity of the peaks is related to the type of atom in the lattice and its occupation number of the atom in the unit cell.

The scattering caused by the electrons in the unit cell results in a complex interference function. The total amplitude of the scattered beam is the sum of the contributions of all electrons, in other words, is proportional to Z (atomic number). These scattering values are the normalized amplitude of electrons number involved in the angle θ = 0 and are the atomic scattering factors. In the case of beams scattered in the direction of incidence, θ = 0, the rays are in phase and amplitude is added. However, when the angle θ is different from zero, the trajectories of the scattered rays are different and the phase difference results in interference. The measure of this phase difference is contained in an exponential factor that defines the change in amplitude as a function of atoms positions. (WILES *et al.*, 1981). The combination of this phase factor and the atomic scattering factor results in the factor of structure (WARREN, 1959) represented by Equation (3.2):

$$F_{hkl} = \sum_{n=1} f_n \exp\left[2\pi i(hx_n + ky_n + lz_n)\right] \tag{3.2}$$

where:
f_n = scattering factor for atom n;

x_n, y_n, z_n = position coordinates of the nth atom;
h, k, l = Müller indice;
$f_n = f_o \exp(-B\, sen^2\theta/\lambda)$, where f_o is the scattering of the temperature absolute zero and *B* is the medium amplitude of normal vibration to the direction of diffraction.

The position of the peaks is related to the interplanar distances of the phase, i.e., with unit cell parameters. As X-rays penetrate only in the atom electrosphere, the oxidation state of the atom influences the intensity of the diffracted beam. Among the factors that affect the intensity, the main ones are: polarization, temperature, atomic scattering, structure and mass attenuation. (KLUG & ALEXANDER, 1954).

The elevation of the background in the range of 2θ of 20 and 50° on the amorphous phase, which can be called a "halo" is not constant over the entire angular range, but prevalent in certain areas. The chemical characteristics of the amorphous phase and its form of development, even at short range, determine the shape of "halo". The analysis of the area, shape and position of the "halo" provides information about the degree of ordering of the amorphous phase, being more open as more desorganized phase is (FLEURENCE, 1968).

According to Borba (2000), through the detailed study of the shape and position of the peak, one can get some information about the crystalline phases, related to the structure, crystallite size, heterogeneity and micro-deformations. The width of the diffraction peak is related to the crystallite size and/or existing micro-deformations in the crystal lattice. The peak broadening of one phase of XRD is the indicative of a small crystallite size. This extension can be perceived in a different way in different reflections, indicating that the crystallites grew preferentially in one direction. The asymmetry at large angles may be an indicative of the presence of residual tension, and this tension can vary with crystallographic orientation. The displacement of the positions of the peaks can be associated with macrodeformation for defects and for changes of network parameters produced by disagreements and segregations of atoms dissolved.

Each crystalline compound has a characteristic diffraction pattern, allowing their identification through the angular positions and relative intensities of the diffracted peaks. The identification of crystalline phases is obtained by comparing the XRD with the diffraction patterns of individual phases provided by the ICDD (*International Center for Diffraction Data*), the former JCPDS (*Joint Committee of Powder Diffraction Standards*). It is also possible to calculate the unit cell parameters, assess the degree of crystallinity, and quantify present phases. The quantification of phases from the X-ray diffraction can be related to the intensities of the peaks of XRD pattern. It also represents the characteristics of the crystalline phases existing in the material, characterizing the ratio of these phases.

3.1.1 Factors that cause changes in the diffraction pattern

Information on the structure of a material can be obtained through the analysis of some features in the diffraction pattern, which can be summarized in:

a. the angular position of diffraction lines, which depends on the geometry of the crystal lattice, indicating the size and shape of the unit cell;
b. intensity of the diffraction lines, which depends on the type of atoms, its arrangement on the crystal lattice and the crystallographic orientation;

c. shape of the diffraction lines, dependent on the instrumental broadening, particle size, and deformation.

Some factors, instrumental or feature from the samples can influence the diffraction pattern of a sample. According to Klug and Alexander (1954), the instrumental factors for a typical X-ray diffractometer, which influence the profile of the diffraction peaks are:

a. geometry of the X-ray source;
b. displacement of the sample;
c. divergence of the axial X-ray beam;
d. transparency of the sample;
e. the purpose of receiving slot;
f. misalignment of the diffractometer.

The most important non-structural factors that affect the widths, shapes and positions in the peaks of diffraction in the geometries of Bragg-Brentano are (KLUG & ALEXANDER, 1974):

a. alignment and collimation of the beam, influencing the width and symmetry;
b. curvature of the diffraction cone, leading to asymmetry of the peaks at high and low angles;
c. flat shape of the sample surface, producing the peak asymmetry at low angles;
d. absorption/transparency of the sample, causing displacement of the diffraction peaks;
e. particle size in the sample and micro-deformations, causing variation in the width and shape of the peaks;
f. intensity of the incident beam (width and shape of the peaks).

4. Quantitative analysis of crystalline phases

Most of the authors refer to the method of Klug and Alexander (1954) as a predecessor of quantitative analysis, since several other methods have been developed based on it. According to Klug and Alexander, the general equation of quantitative analysis is:

$$I_A = \frac{K_A x_A}{\rho_A \left(x_A \left(\mu_A - \mu_M \right) + \mu_M \right)} \tag{4.1}$$

where:
I_A= intensity of line i of unknown phase A;
K_A= constant depending on the nature of phase A and the geometry of the equipment;
ρ_A = density of phase A;
x_A= weight fraction of phase A;
μ_M = mass attenuation coefficient of the matrix;
μ_A= mass attenuation coefficient of phase A.

There are also known methods of standard addition and external standards, with overlapping peaks (KLUG & ALEXANDER, 1974). With the advancement of computer technology, the method of Rietveld (RIETVELD, 1967, 1969), which is based on the simulation of the diffraction profile from the structures of the phase components in a sample, allowed more information to be extracted from the XRD patterns. Analysing the entire diffraction pattern and using the intensities of each individual angular step, the method enabled the refinement of complex crystal structures, and therefore it was applied to the provision of quantitative data accurately recognized.

4.1 Rational analysis

The mineralogical composition is defined by the type and quantity of minerals that constitute the material studied. According to the concept of rational mineralogical analysis, described in Coelho (2002), through a combination of quantitative chemical composition (for example, by X-ray fluorescence) and from the determination of qualitative mineralogical analysis (obtained by XRD), we obtain enough theoretic information to solve the problem of quantitative deduction of mineralogical phase, after the relationship with the chemical composition of the phase. There are two widespread procedures for carrying out the calculations needed to solve the problem: the conventional procedure (HALD, 1952) and the procedure using the method IRTEC (FABBRI *et al.*, 1989). The latter method has possible sources of error such as simplifying the theoretical formula of complex phases and errors due to the presence of two or more phases with the same theoretical formula, besides a wide dispersion (standard deviation) in the solutions obtained when the number of phases is less than the number of oxides.

4.2 Internal standard method

The internal standard method is most suitable to be used in case of a large number of samples, where a component must be determined and the composition of the samples varies widely. The advantage of this method is that any crystalline phase can be analyzed without considering all stages, and it was not necessary to consider the amorphous phase (CULLITY 1978). The intensities of characteristic peaks of the phases of the sample components are related to the internal standard peaks, being generalized in a system of linear equations that allow to use overlapping peaks and links to the proportions of the phases. The analysis is performed by adding a P internal standard to the mixture to be analyzed in known amounts, requiring the presence of one or more individual peaks without overlapping with other peaks, being common the use of crystallized material in the cubic system by presenting simple sructure and few diffracted peaks (BRINDLEY & BROWN, 1980).

In the analysis of a component of a system where the overlapping lines doesn't occur, and, considering the mass attenuation coefficients of the standard (μ_P) and matrix (μ_M) are different ($\mu_P \neq \mu_M$) (general cases), the intensities of line i of conponent A and the line K of pattern P follow the Equations (4.2) and (4.3) (CULLITY, 1978):

$$I_{iA} = \left(\frac{K_1 c_A}{\mu_M} \right) \tag{4.2}$$

$$I_{\kappa P} = \left(\frac{K_2 c_P}{\mu_P} \right) \tag{4.3}$$

where:

c_A e c_P = volume fraction of phase A in the mixture and the standard P;

K_1, K_2, K_3 = constants.

The ratio of two intensties Ii_A/Ik_P leads to the calibration of Ii_A/Ik_P *versus* x_A, where x_A is the fraction of the component to be analyzed according to Equation (4.4):

$$\frac{I_{iA}}{I_{kP}} = K_3 X_A \tag{4.4}$$

4.3 Matrix flushing method by chung

Doneda (2000) in his work, refers to the quantitative methos developed by CHUNG (1974), from Klug and Alexander equations (1954), called *matrix flushing,* in which a mixture is prepared in a 1:1 ratio of phase to be measured and the corundum standard phase. Chung (1974) elected the corundum as a standard (flushing agent); however, it is also possible to use any other phase not present in the sample. The concentration of a x_A phase is obtained from the equation:

$$x_A = \left(\frac{x_C}{k_C}\right)\left(\frac{I_A}{I_C}\right) \tag{4.5}$$

where: x_c is the mass fraction of corundum in the sample; I_A e I_C are the intensities of the hkl plane of the A phase and the corundum respectively and k_C is the constant obtained from the JCPDS.

Equation (4.5) shows that the relations are independent of matrix effects, therefore, to draw a graph I_A/I_C by x_A, the slope of the line would be x_C/k_C. The amorphous phase is determined by the difference between the crystalline phases quantified and total phases in the sample.

4.4 External standard method

The external standard method is to prepare a series of mixtures containing the phase to be measured at an increasing rate. The values of intensity of a peak characteristic of various mixtures allow determining the concentration of the phase to be measured (FLEURENCE, 1968).

A condition for the use of this method is the choice of a standard substance P, which has the same characteristic of pure or mixed diffraction. I_{po} and I_p are the intensities of P pure or in the mixture, respectively, measured under the same conditions, μ_p and μ_M are the mass attenuation coefficients of the pure substance P and the mass attenuation coefficient of the matrix where P is inserted, respectively. The Equation (4.6) relates the intensities of the peaks with the amount of P:

$$\frac{I_p}{I_{Po}} = x_p \frac{\mu_P}{\mu_M} or \frac{I_p}{I_{Po}} = \alpha . x_p \tag{4.6}$$

where: x_p is the proportion of P in the mixture e $\alpha = \mu_p / \mu_M$ (literature).

4.5 Addition method

The addition method developed by Bragg and Copeland and applied by Fleurence (1968) and Alegre (1965), consists of adding known amounts of a pure phase A to the mixture. This phase A, to be measured, belongs to the system. The methodology consists in measuring the intensities of the peaks of the phases A and B, where B is another phase of the system that will serve as reference for the diffrent samples with increasing amounts of A. The curve I_A/I_B in the function of α is generated from the Equation (4.7):

$$\frac{I_A}{I_B} = \left(\frac{\kappa_A \rho_B}{\kappa_B \rho_A x_B}\right)\alpha + \left(\frac{\kappa_A \rho_B x_A}{\kappa_B \rho_A x_B}\right) \quad (4.7)$$

where: I_A = peak intensity of phase A to be measured; I_B = amount of peak of phase B; belonging to the sample and acting as a reference; α = added amount o phase A; ρ_A e ρ_B = density of A and B, respectively; and x_A and x_B = concentrations, in weight, of A and B, respectively.

According to Borba (2000), as the quantity α increases, the ratio I_A/I_B varies linearly, with k, ρ e x constants. The graph obtained by the ratio I_A/I_B for α can be represented by a straight linear coefficient proportional to x_A. This line intercepts the x-axis at a distant point of origin, so that, when extended, x_A, ie, the concentration is obtained by graphical extrapolation. This method is similar to the internal standard method, but the added A phase is one of phases of the mixture.

5. Quantification of crystalline phases in materials by Rietveld method

Unlike other methods based on the integration of the intensity of characteristic peaks of phases, the method developed by Hugo Rietveld (RIETVELD, 1967, 1969), is applied to the total angular range of the diffraction pattern, increasing the accuracy of the obtained data. The problem of overlapping peaks is minimized, allowing maximum extraction of information from the diffraction pattern.

The Rietveld method involves a refinement of crystal structures, using data from X-ray and neutron diffraction by powder. The term refinement in the Rietveld Method refers to the process of adjusting the model of parameters used in the calculation of a diffraction pattern, to the closest observed. The observed XRD pattern should be obtained in a scanning process step by step with constant growth of $\Delta 2\theta$ (PAIVA-SANTOS, 2001). The differences between the two diffraction patterns are calculated by the method of least squares, and this minimized the difference as the theoretical model approximates the characteristics of the structure (YOUNG, 1995).

To use the method, you must know the structure of the phases in the mixture with a good degree of approximation and have information such as: type of crystal structure, atomic coordinates, occupation number, oxidation state of atoms, symmetry points, values of factors of isotropic and anisotropic temperature. The basic requirements for the Rietveld refinement are: accurate measurements of intensities given at intervals of 2θ, an initial model close to the actual structure of the crystal and a model that describes the shape, width and systematic errors in the positions of the Bragg peaks (PAIVA-SANTOS, 1990).

The calculated standard, to fit the observed pattern, provides data on the structural parameters of the material, as well as the parameters of the diffraction profile. The parameters specific to each phase, which vary during the refinement are (YOUNG, 1995):

a. structural: atomic positions, unit cell parameters, occupancy factors, scale factor, thermal vibration parameters (isotropic and anisotropic) and isotropic thermal parameter general;
b. non-structural: parameters of the half width (U, V, W), asymmetry, 2θ zero, preferred orientation and coefficients of the background radiation.

The Rietveld method allows, simultaneously, perform unit cell refinement, refinement of crystal structure, microstructure analysis, quantitative analysis of phases and determination of preferred orientation (RIELLO *et al.*, 1995).

With the advancement of computer technology, the Rietveld method has allowed more information to be extracted from XRD patterns. From the analyzys of all the diffraction pattern and using the individual intensities of each angular step, the method enabled the refinement of complex crystal structures, and therefore can be applied to the provision of quantitative data accurately recognized (KNIESS *et al.*, 2007).

Over the past three decades, the version of the computer program originally developed by RIETVELD (1967, 1969) has been extensively modified. The program DBWS (WILES & YOUNG, 1981) was probably the most widely used until 1995. In the version DBWS 9411 (YOUNG *et al.*, 1998), the entry data of theoretical model can be given through various crystallographic databases, such as: *Inorganic Crystal Structure Database* (ICSD), *Power Diffraction File* (PDF) - International Centre for Diffraction Data (ICDD), *Structure Reports*, *Cambridge Structure Data Base* (CSD) e *Metals Crystallographic Data File* (CRYSTMET).

The GSAS program (General Structure Analysis System), developed by Larson and Von Dreele (1998) at Los Alamos National Laboratory, has great flexibility for single crystal data, powder diffraction, neutron diffraction. It is widespread in the scientific community and constant updated.

The commercial programs available usually refer to the X-Ray diffraction equipment suppliers. The best known are the High Score Plus (Panalytical) and the Topas (Bruker AXS GmbH - Germany). The Siroquant is from CSIRO, Research Centre in Brisbane (Australia).

Young (1995) lists the most commonly used programs available at universities for the refinement of crystal structures by the Rietveld method, as shown in Table 5.1.

Computer Program	Reference
Rietveld	Rietveld (1969)
Rietveld	Hewat (1973)
PFLS	Toraya e Marumo (1980)
DBW	Wiles e Young (1981)
X-ray Rietveld System	Baerlocher (1952)
LHPM1	Hill e Howard (1986)
GSAS	Larson e Von Dreele (1988)
FullProf	Rodrigues-Carvajal (1990)

Table 5.1. Most widely used programs in universities to the refinement of crystal structures.

5.1 Refinement methodology

Obtaining data suitable for the Rietveld refinement requires attention to the choice of material, which must have small particle size and a minimum of preferred orientation. The sample should be prepared so that the surface is smooth and homogeneous, to avoid the effect of surface roughness.

In order to obtain optimal results , the conditions of data collection should be determined prior to the refining. The main factors to determine are the wavelenght, the collimation of the beam, the angular range and angular distance between the steps and counting time. The choice of appropriate values for the count time T (which defines the intensity and for angular interval of step (which in a given interval determines the number of steps N) depends on the experimental conditions and characteristics of the studied material.

With the choice of a theoretical model of the structure through a database, and subsequent entry of the theoretical data in the program, followed by the step of refining the experimental parameters. The variables contain in the input the data necessary for the construction of the calculated diffraction pattern, ie, data on the crystal structure of the material. The main data are: 2θ limits, wavelenghs of radiation used, specification of the background radiation, the space group symbol, symbol and valence of each atom (used for entering tables scattering factors) and number of phases.

The main parameters that can be adjusted simultaneously in the refinement are (CARVALHO, 1996):

a. scale factor: is the correction of proportionality between the calculated and observed diffraction patterns. The refinement of the scale factor is directly related to the amount of phase;
b. baseline (background): is corrected from data collected in the same XRD pattern and interpolation between these points. It is important to understand the behavior of the baseline, since this provides information about the presence of amorphous phases in the sample and can be included in a routine of quantification of involved phases. The fit of the background equation uses a polynomial up to fifth grade and instrumental aberrations can be considered during the refinement;
c. peak profile: set of analytical functions that are modeled effects related to the profile. Some analytical equations are proposed to correct these effects, such as the Gaussian and the Lorentzian equation and the equation that corrects the asymmetry. The width and position of the peaks are related to the characteristics of crystallite size and cell, respectively;
d. cell parameters: the cell parameters can be fited by Bragg's Law, where the interplanar distance of successive planes of the crystal (d) is related to the Miller indices and therefore the parameters of cell (a, b, c, α, β, γ). The indexing of the peaks is made taking into account the parameters of the cell and the intensity calculated, which shows a certain advantage over convertional techniques, for all the parameters that influence the discrepancy in the values of "d", are handled jointly with the intensities;
e. factor of structure: the variable parameters of this factor are the atomic positions, the isotropic or anisotropic temperature factors and the occupation number;
f. offset: parameters for correction of displacements due to leakage of the focal point of the optical diffractometer;

g. factors of temperature: can absorb deficiencies in the model for the background radiation, absorption and surface roughness, and show discrepancies with the values determined by single crystal diffraction experiments. Therefore, the factors of temperature can be set to values obtained from the literature and an overall temperature factor refined;
h. preferential orientation: correlation of problems generated in the sample preparation.

In the sudy by Post and Bish (1989), the authors suggest the refining steps for any crystalline sample. If the analyzed sample has several crystalline phases, models of atomic structures should be entered for each phase. The study shows that the first few cycles of least squares should be performed with the coefficients of the baseline and the scale factor set, and then several cycles with the inclusion of other parameters should be performed. During the refinement, it is essencial to observe the differences between the spectra of the calculated and observed patterns, so that problems can be detected, like background settings and peak of the profile irregularities. The spectral differences are also important for the verification phase that may not have been included in the refinement. Post and Bish (1989) and Young (1995) consider the control chart refinement important for checking the quality of refinement.

5.2 Output file and refinement evaluation

Through the Equation (5.1) one can calculate the X_B concentration of a given B phase, after the refinement of all *i* phases.

$$x_B = \frac{F_B(ZMV)_B}{\sum_i [S_i(ZMV)_i]} \tag{5.1}$$

where: F is the scale factor refined by the program; Z is the number of formula units per unit of unit cell, M is the mass of the formula unit, V is the volume of the unit cell.

The refinement can be assessed by verification of the structural parameters and profile obtained, the comparison of results with those obtained for single crystals and the observation of plot of calculated and observed patterns, as well as the residuals obtained.

The quality of refinement is verified by numerical statistical indicators that are used during the iterative process (calculations) and after its termination, in order to check whether the refinement proceeds satisfactorily. Table 5.2 presents the metrics most commonly used in the refinements using the Rietveld method (POST E BISH, 1989; YOUNG, 1995; GOBBO, 2003).

Equation	Indicator
$R_F = \Sigma\|(I_K(\text{"OBS"}))^{1/2} - (I_K(\text{CALC}))^{1/2}\| / \Sigma(I_K(\text{"OBS"}))^{1/2}$	R - Structure Factor
$R_B = \Sigma\|I_K(\text{"OBS"}) - I_K(\text{CALC})\| / \Sigma(I_K(\text{"OBS"})$	R - Bragg
$R_P = \Sigma\|y_i(\text{OBS}) - y_i(\text{CALC})\| / \Sigma y_i(\text{OBS})$	R Profile
$R_{WP} = \{\Sigma w_i (y_i(\text{OBS}) - y_i(\text{CALC}))^2 / \Sigma w_i (y_i(\text{OBS}))^2\}^{1/2}$	R - Profile Adjust
$S = [S_y/(N-P)]^{1/2} = R_{WP}/R_{EXP}$	Goodness of Fit = GOF = S
$R_{EXP} = [(N-P)/\Sigma W_i y^2_{io}]^{1/2}$	R - Expected

Table 5.2. Statistical indicators most frequently used in the refinements based on the Rietveld method.

1. I_K is the intensity of Bragg K reflection at the end of each cycle of refinement. In the expressions for R_F and R_B, the "obs", an observed, is placed between quotation marks because the I_K is computed as Rietveld (1969);
2. N = number of parameters being refined; P = number of observations
   Source: GOBBO, (2003).

The quality of refinement is verified by two numerical statistical indicators R_P and R_{WP}, comparative parameters between theoretical and experimental XRD patterns, which can be used to monitor the convergence of the model. R_P and R_{WP} should reach a value of R_{EXP} to consider the model acceptable. The residue R_{WP} considers the error associated with each intensity value by the number of counts, using the weighting factor $w\ (2\theta)$. The value of R_{WP} for good results is 2-10%, while the typical values obtained range from 10-20%. To evaluate the goodness of fit, we compare the final values of R_{WP} with the expected value of the error (R_{EXP}). The expected error is derived from the statistical error associated with the measured intensities. R_{EXP} is related to the quality of the experimental diffractogram, this value being smaller is better (BORBA, 2000). In practice, differences of up to 20% between R_{EXP} and R_P are acceptable. R_{WP} is a statistical indicator that represents the best approach, since the numerator is the residual minimized in the least squares procedure. The factors that modify the R_{WP} are differences in the peaks (as the width) and the background radiation.

From the mathematical point of view, Rwp is one of the indices that best reflect the progress of refinement, because the numerator has the residue that is minimized. Rwp is the index that should be analyzed to see if the refinement is converging. If Rwp is decreasing, then the refinement is successful. At the end of the refinement it should not be more varied, meaning that the minimum has been reached. If Rwp is increasing, then some parameter (s) is (are) diverging from the actual value and refinement should be discontinued for a more detailed analysis of the parameters being refined.

S is called the "*goodness-of-fit* (GOF)" (Equation 5.2) and should be close to 1,0 at the end of refinement, meaning that nothing else can be improved because the Rwp has reached the limit that can be expected to those measured diffraction data. The *goodness-of-fit* should be equivalent to 1,0 in a perfect refinement. In practice less than five figures reflect an optimized refinement.

$$S = R_{WP}/R_{EXP} \tag{5.2}$$

where: R_{EXP} is the value statistically expected to R_{WP}.

All these indices provide subsidies to the user to judge the quality of refinement. However, none of them is related to the crystal structure but only to the profile of the XRD pattern. To assess the quality of the refined structural model, one should calculate the R_{BRAGG}, which is described as a function of the integrated intensities of the peaks (PAIVA-SANTOS, 2005).

$$R_{BRAGG} = 100[(\Sigma\,|\,\text{"Io"} - \text{Ic}\,|)/(\Sigma\text{"Io"}) \tag{5.3}$$

where: I_0 is the intensity observed in the angular position; I_C is the intensity calculated in the angular position.

As the integrated intensity is related to the crystal structure (types of atoms, atomic positions and displacement), this index is to be considered when evaluating the quality of the refined model of the crystal structure.

6. Applications of Rietveld method: Case study

Several studies in the literature addressing the application of the Rietveld method in the quantification of crystalline phases of materials, such as: Borba (2000); Paiva-Santos (2001); Gobbo (2003); Kniess, *et al.* (2009a); Kniess, *et al.* (2010).

In this part of the chapter, we will discuss two case studies in which the Rietveld method was used in the refinement and quantification of crystalline phases of materials.

The first case described in the work of Kniess, *et al.* (2007), refers to the refinement and quantification of crystalline phases of mineral coal bottom ash, residue resulting from combustion of coal in thermoelectric plants to produce electricity.

The second case described in Kniess, *et al.* (2009b), deals with the study of the structural refinement and quantification of crystalline phases of ceramic bodies materials.

6.1 CASE 1 – By-product mineral coal bottom ash

6.1.1 Characterization of the problem

The Jorge Lacerda thermoelectric plant, located at Santa Catarina State, Brazil, uses mineral coal, extracted from the Criciuma region (Santa Catarina State). The burning of coal mineral results in two types of residues: fly and bottom ashes. The first residue has been used as by-products in the cement industries while the second one, which represents almost 50% of the total residues generated by the thermoelectric plants, up to now has no industrial applications causing a serious problem for the environment.

The physical, chemical and mineralogical properties of coal ash are compatible with various raw materials used in ceramic industries coating, which indicates a possibility of partial or full utilisation for this residue. The mineral coal ash has the market value of at least four times smaller than many commercial minerals commonly used as raw materials. The production cost can be reduced both in terms of raw material is a residue of low cost, but also in its physical presentation, whereas the particle size distribution of the residual range of about 5-200μm.

The mineral coal bottom ash is composed mainly of SiO_2 and Al_2O_3 (about 80% by weight). The measured X-ray diffraction (XRD) pattern for coal bottom ash has shown that it is composed majority by low quartz (SiO_2) and mullite ($3Al_2O_3.2SiO_2$).

6.1.2 Mineralogical and chemical characterization of mineral coal bottom ash

The mineral coal bottom ash used in the study was submitted to drying in an oven as 110º C during 24 hours. The percentage of moisture in this particular batch was 20%. Removing the fraction with detectable levels of carbonaceous material macroscopically, because the presence of carbon generates gases that can remaind attached, making the sample homogeneity.

Chemical analysis of industrial by-product was carried out in a XRF spectrometer Philips PW 1400 with Rh tube. To obtain the sample, glass was prepared by using a mixture of lithium tetraborate and lithium metaborate as a flux. Chemical analysis of coal bottom ash is presented in Table 6.1. Some trace elements such as Ba, Pb, Cu, Cr and Ni were determined by atomic absorpion spectrometry, Table 6.2.

Constituents	Coal Bottom Ash (%)
SiO_2	54,04
Al_2O_3	25,19
Fe_2O_3	4,61
CaO	2,26
MnO	0,03
MgO	1,41
TiO_2	0,91
Na_2O	0,86
K_2O	0,95
P_2O_5	0,22
Losses in fire	8,52

Table 6.1. Chemical analysis in oxides, of the mineral coal bottom ash.

Constituentes	Coal Bottom Ash ppm)
Ba	299
Nb	27
Zr	286
Sr	168
Rb	71
Pb	27
Zn	32
Cu	34
Ni	48
Cr	224

Table 6.2. Chemical analysis of mineral coal bottom ash. (trace elements).

The mineralogical analysis of coal ash was performed on its powder using the a difractometric method. The equipment used was a Philips diffractometer, model X'Pert, Copper K_α radiation (λ = 1,54056 Å), power of 40 kV e 30 mA, nickel filter in the secondary optic. The byproduct was crushed in a mortar, sieved and separated fractions withparticle size less than 45 μm. The analysis conditions were: 0,02° for the step, step time 2s and measuring range in 2θ, of 10 to 90°.

The figure 6.1 shows the X-ray diffractogram of bottom coal ash. The presented crystalline phases were identified using the JCPDS (1981) and ICSD (1995) database. The analysis of the XRD pattern shows that the bottom ash is formed by the crystalline phases quartz (SiO_2 - JCPDS 5-490), mullite ($Al_2Si_6O_{13}$ - JCPDS 15-776), magnetite (Fe_3O_4- JCPDS 19-629) and hematite (Fe_2O_3 - JCPDS 13-534).

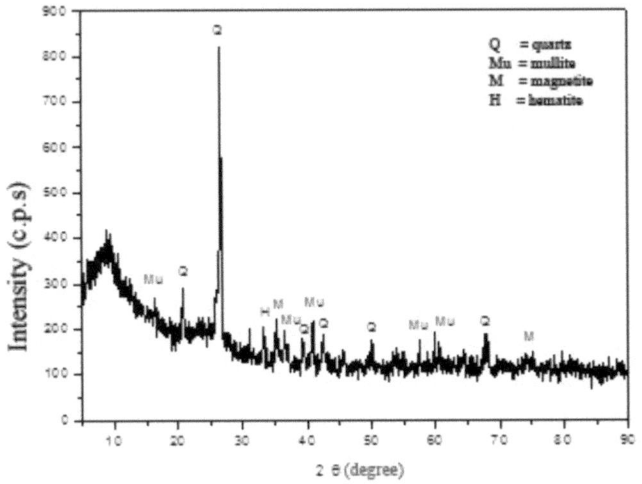

Fig. 6.1. X-ray diffractogram of mineral coal bottom ash.

6.1.3 Structural refinement and quantification of crystalline phases by the Rietveld method

The crystalline phases identified from mineral coal bottom ash were quantified using the Rietveld Method (RIETVELD, 1965, 1965). The software used was DB8K98. The graphic display of the plots of the XRD patterns (simulated and experimental) was obtained through the program DMPLOT.

In order to estimate the relative amount of each crystalline phase, the XRD experimental spectrum was simulated using the Rietveld method for structure refinement. Crystallographic data (atomic coordinates, space group, cell parameters and temperature factor) used were those available in the archives ICSD numbers 174, 23726, 20596 e 15840 for all phases identified, presented in Table 6.3. These are the input information for the refinement, which consists in comparing the structure of theoretical models and the observed spectrum.

Table 6.4 present the refined crystallographic data for the crystalline phases present in mineral coal bottom ash. It is observed that the values of network parameters were refined and the atomic positions and isotropic temperature factors of atoms remained constant after the refinement.

The intense background in the diffractogram of coal bottom ash (Figure 6.1), which was set in the simulation proces, was not regarded as an amorphous phase, since it is difficult to measure it separately. The curves of experimental and simulated pattern (Figure 6.2) showed a good agreement, resulting in the following percentages of relative phases: quartz (34,94%), mullite (64,22%), magnetite (0,38%) and hematite (0,46%).

Phase	Space Group	Lattice Parameters (Å)	Temperature Factor (B_o)	Atomic Positions
Quartz (SiO_2) ICSD # 174 PDF # 05-490	P 32 2 1 S (No. 154)	a = 4.913 b = a c = 5.405	B_o = 0 B_o = 0	Si (3a), x = 0.4698, y = 0.0, z = 0.0 O (6c), x = 0.4141, y = 0.2681, z = 0.1188
Mullite ($Al_{2.35}O_{4.82}Si_{0.64}$) ICSD # 23726 PDF # 15-776	P B A M (No. 55)	a = 7.566 b = 7.682 c = 2.884	B_o = 0.43 B_o = 0.51 B_o = 0.49 B_o = 0.49 B_o = 0.97 B_o = 0.92 B_o = 1.40 Bo = 0.84	Al (2a), x = y = z = 0.0 Al (4h), x = 0.2380, y =0.2945, z = 1/2 Al (4h), x = 0.3512, y = 0.1590,z = 1/2 Si (4h), x =0.3512, y = 0.1590, z = 1/2 O (4g), x =0.3929, y = 0.2808, z = 0.0 O (4h), x =0.1420, y = 0.0777, z = 1/2 O (2d), x =0.0, y = 1/2, z = 1/2 O (4h), x =0.0509, y = 0.4482, z = ½
Magnetite (Fe_3O_4) ICSD #20596 PDF No. 19-629	F D 3 M (No. 227)	a=b=c=8.400	B_o = 0 B_o = 0 B_o = 0	Fe (8a), x = y = z = 1/8 Fe (16d), x = y = z = 1/2 O (32e), x = y = z = 0.258
Hematite (Fe_2O_3) ICSD # 15840 PDF No. 13-534	R -3 C H (No. 167)	a= 5.038 b = a c = 13.772	B_o = 0 B_o = 0	Fe (12c), x = y = 0.0, z = 0.3553 O (18e), x = 0.3059, y = 0.0, z = ¼

Table 6.3. Theoretical crystalloghaphic data for the crystalline phases present in mineral coal bottom ash.

Phase	Space Group	Lattice Parameters (Å)	Temperature Factor (B_o)	Atomic Position
Quartz (SiO_2)	P 32 2 1 S (No. 154)	a= 4.919 b = a c = 5.414	B_o = 0 B_o = 0	Si (3a), x = 0.4698; y = 0.0, z = 0.0 O (6c), x = 0.4141; y =0.2681, z = 0.1188
Mullite ($Al_{2.35}O_{4.82}Si_{0.64}$)	P B A M (No. 55)	a = 7.563 b = 7.706 c = 2.890	B_o = 0.43 B_o = 0.51 B_o = 0.49 B_o = 0.49 B_o = 0.97 B_o = 0.92 B_o = 1.40 B_o = 0.84	Al (2a), x = y = z = 0.0 Al (4h), x = 0.2380; y =0.2945, z = 1/2 Al (4h), x = 0.35120, y = 0.1590, z = 1/2 Si (4h), x =0.35120, y = 0.1590, z = 1/2 O (4g), x=0.3929, y = 0.2808, z = 0.0 O (4h), x =0.1420, y = 0.0777, z = 1/2 O (2d), x =0.0, y = 1/2, z = 1/2 O (4h), x =0.0509, y = 0.4482, z = ½
Magnetite (Fe_3O_4)	F D 3 M (No. 227)	a = 8.401 b = a c = a	B_o = 0 B_o = 0 B_o = 0	Fe (8a), x = y = z = 1/8 Fe (16d), x = y = z = 1/2 O (32e), x = y = z = 0.258
Hematite (Fe_2O_3)	R -3 C H (No. 167)	a = 5.048 b = a c = 13.793	B_o = 0 B_o = 0	Fe (12c), x = y = 0.0 z = 0.3553 O (18e), x = 0.3059, y = 0.0, z = 1/4

Table 6.4. Refined crystallographic data for the crystalline phases present in mineral coal bottom ash.

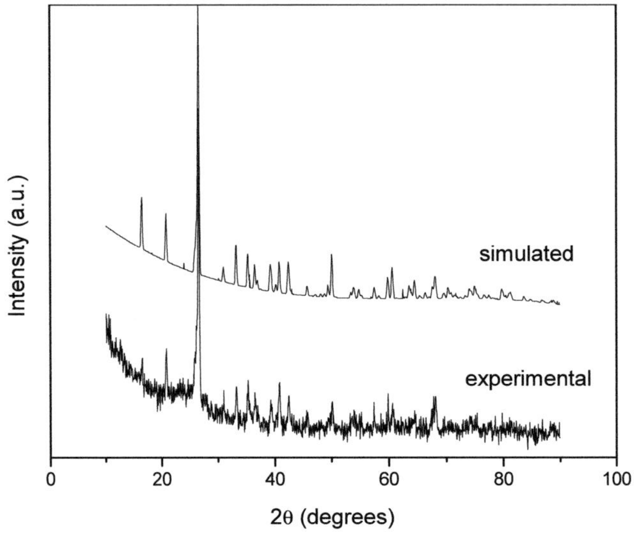

Fig. 6.2. XRD patterns of mineral coal bottom ash: experimental and simulated by the Rietveld Method.

An estimate of the crystallinity of the bottom ash can be performed from the analysis of the integrated intensity areas contained in the diffraction pattern. Subtracting the value of "amorphous area" of the value of "total area" is possible to estimate the value of "crystalline area". According to this methodology, the bottom ash under study has a crystallinity of 43,10%, or 58,90% of amorphous phase. These results are consistent with results found in the literature (KNIESS, 2005).

6.2 Case 2 – Ceramic materials

6.2.1 Obtaining and characterization of materials

The materials characterized in this study are ceramics materials for covering with addition of clay raw materials (argilominerals) and mineral coal bottom ash, sinterized at 1150º C to 2 hours.

X Ray diffraction analysis of the developed material were obtained with a Philips X'Pert equipment, (= 1,54 Å) through the powder method. Analyses were done with 0,02º/ 2s and 2θ from 10 to 90º.

Crystalline phases were identified based on JCPDS database.

In order to obtain the crystallographic data, necessary to the structural refinement through the Rietveld Method, the ICSD was used. Input data to the refinement by the Rietveld Method are presented in Table 6.5

The used refinement program was the DBWS 98. DMPLOT program made possible the comparison between the theoretical spectrum and the refined one.

6.2.2 Characterization of ceramic materials - Crystalline phases identification

The Figure 6.3 presents the X ray patterns of ceramic material studied: MA, MB, MC, MD. These materials were obtained in the paper described by Kniess *et al.* (2006). Identified crystalline phases (in ceramic materials) are also presented in Figure 6.3.

The Table 6.5 shows theoretical crystallographic data of crystalline phase present in sinterized ceramic materials.

Phase	Lattice parameters (Å)	Atomic Position	Occupation Number	Thermal Isotropic Factors (B_o)
Quartz (α-SiO_2) ICSD 29210 PDF 05-490 P 32 2 1 S (154)	a = b = 4,913 c = 5,405 $\alpha = \beta = 90$ $\gamma = 120$	Si (3a), x = 0,469, y = 0,0, z = 0,0 O (6c), x = 0,403, y = 0,253, z = 0,122	Si = 1,0 O = 1,0	B_o (Si) = 0 B_o (O) = 0
SiO_2 ICSD 34889 PDF 76-0912 P 43 21 2 (96)	a = b = 7,456 c = 8,604 $\alpha = \beta =$ $\gamma = 90$	Si (8b), x = 0,326, y = 0,120, z = 0,248 Si (4a), x = 0,410, y = 0,410, z = 0,0 O (8b), x = 0,445, y = 0,132, z = 0,400 O (8b), x = 0,117, y = 0,123, z = 0,296 O (8b), x = 0,334, y = 0,297, z = 0,143	Si (8b) = 1,0 Si (4a) = 1,0 O (8b) = 1,0 O (8b) = 1,0 O (8b) = 1,0	Si (8b) = 2,39 Si (4a) = 2,39 O (8b) = 2,39 O (8b) = 2,39 O (8b) = 2,39
Tridimita (SiO_2) ICSD 29343 PDF 75-0638 P 63 2 2 (182)	a = b = 5,01 c = 8,18 $\alpha = \beta = 90$ $\gamma = 120$	Si (4f), x = 0,333, y = 0,667, z = 0,47 O (2c), x = 0,333, y = 0,667, z = 0,25 O (6g), x = 0,425, y = 0,0, z = 0,0	Si (4f) = 1,0 O (2c) = 1,0 O (6g) = 1,0	B_o (Si) = 0 B_o (O) = 0
Mullite ($Al_{2,35}Si_{0,64}O_{4,82}$) ICSD 23726 PDF 15-776 P B A M (55)	a = 7,566 b = 7,682 c = 2,884 $\alpha = \beta =$ $\gamma = 90$	Al (2a), x = y = z = 0,0 Al (4h), x = 0,2380, y =0,2945, z = 1/2 Al (4h), x = 0,3512, y = 0,1590,z = 1/2 Si (4h), x =0,3512, y = 0,1590, z = 1/2 O (4g), x =0,3729, y = 0,2808, z = 0,0 O (4h), x = 0,1420, y = 0,0777, z = 1/2 O (2d), x =0,0, y = 1/2, z = 1/2 O (4h), x = 0,0509, y = 0,4482, z = ½	Al (2a) = 1,0 Al (4h) = 0,34 Al (4h) = 0,34 Si (4h) = 0,33 O (4g) = 1,0 O (4h) = 1,0 O (2d) = 0,41 O (4h) = 0,21	Al (2a) = 0,43 Al (4h) = 0,51 Al (4h) = 0,49 Si (4h) = 0,49 O (4g) = 0,97 O (4h) = 0,92 O (2d) = 1,4 O (4h) = 0,84
Hematite (Fe_2O_3) ICSD 15840 PDF 13-0534 R -3 C H (167)	a = b = 5,038 c = 13,772 $\alpha = \beta = 90$ $\gamma = 120$	Fe (12c), x = 0,0, y = 0,0, z = 0,3553 O (18e), x = 0,3059, y = 0,0, z = 0,25	Fe = 1,0 O = 1,0	B_o (Fe) = 0 B_o (O) = 0

Table 6.5. Crystallographic theoretical data of crystalline phase present in sinterized ceramic materials.

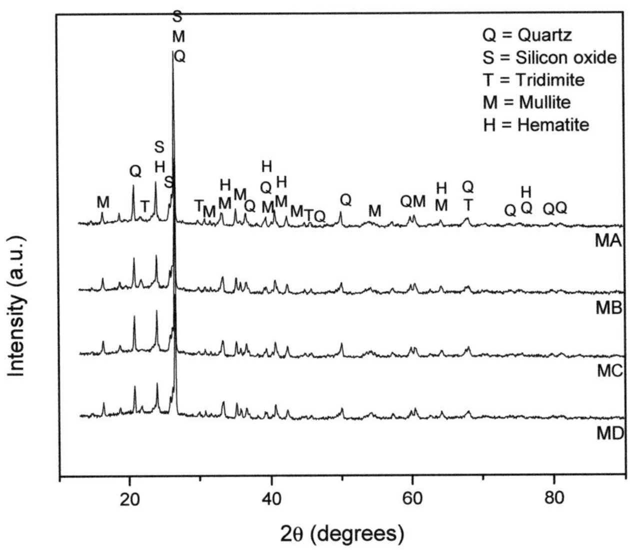

Fig. 6.3. X Ray patterns of ceramic materials, sinterized under the temperature of 1150ºC.

6.2.3 Crystalline phase quantification by Rietveld method

The Figure 6.4 presents the comparison between the experimental pattern and MA, MB, MC and MD samples simulation, through the Rietveld Method. Materials MA, MB, MC and MD spectrum plotting presented a good approach to the diffraction pattern simulated and the observed one, with a good definition to intensities and peak positions.

Percentage related to crystalline phases, obtained through the Rietveld Method, is presented in Table 6.6, which also presents statistical numeric indicators R_P, R_{WP} e R_{EXP}.

Crystalline and amorphous phases' characteristics are considered very important factors, which influence mechanical properties of ceramic materials (KNIESS, *et al* 2006).Quartz and mullite were identified as major crystalline components in all four samples. It is possible to observe that the MA material presented the highest percentage of residual quartz after sintering (54,89%). MC material presents the highest percentage of tridymites phase in comparison with the other materials obtained (8,39%). MD material presented the highest percentage of mullite phase (22,45%) and hematite phase (14,49%).

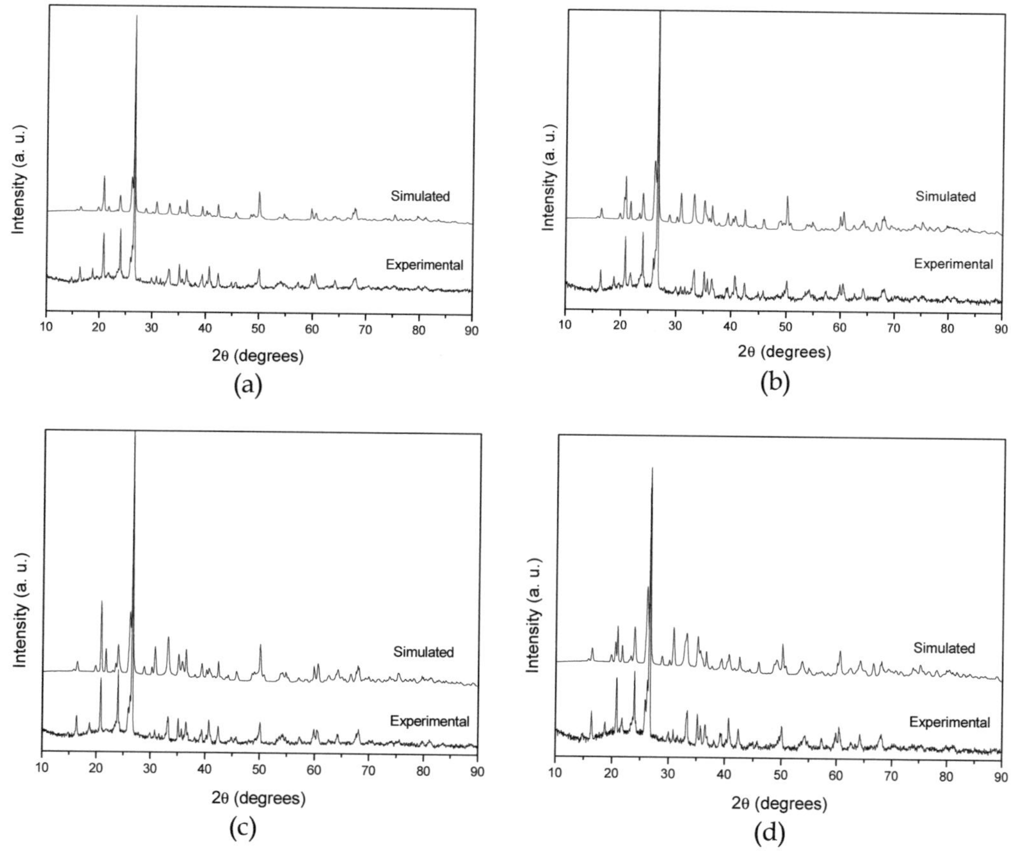

Fig. 6.4. Materials MA (a), MB (b), MC (c) and MD (d) patterns, experimental and simulated by the Rietveld Method.

Sample	R_P* (%)	R_{W-P}* (%)	R_{EXP}* (%)	Relative percentage of the crystalline phases calculated by Rietveld Method				
				α -Quartz	Silicon oxide	Tridimite	Mullite	Hematite
MA	7,41	9,97	3,32	54,89	21,32	3,31	15,91	4,58
MB	7,27	10,23	3,19	39,90	19,92	7,20	23,25	9,73
MC	8,92	12,39	3,22	37,76	18,66	8,39	19,95	15,24
MD	8,66	13,0	3,18	29,11	21,11	6,05	26,56	17,14

Table 6.6. Statistical numeric indicators R_P, R_{WP} e R_{EXP} and related percentage calculated through Rietveld of crystalline phases.

The structural refinement of MA material presented the lowest of R_{W-P} (9,97%). Convergence was verified through indexes R_P e R_{WP}. In despite of the fact that R_P e R_{WP} are distant for more than the recommended 20%, experimental and simulated spectrum curves presented a good correspondence (Figure 6.4a), and the R_{WP} value is within the recommended range for good results ($2 \leq R_{WP} \geq 10$).

Pattern of MB material presented a good approaching between the simulated diffraction pattern and the observed one, with a good definition to the intensities and peak positions, as shown in Figure 6.4b. Refinement quality indicators are R_P =7,27 %, R_{WP} = 10,23 % and R_{EXP} =3,19 %.

It is possible to observe that, to MC material, there is also a larger difference between peaks intensities than in the pattern, experimental and simulated in the lower angles region, where more intense peaks are placed (Figure 6.4c). Refinement quality was evaluated through R_P e R_{WP} indexes, equals to 8,92% and 12,39 %, respectively, while R_{EXP} was 3,22%. In despite of the difference between entre $R_{W\text{-}P}$ and R_{EXP} is higher than the recommended 20%, simulated and experimental spectrum curves present good concordance, and the $R_{W\text{-}P}$ value is within the recommended range in literature (Mccusker, & Von Drelle, 1990).

Crystalline phases structural refinement of MD material was the one that presented the highest R_{WP} value (13,0%) compared with the structural refinement done in the other materials. Convergence Indexes R_P e R_{EXP} were equal to 8,66%, and 3,18% respectively.

7. Conclusion

The methodology for quantification of the crystalline phases by Rietveld Methods was aaplied to ceramics materials and raw materials. The crystalline phases formation is related with the sintering process. The Rietveld method allows, simultaneously, perform unit cell refinement, refinement of crystal structure, microstructure analysis, quantitative analysis of phases and determination of preferred orientation. With the advancement of computer technology, the Rietveld method has allowed more information could be extracted from XRD patterns. Analyzing all the diffraction pattern and using the individual intensities of each angular step, the method enabled the refinement of complex crystal structures, and later applied to the provision of quantitative data accurately recognized.

Rietveld Method, due to the fact of using all X-ray diffraction profile in calculations, overcomes several compounds peaks superposition problem and turns possible to obtain results from all crystalline phases simultaneously, without the need of pattern samples and calibrations curves. It means an expressive gain in relation to other techniques to multiphase systems crystalline phases quantification through X-ray diffraction.

8. References

Alégre, R. (1965). Généralisation de la méthode d´addition pour l analyse quantitative par diffraction X. *Bull. Soc. Franc. Miner. Crist.* v. 88, p. 569-574.

Borba, C. D. G.(2000). *Obtenção e caracterização de vitrocerâmicos de nefelina: medição de tamanho de cristalito e quantificação de fases por difração de raios-X*. Florianópolis, SC, Abr, 137p. Tese de Doutorado em Ciência e Engenharia de Materiais. Universidade Federal de Santa Catarina.

Brindley, G. W., Brown, G. (1980). *Crystal structures of clay minerals and their X-ray identification. London*: Mineralogical Society, 495 p.

Callister, W. D. (2002). *Materials Science and Engineering, an Introduction.* New York, John Wiley & Sons, Inc.

Carvalho, F. M. S. (1996). *Refinamento da estrutura cristalina do quartzo, coríndon e criptomelana utilizando o método de Rietveld.* São Paulo, 73 p. Dissertação de mestrado. Instituto de Geociências da Universidade de São Paulo.

Chung, F. H. (1974). Quantitative interpretation of X-ray diffraction patterns of mixtures. I. Matrix flushing method for quantitative multicomponent analysis. *Journal of Applied Crystallographic.* v.7, p.519-525.

Coelho, C. (2002). *Quantificação de fases mineralógicas de matérias-primas cerâmicas via numérica.* Florianópolis, SC, Mai, 91p. Tese de Doutorado em Ciência e Engenharia de Materiais. Universidade Federal de Santa Catarina

Cullity, B. C. (1978). *Elements of X-Ray difracction.* New York: Addilson-Wesley, Ed.2, 555p.

Fabbri, B., Fiori, C., RavaglioII, A. (1989). *Materie prime ceramiche: tecniche analitiche e indagini di laboratório.* Faenza Editrice, v.3, 531p. 1989.

Fleurence, A. (1968). *Analyse diffractométrique aux rayons X. Intrustrie Céramique.* n. 605, p. 203-211.

Fonseca, A. T. (2000). *Técnicas de processamento cerâmico.* Lisboa: Universidade Aberta, 554p.

Gobbo, L. A. (2003). *Os compostos do Clínquer Portland: sua Quantificação por Difração de Rios X e Quantificação por Refinamento de Rietveld.* Dissertação de Mestrado. Instituto de Recursos Minerais e Hidrogeologia. Universidade de São Paulo.

Hald, P. (1952). *Técnica de la cerámica.* Barceloma: Ediciones. Omega, 126p.

ICSD (Inorganic Crystal Structure Database) (1995). Gmchin-Intitut fur Anorganishe Chemie and Fachinformationzentrum FIZ. Karlsruhe, Germany.

International tables for x-ray crystallography (1997). International Union of Crystallography (IUCr).

JCPDS (Joint Committee of Powder Diffraction Standards) (1981). International Centre for Diffraction Data. Pennsylvania, USA.

Kingery, W. D., Bowen, H. K., Uhlmann, D. R. (1976). *Introduction to ceramics.* New York: John Wiley & Sons, Ed.2.

Klug, H. P., Alexandder, L. E. (1954). *X-ray diffraction procedures for polycrystalline and amorphous materials.* Ed 2., 966p.

Kniess, C.T. (2005). *Desenvolvimento e caracterização de materiais cerâmicos com adição de cinzas pesadas de carvão mineral.* 285p. Tese de Doutorado em Engenharia e Ciência dos Materiais. PGMat / Universidade Federal de Santa Catarina, Florianópolis.

Kniess, C. T.; Prates, P. B.; Lima, J. C.; Kuhnen, N. C.; Riella, H . (2006). Influência da adição de cinzas pesadas de carvão na resistência mecânica à flexão de revestimentos cerâmicos. In: 50º *Congresso Brasileiro de Cerâmica,* Blumenau. Anais do 50º Congresso Brasileiro de Cerâmica.

Kniess, C. T.; Prates, P. B.; Lima, J. C.; Kuhnen, N. C.; Riella, H . (2007). Dilithium dialuminium trisilicate phase obtained using coal bottom ash. *Journal of Non-Crystalline Solids,* v. 353, p. 4819-4822.

Kniess, C. T.; Prates, P. B.; Kuhnen, N. C.; Riella, H; Lima, J. C.; Franjdlich, E. U. (2009a). Determination of the Amorphous and Crystalline Phases of the Ceramic Materials. In: *11 th International Conference on Advanced Materials* - ICAM 2009, Rio de Janeiro, Brazil.

Kniess, C. T.; Prates, P. B.; Milanez, K.; Kuhnen, N. C.; Riella; Lima, J. C.; Maliska, A. M. (2009b). Quantitative Determination of the Crystalline Phases of the Ceramic

Materials Utilizing the Rietveld Method. In: *Seventh International Latin American Conference on Powder Technology (PTECH)*, Atibaia, Brazil.

Kniess, C. T.; Riella, H; Franjdlich, E. U; Durazzo, M.; Saliba-Silva, A; Prestes, L. (2010). Determination of Crystalline Phases in the Uranium Silicide by X-ray Diffraction. In: International Meeting on Reduced Enrichment for Research and Test Reactors - RERTR 2010; 32nd, Lisboa, Portugal.

L. D. Mccusker, R. B. Von Drelle, D. E. Cox, D. Louer, P. Scardi. (1990). *Journal of Applied Crystallographic.* v. 32, p. 36-50.

Navarro, J. M. F. (1991). *El Vidrio.* Madrid: CSIC (Consejo Superior de Investigaciones Científicas).

Padilha, A. F. (1997). *Materiais de Engenharia - Microestrutura e propriedades.* São Paulo: Hemus.

Paiva-Santos, C.O. (1990). *Estudos de Cerâmicas Piezelétricas pelo método de Rietveld com dados de difração de raios X.* São Carlos, SP. Tese de doutorado. Instituto de Física e Química de São Carlos - Universidade de São Paulo.

Paiva-Santos, C. O. (2001). *Caracterização de materiais pelo método de Rietveld com dados de difração por policristais.* São Paulo: Instituto de Química, UNESP, 46p.

Paiva-Santos, C. O. (2005). *Aplicações do Método de Rietveld.* São Paulo: Instituto de Química, UNESP, 2005.

Post, J. E., Bish, D. L. (1989). Rietveld refinement of crystal structures using power X-ray diffraction data. Modern Power Diffraction. *Mineralogical Society of America.* v.20. p. 277-308, 1989.

Riello, P., Faguerazzi, G., Canton, P., Clemente, D., Signoretto, M. (1995). Determining the degree of crystallinity in semicrystalline materials by means of the Rietveld analysis. *Journal of Applied Crystallographic.* v. 28, p.121-126.

Rietveld, H. M. (1967). Line profiles of neutron powder-diffraction peaks for structure refinement. *Acta Crystallographica* . n.22, p.151-1152, 1967.

Rietveld, H. M. (1969). A profile refinement method for nuclear and magnetic structures. *Journal of Applied Crystallographic*, v.2, p.65-71, 1969.

Warren, E. B. (1959). *X ray Diffraction.* London: Addison-Wesley Pub. Company.

Wiles, D. B., Young, R. A. (1981). A new computer program for Rietveld analysis of X-ray powder diffraction pattern. *Journal of Applied Crystallographic.* v.14, p.149-151.

Young, R. A. (1995). *The Rietveld Method.* New York: Oxford University Press, 298p.

Young, R.A., Larson, A.C, Paiva-Santos, C.O. (1998). *Rietveld analysis of x-ray and neutron powder diffraction patterns.* Atlanta: School of Physics, Georgia Institute of Techonology.